Biomining Technologies

David Barrie Johnson •
Christopher George Bryan • Michael Schlömann •
Francisco Figueroa Roberto
Editors

Biomining Technologies

Extracting and Recovering Metals from Ores and Wastes

Springer

Editors
David Barrie Johnson
School of Natural Sciences
Bangor University
Bangor, UK

Christopher George Bryan
BRGM
Orléans, France

Michael Schlömann
Institute of Biosciences
Technische Universität Bergakademie Freiberg
Freiberg, Germany

Francisco Figueroa Roberto
Technical Services Processing and Metallurgy
Newmont USA Limited
Denver, CO, USA

ISBN 978-3-031-05381-8 ISBN 978-3-031-05382-5 (eBook)
https://doi.org/10.1007/978-3-031-05382-5

This Springer imprint is published by the registered company Springer Nature Switzerland AG
The registered company address is: Gewerbestrasse 11, 6330 Cham, Switzerland

This book is dedicated to the memory of Professor Douglas E. Rawlings, the pioneer of molecular studies of biomining microorganisms and who also played a key role in deciphering their roles in mineral processing operations. Doug Rawlings was also the editor/co-editor of previous texts in this field.

Contents

Editors and Contributors

About the Editors

Barrie Johnson is a research professor at the Universities of Bangor and Coventry (UK) and a Research Associate of the Natural History Museum (London) with >30 years' experience with using specialised microorganisms to extract and recover metals. He is an elected member of the Learned Society of Wales, is an Industrial Research Fellow of the Royal Society, has held honorary professorships at Exeter University and Central South University (China), has been awarded research fellowships by the US Department of Energy and Bureau of Mines, and has, on two occasions, acted on behalf of the International Atomic Energy Agency (in Peru). He has worked with many of the leading global mining companies (Rio Tinto, BHP Billiton, Newmont, Vale, etc.) and collaborates with many research groups throughout the world on fundamental and applied aspects of biohydrometallurgy. He has published over 380 articles in journals, books, and conference proceedings and has worked on over 70 funded (research council, EU, and industry) research projects. His research was highlighted by *Scientific American* as one of "10 world changing ideas" in their December 2011 edition. He co-edited (with Professor Doug Rawlings) the second edition of *Biomining*, which was published by Springer-Verlag in 2007.

Chris Bryan is head of the Geomicrobiology and Environmental Monitoring unit at BRGM—the French Geological Survey, leading a team of microbiologists and geochemists providing expertise and technological assistance for environmental protection and monitoring. This includes the remediation and restoration of mine wastes and post-mining environments and the development of ecotechnologies for the sustainable exploitation of primary and secondary raw materials. Prior to joining BRGM, he was Senior Lecturer in Sustainable Mining at the Camborne School of Mines (University of Exeter, UK). He is an environmental and molecular microbiologist, with >20 years' experience in the microbiology of biohydrometallurgical systems (both tank and heap leaching) and acidogenic mine wastes.

Michael Schlomann is full professor for environmental microbiology at Technische Universität Bergakademie Freiberg. Based on a background in the degradation of organic contaminants, molecular ecology, and white biotechnology, after having been appointed in Freiberg he turned more and more to mining-related microbiology. Several projects with regional companies and other partners dealt with microbial treatment of acid mine drainage. More recently, projects focused on leaching of ores and included partners mainly in Chile, but also Sweden and South Africa. He is the initiator and speaker of the Biohydrometallurgical Centre Freiberg. He has published ca. 140 papers in peer-reviewed journals or extended conference contributions. He has served on the editorial board of *Applied Microbiology* and *Biotechnology* as well as of *Bioresource Technology* and has published the textbook *Umweltmikrobiologie* with Prof. Reineke. For more than 15 years, he has served on the selection committee of the German Environmental Foundation (DBU) for doctoral fellowships.

Frank Roberto is a director of processing at Newmont Corporation. He joined Newmont in 2012 and held senior and chief metallurgist positions in the Newmont Metallurgical Services corporate laboratory before moving to his current role in the Technical Services process development team. He is the corporate subject matter expert (SME) for biohydrometallurgy. He is responsible for identifying and developing transformational technologies for processing, tailings, and water treatment and R&D advisor to the corporate energy and decarbonisation team. Prior to joining Newmont, he directed and executed biotechnology R&D programmes at the US Dept. of Energy's Idaho National Laboratory for 23 years, in areas ranging from bioleaching to biodefense. He was elected to the US National Academy of Engineering in 2020, is a co-inventor for 14 US patents, and has published over 75 technical articles, including a chapter with Barrie Johnson in the first edition of *Biomining* published in 1997.

Contributors

Patricio Martínez Bellange Santiago, Chile

Aleksandr Vasilyevich Belyi JSC "Polyus Krasnoyarsk", Krasnoyarsk, Russia

Christopher George Bryan BRGM, Orléans, France

Craig Bradley van Buuren Gold Technology Group, Metso Outotec, Centurion, Gauteng, South Africa

Miao Chen CSIRO Mineral Resources, Clayton South, VIC, Australia

David William Dew University of Exeter, Cornwall, UK

David Grant Dixon Department of Materials Engineering, University of British Columbia, Vancouver, BC, Canada

Mark Dopson Centre for Ecology and Evolution in Microbial Model Systems (EEMiS), Linnaeus University, Kalmar, Sweden

Mariekie Gericke Biometallurgy Division, Mintek, Randburg, South Africa

Anne-Gwénaëlle Guezennec BRGM, Orléans, France

Susan Therese Largier Harrison Department of Chemical Engineering & DVC: Research and Internationalisation, Centre for Bioprocess Engineering Research (CeBER), University of Cape Town, Cape Town, South Africa

Sabrina Hedrich TU Bergakademie Freiberg, Institute of Biosciences, Freiberg, Germany

Agathe Hubau BRGM, Orléans, France

David Barrie Johnson School of Natural Sciences, Bangor University, Bangor, UK
Faculty of Health and Life Sciences, Coventry University, Coventry, UK
Natural History Museum, London, UK

Anna Henriikka Kaksonen Commonwealth Scientific and Industrial Research Organisation (CSIRO), Land and Water, Floreat, Western Australia, Australia

Päivi Hanna-Maarit Kinnunen VTT Technical Research Centre of Finland Ltd, Tampere, Finland

Hamer Arévalo Lara Minera Peñasquito, Carretera Mazapil, Zacatecas, Mexico

Alvaro Videla Leiva Departamento de Ingeniería de Minería, Pontificia Universidad Católica de Chile, Santiago, Chile

Xueduan Liu Central South University, Changsha, People's Republic of China

Hector Mark Lizama Technical Services Trail, Teck Metals Ltd., Trail, BC, Canada

Pasi Määttä Elementis Minerals B.V. Suomen Sivuliike, PL, Kajaani, Finland

John William Neale Biometallurgy Division, Mintek, Randburg, South Africa

Jan Albert van Niekerk Gold Technology Group, Metso Outotec, Centurion, Gauteng, South Africa

Naoko Okibe Faculty of Engineering, Department of Earth Resources Engineering, Kyushu University, Fukuoka, Japan

Johan Waldemar Olivier Gold Technology Group, Metso Outotec, Centurion, Gauteng, South Africa

Jochen Petersen Department of Chemical Engineering, University of Cape Town, Rondebosch, Cape Town, South Africa

Guanzhou Qiu Central South University, Changsha, People's Republic of China

Raquel Quatrini Fundación Ciencia & Vida, Santiago, Chile
Facultad de Medicina y Ciencia, Universidad San Sebastián, Providencia, Santiago, Chile
ANID – Millennium Science Initiative Program – Millennium Nucleus in the Biology of the Intestinal Microbiota, Santiago, Chile

Francisco Figueroa Roberto Technical Services Processing and Metallurgy, Newmont USA Limited, Denver, CO, USA

Gary Vernon Rorke Prudentia Process Consulting, Spring Hill, QLD, Australia

Ana Laura Santos School of Natural Sciences, Bangor University, Bangor, UK

Axel Schippers Federal Institute for Geosciences and Natural Resources (BGR), Hannover, Germany

Michael Schlömann Technische Universität Bergakademie Freiberg, Institute of Biosciences, Freiberg, Germany

Olga Vladimirovna Tupikina Nordgold, Moscow, Russia

Mario Vera Véliz Departamento de Ingeniería Hidráulica y Ambiental. Escuela de Ingeniería, Instituto de Ingeniería Biológica y Médica Escuelas de Ingeniería, Ciencias Biológicas y Medicina, Pontificia Universidad Católica de Chile, Santiago, Chile

Elizabeth Lindsay Jane Watkin Curtin Medical School, Curtin University, Perth, WA, Australia

Ruiyong Zhang Key Laboratory of Marine Environmental Corrosion and Bio-fouling, Institute of Oceanology, Chinese Academy of Sciences, Qingdao, People's Republic of China
Open Studio for Marine Corrosion and Protection, Pilot National Laboratory for Marine Science and Technology (Qingdao), Qingdao, People's Republic of China

Chapter 1
Evolution and Current Status of Mineral Bioprocessing Technologies

David Barrie Johnson and Francisco Figueroa Roberto

Abstract Metals have been used for thousands of years and have had pivotal roles in the development of human civilisation. Both the scale and range of metals that are used in modern and emerging technologies, and industrial and domestic applications have increased vastly in recent years. Harnessing microorganisms to facilitate the extraction and recovery of metals from mineral ores and waste materials has been often promulgated as a more environmentally benign approach than conventional methods, such as pyrometallurgy, yet "biomining" remains a niche technology, used primarily to bioleach copper ores and biooxidise refractory gold ores. This chapter charts the development of mineral bioprocessing technologies since the discovery of the first bacteria that were shown to accelerate the dissolution of sulfide minerals, and highlights their perceived strengths and weaknesses. The diverse engineering approaches used in biomining, and the role of microbial consortia in liberating metals from sulfide ores, are highlighted.

Keywords Acidophiles · Biomining · Bioheaps · Bioleaching · Biooxidation · Microbial consortia · Stirred tank bioreactors

D. B. Johnson (✉)
School of Natural Sciences, Bangor University, Bangor, UK

Faculty of Health and Life Sciences, Coventry University, Coventry, UK

Natural History Museum, London, UK
e-mail: d.b.johnson@bangor.ac.uk

F. F. Roberto
Technical Services Processing and Metallurgy, Newmont USA Limited, Denver, CO, USA
e-mail: frank.roberto@newmont.com

D. B. Johnson et al. (eds.), *Biomining Technologies*,
https://doi.org/10.1007/978-3-031-05382-5_1

1.1 Metals, Minerals, and Human Civilisation: The Context and Early History of Biomining

Humans have a long and sometimes fraught relationship with metals and metal mining. Major changes in progression of civilisations have, in the past, been driven by our ancestors discovering how to obtain and use "new" metals and alloys. The transformation of civilisations from Neolithic (New Stone Age) to the Bronze Age (the first period of the "Metal Age") is estimated to date from about 4300 years BP in Indo-Europe. This was underpinned by the discovery and production of copper–tin alloy, which had many superior properties to the two metals individually, and was accompanied by major social and technological advances. Archaeological evidence for bronze age mining of copper and tin is widespread, stretching from the southeast (e.g., Cyprus) to the northwest (e.g., Wales) of Europe. The following era in human civilisation, the iron age, estimated to extend from ca. 2800 to 1900 years BP was defined by metalworking being dominated by ferrous metallurgy. Since iron is harder and more abundant than copper and tin, this became the metal of choice for many applications. The method used then (and mostly today) to obtain metals involved roasting with materials such as charcoal to provide the reductant for reducing the metal to zero-valent iron. As in modern times, metal mining in pre-antiquity would have doubtless provided opportunities for microorganisms to colonise exposed ore bodies and degrade minerals. There is, however, at least one reference from the Renaissance period to what appear to be bioleach liquors in *De re metallica* ("On the nature of metals") a text written in Latin by Georg Bauer (Georgius Agricola) in 1556. In western Europe, China and probably elsewhere in the middle ages, miners learned that, by periodically allowing the deeper mines and adits to flood and then draining the waters into lagoons and ponds, it was possible to obtain crude copper without roasting the ore bodies. This was done by adding scrap iron, which dissolved concurrent with the appearance of "cement copper", a phenomenon that was perceived as alchemy (the transformation of one metal into another). Interestingly, this approach for extracting copper (in situ (bio)mining) and its electrochemical recovery has persisted well into the modern era of biomining.

While *Homo sapiens* has been exploiting metal ores for thousands of years, this is dwarfed by the millions of years since prokaryotic microorganisms first developed intimate associations with metal-containing (and other) minerals. The first primitive single-celled life forms are thought to have emerged relatively soon (ca. 4000 million years ago; Mya) after the planet coalesced (ca. 4600 Mya). There is considerable debate about what forms of energy (electron donors) and electron acceptors were used by primitive prokaryotes, though it seems highly probable that inorganic materials are prime candidates for both roles. This may be observed today with some acidophilic archaea and bacteria that can colonise biomining environments, e.g., *Acidianus* spp. that couple oxidation of hydrogen to the reduction of sulfur. The "great oxidation event", which caused the anoxic planet to become transformed with an oxygen-enriched atmosphere, and which is thought to have initiated somewhere between 2000 and 2400 million years Mya, would have been a major game changer for bacteria and archaea, allowing them to access, indirectly, the energy released by transforming sulfide minerals such as pyrite (FeS_2) using molecular oxygen as an electron acceptor. This is essentially the same fundamental microbial metabolism that is harnessed in all current full-scale biomining operations.

There are two major differences between historic and modern-day mining of metals and how they are used. One is in scale. It has been estimated that humans have mined as much metal from the lithosphere in the past 30 years as in all previous times combined, with ore grades steadily decreasing, and consequently more waste produced. Second, the range of metals required for modern applications, technologies, and consumables, is far greater than in the past, and includes a number of rare earth as well as transition metals. The demand for some metals which had only minor use in past times (e.g., cobalt, which had been used primarily as a pigment) has dramatically escalated in more recent times, causing their commodity prices to rocket and increasing effort to find and exploit new sources, both primary ores and waste materials. The list of critical raw materials compiled by the European Union, the majority of which are metals, increased from 14 in 2011 to 30 in 2020. The current trend in switching from fossil fuel-driven to electric vehicles, and in generating energy from renewable sources, will doubtless result in major increases in demand for metals in general and for those used for generating and storing electric energy, in particular, in the twenty-first century. The question of whether accessible reserves of many of these metals are adequate for meeting the projected demands is something that has not always been considered sufficiently.

1.2 The Modern Era: Development and Application of Engineering Designs of Full-Scale Biomining Operations

The modern era of commercial-scale mineral bioleaching (i.e., post Antonie van Leeuwenhoek, who is credited with the discovery of bacteria in 1676) began within 20 years of the isolation of a bacterium (then named as *Thiobacillus ferrooxidans*) from an acidic ferruginous coal mine drainage stream in West Virginia. Since then, understanding how microorganisms can liberate or make accessible metals in sulfidic ore bodies, their diversities and interactions, and the engineering design options that can be used for bioprocessing sulfidic ores and concentrates has expanded greatly. A number of review articles have been written on this topic in scientific journals and books, including Brierley (2008a, b), Brierley and Brierley (2001, 2013), Rawlings (1997, 2002, 2005), Rawlings and Silver (1995), Rawlings et al. (2003), Rawlings and Johnson (2007a), Johnson (2010, 2013, 2014, 2018). In addition, proceedings published from the biennial International Biohydrometallurgy Symposia and Biohydrometallurgy/Biomining conferences are a major repository and resource of information on mineral bioprocessing from fundamental studies to full-scale operations.

Dump leaching was the initial design used for engineering a biomining operation, with the first full-scale system established by the (then) Kennecott Copper Corporation at the Bingham Canyon mine in Utah, and shortly afterwards at the Chino mine in New Mexico. Dump leaching has since been applied in many other countries and is used still to extract, primarily, copper from low grade (<0.1–0.5%) waste rock and run-of-mine-ore. Ungraded material, comprising fine dust particles to large boulders, are stacked in mounds that may exceed 100 m in height, and irrigated with sulfuric acid to stimulate the activities of indigenous acidophilic prokaryotes that degrade the copper sulfide minerals present. The copper-enriched leachates (pregnant leach solutions; PLS) are channelled into vats and copper is recovered, often using cementation as described above. The process is slow and individual dumps may be used over several years or even decades. Although it uses relatively crude engineering and little or no control of microbiology, dump leaching continues to be a major contributor to global copper production. Heap leaching, which has its origins in Chile, Australia, and elsewhere in the 1970s, uses a similar general approach, but with greater refinement and capital investment. Again, heap leaching is used largely to extract copper, though bioheaps have also been used as a pre-treatment for extracting gold from refractory ores, and more recently for processing polymetallic ores. The main upgrades used in heap compared with dump biomining are: (1) ores are usually crushed and graded, and sometimes also agglomerated; (2) materials are transferred to pads that have impermeable high-density plastic liners and pipe networks to collect and transfer PLS, and the heaps are ideally constructed by conveyor stacking rather than truck dumping; (3) heaps are actively aerated to provide the mineral-degrading microorganisms with not only a supply of oxygen, which is required for the oxidative dissolution of sulfide minerals, but also carbon dioxide, as the principal prokaryotes involved are, like green plants, autotrophic; (4) heaps are often inoculated to ensure their exposure to suitable biomining microflora (which may need to include those that operate over very different temperature ranges to cope with those often found in heterogeneous heaps); (5) target metals in PLS are often extracted using solvents (followed by electrowinning to produce high-grade cathodic metals) and the raffinate liquors generated pumped back into the heap circuit (with or without an additional inoculum) using an irrigation network placed on the heap surface, sometimes below a plastic cover that serves both to act as a thermo-insulator and, depending on climatic factors, reduce moisture loss by evaporation or inputs of meteoric water. The life of heaps is typically 1–2 years, after which they can be removed and treated to minimise the ongoing dissolution of remaining reactive minerals. Alternatively, heaps can be stacked progressively upon each other, ultimately forming a large multilayered structure, such as at the Escondida mine in Chile. A modified approach for heap leaching involves agglomerating fine-grain mineral concentrate particles onto coarser rock fragments and regenerating the carrier material once the former have been oxidised. The Geocoat® process (Harvey and Bath 2007) was claimed to reduce the processing time in a bioheap to about 2 months.

A radically different operational design for mineral bioprocessing was initiated in the 1980s in South Africa and has since seen plants established in many countries in different parts of the world. Continuous-flow stirred tank reactors are used primarily to biooxidise refractory gold concentrates but have also been used to bioleach cobalt and nickel from mineral wastes. Large (>1000 m^3) tanks constructed from corrosion-resistant stainless steel and fitted with pipework to facilitate the flows of liquids and air, one or more impeller connected to a motor to maintain fine-grain particles in the mineral slurries in suspension, and a cooling system (the accelerated rate of mineral oxidation compared to dumps and heaps generates a lot of heat) are used either as single units or in series, through which the slurries are transferred. Target metals are recovered from the solution phase in the case of bioleaching, or from the partially processed biooxidised mineral phase by chemical extraction in the case of gold. Stirred-tank bioprocessing is much more rapid than dump and heap operations, typically requiring only 3–6 days to be complete.

A number of worked out uranium mines in Canada were subjected to an end-of-life bioleaching phase (in-place, or in situ, leaching) in the 1970s and 1980s. Controlled blasting was used to fracture the residual buried ore bodies and the collapsed underground structures were allowed to flood. Soluble uranium (VI) was extracted from the acidic leach liquors produced from the solubilised uranite (UO_2), accounting for about 300 t in one mine (the Denison mine) during 1 year of in-place leaching. This was essentially a refined application of the mediaeval practice of in situ bioleaching described above. A development of this approach (deep in situ biomining) wherein ore bodies present deep in the lithosphere are processed without the need for haulage and comminution, is currently being evaluated for its economic viability and acceptance by society as an alternative strategy for mining metals in the twenty-first century (Chap. 17).

Some milestones in the development of biomining engineering and operational designs are given in Table 1.1, and images from some of these are in Fig. 1.1.

1.3 The Biomining Niche: Limitations and Opportunities

In the late twentieth century, by which time biomining had become an established and expanding biotechnology with operations in place in various parts of the world, there was great optimism shared by many researchers working in the field that it would have major, and even revolutionary, impact on the metal mining sector. In reality, however, its impact and uptake have been far more limited. While bioprocessing of metal ores has been estimated to account for between 10 and 20% of global copper production, ~1% of gold, and smaller amounts of other base metals, such as cobalt and nickel, it remains, in essence, a niche technology. There are a number of reasons for this.

Biomining is frequently claimed by its protagonists to be a "green technology" but the actual case for this is not always that strong. While many of the bacteria and archaea involved in biomining operations are autotrophic (i.e., fix CO_2), their contributions to global carbon budgets are minor. Much of the energy demand and carbon footprint of metal mining is associated with excavating, haulage, and

Table 1.1 Some key milestones in the development of biomining technologies

Year	Event
	Dump operations
1960s	Bioleaching of run-of-mine ore in dumps established at two sites (in Utah and New Mexico) operated by (the then) Kennecott Corporation
1979	Recovery of copper from "waste" rock at the Dexing mine (China)
	Bioheap operations
1980	Copper heap bioleaching established in Chile (Lo Aguirre mine)
1992	Copper bioheap established at Mount Gordon, Australia
1998	First copper heap bioleaching operation in Myanmar (Sabetaung and Kyisintaung mine)
1999	First biooxidation heap leach commissioned for processing refractory gold ore (Newmont Corp)
2003	Commissioning of the first heap biooxidation operation using GEOCOAT® technology Agnes mine, South Africa)
2006	Copper bioleached from stacked bioheaps at Escondida, Chile (the world's largest copper mine)
2008	Heap bioleaching of a polymetallic (Ni, Zn, Cu) schist established at Talvivaara, Finland
2013	First cathode produced from 1 million tonne commercial-scale enargite bioleach demonstration at Minera Yanacocha, Peru
	Continuous stirred-tank operations
1986	First commercial biooxidation reactor for refractory gold ore (Fairview, South Africa); BIOX® technology
1994	First full-scale operation using BacTech technology to process refractory gold ores (Youanmi, Australia)
1999	Stirred-tank bioleaching of pyritic waste to extract and recover cobalt (Kasese, Uganda)
2003	Large-scale (<8000 t/d) ore treatment of refractory gold concentrate, Olimpiada, Polyus, Russia
2015	Stirred-tank bioleaching of a nickel sulfide concentrate (by-product of talc extraction; Mondo Minerals, Finland)

Referenced from: Olson et al. (2003), Morin and d'Hugues (2007), Brierley (2008b), Wu et al. (2008), Brierley and Brierley (2013), Riekkola-Vanhanen (2013), and Chap. 12. More comprehensive lists of earlier bioheap and stirred-tank bioleaching operations can be found in Watling (2006) and Brierley (2008a).

comminution (rock grinding and sorting), which is often followed by generating mineral concentrates for final processing (Curry et al. 2014). Most stirred-tank operations require all of these up-front processing steps, and it is only the concentrate bioprocessing stage that can justifiably be regarded as relatively "green". Dump and heap leaching do not require all of the preprocessing stages, while the bioleaching of mineral tailings and other wastes uses materials that have already been subjected to haulage etc., and are therefore among the most environmentally benign applications of biomining. Combined with this, bio- and subsequent processing can generate more secure secondary mineral wastes that could be used for other purposes, fulfilling an objective of a circular economy.

The general area of hydrometallurgy includes biomining technologies, as well as others (e.g., chemical leaching) that do not involve biological systems. The fact that

Fig. 1.1 Images from sites where different approaches are used to bioleach or biooxidise sulfidic ores and concentrates to recover base and precious metals. (**a**) precipitation pond, used to recover copper from in situ bioleaching (Mynydd Parys, Wales); (**b**) run-of-mine and crushed ore bioleach heaps (Yanacocha mine, Peru); (**c**) pond receiving copper-rich pregnant leach solution from a trial heap leach (Bingham Canyon mine, Utah); (**d**) bioleach aeration blowers and distribution plenum (Yanacocha mine, Peru); (**e**) copper cathode produced by SX-EW (Phoenix mine, Nevada); and (**f**) continuous stirred tanks used to process refractory gold ore (Suzdal mine, Kazakhstan)

biotechnologies are regarded by some, and possibly many, in the mining industry as being not sufficiently robust, has impeded their acceptance. Bioprocessing ores and concentrates has to compete with alternative approaches, such as pressure leaching, which are also continuing to make significant technological advances, as well as with pyrometallurgy. Smelters represent major investments for mining companies, with constructing and commissioning a single smelter costing in the order of $1 billion. However, bioprocessing mineral ores and concentrates is invariably much slower than competing technologies, and this is a significant detraction. This latter downside, throughput, is probably the biggest reason bioprocessing is not applied by mining companies on a similar scale to competitor technologies like pressure oxidation. Ultimately, time under leach is a negative for bioleach/whole-ore biooxidation compared to chemical leaching—sulfuric acid (perhaps with ferric iron) for copper, and cyanide for gold.

There are, however, niche areas where biomining can compete with alternative technologies, some of which, such as for processing low-grade/run of mine ores, reprocessing mine wastes, and biooxidising refractory gold ores, have already been touched upon. In some situations, it has not been found possible to produce a high-grade mineral concentrate from an ore (such as the polymetallic ore body at Talvivaara/Terrafame in Finland, which contained ~10% graphite) and bioleaching rather than smelting ground ore was therefore considered to be preferable. The elevated arsenic content of some mineral ores and concentrates can preclude processing in smelters due to legislative restrictions, though companies often blend high and low arsenic-containing materials to get around this barrier. New technology is emerging at the world's only smelter currently taking high-arsenic concentrates (Tsumeb in Namibia) to incorporate up to 20% arsenic (by weight) in glass. Bioprocessing, like other hydrometallurgical approaches, has the advantage that solubilised arsenic is retained in solution rather than emitted in flue gases, and can be precipitated from liquid wastes as a relatively stable mineral (such as scorodite or ferric arsenate) and stored securely.

1.4 The Microbiological Context of Biomining

Biomining environments are typified by being acidic (sometimes extremely so), rich in soluble metals and other solutes, such as sulfate, and having widely varying temperatures. Knowledge of microbial species that contribute to biomining processes and understanding of how these interact both with minerals and with each other has increased markedly since the early days when the sole bacterium thought to have a direct role in accelerating the oxidative dissolution of sulfide minerals was *(Acidi) thiobacilllus ferrooxidans*. Since all current commercial-scale biomining operations operate at low pH, active microbial populations are limited to acidophiles. An account of how this area has expanded since the discovery of the very first acidophile (the sulfur-oxidiser *(Acidi)thiobacillus thiooxidans*, in the early 1920s) can be found in Johnson and Quatrini (2020). Several novel isolates, representative of genera and species that would subsequently be identified as prokaryotes that have widespread and major roles in biomining operations, were described in the 1970s. These include: (1) *Leptospirillum ferrooxidans*, isolated by Markosyan (1972) from a copper mine in Armenia, though its related thermo-tolerant relative *Leptospirillum ferriphilum* is now recognised as having a more important role in commercial biomining operations, and has often been identified as the dominant iron-oxidising bacterium present; (2) the first thermo-acidophilic archaeon (a member of the order Sulfolobales; Brierley and Brierley 1973); (3) the first mixotrophic mineral-oxidising bacterium (the thermo-tolerant Firmicute, *Sulfobacillus*; Golovacheva and Karavaiko 1979). Many new species of acidophilic prokaryotes have since been described, which has been greatly aided by the advent of molecular microbiology techniques. For example, in the past all rod-shaped mesophilic acidophiles that could oxidise both iron and sulfur tended to be classified as strains of *At. ferrooxidans*, the iron-oxidising acidithiobacilli currently comprise five distinct species. Not all acidophilic microorganisms can thrive in bioleach liquors, however, as other factors, particularly elevated concentrations of transition metals, may preclude this.

Probably as important as the isolation and identification of species that can mediate sulfide mineral dissolution was the recognition, particularly over the past 20 years, that these both co-exist with, and interact with other microorganisms both in natural and anthropogenic environments, such as "biomines" (Rawlings and Johnson 2007b). Biomining systems from dumps to stirred tanks all necessarily operate as open, non-sterile environments where microorganisms can be introduced from a number of sources, such as atmospheric deposition or the ore/concentrate itself. Likewise, it is not possible to totally preclude microorganisms in a mine site from migrating to the wider environment, which is why it is unlikely that genetically modified acidophiles will find an application in industrial-scale biomining operations. Although sulfide minerals can be degraded by pure cultures of acidophilic iron-oxidising bacteria such as *Leptospirillum* spp. and some *Acidithiobacillus* spp. in the laboratory, microbial consortia are both more effective and robust, and invariably found in actual biomining operations. While the individual species may differ from site to site, and especially with temperature, the presence of the same three functional groups appears to be universal in dumps, tanks, and heaps. These are: (1) iron-oxidisers, which catalyse the initial oxidative dissolution of sulfide minerals by their continuous regeneration of the oxidant, ferric iron; (2) sulfur-oxidizers, which oxidise sulfur oxyanions and zero-valent sulfur, generating sulfuric acid and thereby maintaining conditions that are conducive both to the iron-oxidisers and also for retaining the cationic metals released from minerals in solution, facilitating their downstream recovery; (3) organic carbon degraders (heterotrophic and mixotrophic acidophiles) that metabolise the soluble organic carbon compounds released from active and moribund/dead iron/sulfur-oxidisers and which may otherwise build up to concentrations that inhibit the latter. Many of the third group can also oxidise iron and/or sulfur, such as *Sulfobacillus* and *Ferroplasma* spp., and some, at least, of the CO_2 they generate is used by bacteria in groups (1) and (2), which are primarily autotrophic.

While the generic composition of microbial consortia required for efficient bioleaching may be essentially the same for all engineering configurations, there are important differences between bioheaps and stirred tanks that have major influences on formulating the compositions of microbial inocula (Rawlings and Johnson 2007b). Bioheaps can be highly heterogeneous in terms of temperature and chemical microenvironments, and major selective pressure on the indigenous microflora is their ability to attach to mineral particles (as biofilms) in the heap to minimise or avoid washout. In contrast, stirred tanks are homogeneous, providing constant conditions for microbial growth though the relatively fast throughput selects for faster-growing consortia. As a consequence, microbial populations in stirred tanks tend to be dominated by relatively few (3–4) species of acidophiles, whereas far greater biodiversity is found in heaps, and these are also subject to temporal changes as the physico-chemistry of heaps evolves during leaching. Other selection pressures will apply in certain situations, e.g., for mineral-oxidising prokaryotes that are able to tolerate elevated concentrations of salt (NaCl) as well as extreme acidity and transition metals.

1.5 Commercial Bioleach and Biooxidation Operations in 2020

Biomining was described above as a niche technology, but the encouraging reality is that within the mining industry, bioleaching is considered a viable alternative hydrometallurgical process for extraction of base metals, and biooxidation competes with pressure oxidation or roasting for pre-oxidation of refractory gold-bearing ores and concentrates. A section of the *SME Mining Reference Handbook* has devoted a chapter to bioleaching since 2002 (Briggs 2002) and major metallurgical conferences (CIM) interleave biohydrometallurgical papers among other hydrometallurgical presentations. The comprehensive *SME Mineral Processing and Extractive Metallurgy Handbook* published in 2019 (Dunne et al. 2019) dedicated two chapters to bioleaching and agitated bioleach reactors within the section devoted to hydrometallurgy. Therefore, there can be no question that bioleaching and biooxidation are accepted unit processes in mineral processing, though they have a limited range of applications compared to more common unit processes such as conventional heap leaching with sulfuric acid (for copper) and high-temperature oxidation (for gold). Ultimately the mineralogy and deportment of the base or precious metal, and the economics relative to the location and grade of the deposit, will drive the decision of which process(es) to build a mine around.

More than a decade since the last edition of *Biomining*, there have been several notable long-term studies of commercial copper bioleach operations around the world. In Chile, a focused effort to develop a logic-based control system for heap bioleaching at the Escondida mine incorporated microbiological, genetic, and production information with machine learning to optimise the bioleach component of the overall operation (Demergasso et al. 2018). Heap aeration, specialised material handling, and inoculation were key components of the Bio-Leach Sulfide Project. The Escondida mine, a joint venture of BHP Billiton, Rio Tinto, and JECO Corporation is the world's largest copper mine, produced nearly 1.2 million tonnes of copper in 2020, or about 6% of the world's copper and over 20% of Chile's national production (2020 preliminary data). For nearly 15 years, Biosigma, a joint venture between Codelco and JX Nippon Mining and Metals operated in Chile to develop microbiological improvements in copper heap leaching, including dozens of Chilean and US patents. JX Nippon Mining and Metals sold its stake to Codelco in 2015, after which Biosigma became a part of Codelco Tech SpA, and was disbanded in 2017. At Codelco's Radomiro Tomic mine, Biosigma implemented a novel fluidised bed bioreactor to cultivate microorganisms for heap inoculation that was tested at kiloton scale. At the Zijinshan mine in China, a body of work has described the directed manipulation of environmental conditions to influence microbial populations and activities in order to improve heap leach performance that benefits both copper and gold production. This practice is being transferred to other mines operated by Zijin Mining Company including the Monywa mine in Myanmar (Chen et al. 2020). These refinements and advances are further described in Chaps. 8–10.

In refractory gold ore biooxidation, BioMin (formerly GoldFields) was acquired by mining industry support giant Outotec (now Metso Outotec) in 2016, and the BIOX® process is now another product offering in that company's suite of hydrometallurgical process options that include the related ASTER™ thiocyanate biodegradation process and MesoTHERM® elevated temperature BIOX®. Coupled with Metso Outotec's strong capabilities in large-scale, stirred-tank reactor systems, it can be anticipated that the BIOX® process will continue to be refined and improved to increase its competitiveness with roasting and pressure oxidation for pre-oxidation of refractory gold ore concentrates. The BIOX® process is currently operating on 3 continents at 7 different mines and accounts for 1% (approximately 1.1 million oz. in 2020) of global gold production and will be further described in Chaps. 4 and 11. According to Metso Outotec, several new projects are in the pipeline for North America.

1.6 Scope of the Current Textbook

This book is the successor to two previous texts on biomining that became firmly established as major references for this area of biotechnology: *Biomining: theory, microbes and industrial practices* (Rawlings 1997) and *Biomining* (Rawlings and Johnson 2007a). It provides both an update on the topic and projects on how the technology is developing and expanding into potential new areas, with contributions from experts and leading authorities from industry, government agencies, and academia from around the world. The book comprises six parts. The first (this chapter) describes the context and development of biomining, while Part II has three chapters that describe the engineering designs and operation of biomining systems (bioheaps and stirred tank systems) and an up-to-date account of the bioprocessing of refractory gold ores. Part III focuses on the microbiology of biomining, with individual chapters covering the biodiversity of acidophiles and how they mediate mineral dissolution, the cultivation and molecular techniques available to study them, and the microbial ecology of bioheaps, stirred tanks, and abandoned mine wastes. Part IV highlights commercial mineral bioprocessing operations carried out in different parts of the world (China, Chile, Peru, Russia, Kazakhstan, and Finland), while Part V describes four areas of biohydrometallurgy that are emerging as potential new areas: bioleaching in the presence of elevated salt concentrations, bioprocessing electronic wastes, reductive bioleaching of oxidised ores, and the use of microorganisms to recover metals from acidic waste and process waters. Finally, Part VI considers, in a concluding chapter, how biomining technologies may develop and be applied in the twenty-first century, in the context of ever-expanding demand for metals and the need for sustainability.

References

Brierley CL (2008a) How will biomining be applied in future? T Nonferrous Met Soc China 18: 1302–1310. https://doi.org/10.1016/S1003-6326(09)60002-9

Brierley JA (2008b) A perspective on developments in biohydrometallurgy. Hydrometallurgy 94: 2–7. https://doi.org/10.1016/j.hydromet.2008.05.014

Brierley CL, Brierley JA (1973) A chemoautotrophic and thermophilic microorganism isolated from an acid hot spring. Can J Microbiol 19:183–188. https://doi.org/10.1139/m73-028

Brierley JA, Brierley CL (2001) Present and future commercial applications of biohydrometallurgy. Hydrometallurgy 59:233–240. https://doi.org/10.1016/S0304-386X(00)00162-6

Brierley CL, Brierley JA (2013) Progress in bioleaching: part B: applications of microbial processes by the minerals industries. Appl Microbiol Biotechnol 97:7543–7552. https://doi.org/10.1016/S0304-386X(00)00162-6

Briggs A (2002) Bacterial oxidation. In: Lowrie RL (ed) SME mining reference handbook. Society for Mining, Metallurgy and Exploration, Englewood, Colorado, pp 319–323

Chen J, Zhong S, Tang D, Kuang C (2020) Practical experience in large-scale development of Zijinshang low-grade gold-copper mine. Min Metall Explor 37:1339–1347. https://doi.org/10.1007/s42461-020-00237-2

Curry JL, Ismay MJL, Jameson GJ (2014) Mine operating costs and the potential impacts of energy and grinding. Miner Eng 56:70–80. https://doi.org/10.1016/j.mineng.2013.10.020

Demergasso CS, Veliz R, Galleguillos P, Marin S, Acosta Grinok M, Zepeda Alvarez VJ, Zeballos J, Henriquez F, Pizarro R, Bekios-Calfa J (2018) Decision support system for bioleaching processes. Hydrometallurgy 181:113–122. https://doi.org/10.1016/j.hydromet.2018.08.009

Dunne RC, Kowatra K, Young CA (eds) (2019) SME mineral processing and extractive metallurgy handbook. Society for Mining, Metallurgy and Exploration, Inc., Englewood, Colorado, pp 1267–1282, & 1283–1292

Golovacheva RS, Karavaiko GI (1979) *Sulfobacillus* – a new genus of spore forming thermophilic bacteria. Microbiology (English translation of Mikrobiologiya) 48:658–665

Harvey TJ, Bath M (2007) The GeoBiotics GEOCOAT^{R} technology – progress and challenges. In: Rawlings DE, Johnson DB (eds) Biomining. Springer, Heidelberg, pp 97–112

Johnson DB (2010) The biogeochemistry of biomining. In: Barton L, Mandl M, Loy A (eds) Geomicrobiology: molecular and environmental perspective. Springer, Heidelberg, pp 401–426

Johnson DB (2013) Development and application of biotechnologies in the metal mining industry. Environ Sci Pollut Res 20:7768–7776. https://doi.org/10.1007/s11356-013-1482-7

Johnson DB (2014) Biomining – biotechnologies for extracting and recovering metals from ores and waste materials. Curr Opin Biotechnol 30:24–31. https://doi.org/10.1016/j.copbio.2014.04.008

Johnson DB (2018) The evolution, current status and future prospects for using biotechnologies in the mineral extraction and metal recovery sectors. Minerals 8:343. https://doi.org/10.3390/min8080343

Johnson DB, Quatrini R (2020) Acidophile microbiology in space and time. Curr Issues Mol Biol 39:63–76. https://doi.org/10.21775/cimb.039.063

Markosyan GE (1972) A new iron-oxidizing bacterium, *Leptospirillum ferrooxidans* gen. et. sp. nov. Biol J Armenia 25:26–29. (in Russian)

Morin D, d'Hugues P (2007) Bioleaching of a cobalt-containing pyrite in stirred reactors: a case study from laboratory scale to industrial application. In: Rawlings DE, Johnson DB (eds) Biomining. Springer, Heidelberg, pp 35–55

Olson GJ, Brierley JA, Brierley CL (2003) Bioleaching review part B: progress in bioleaching: applications of microbial processes by the minerals industry. Appl Microbiol Biotechnol 63: 249–257. https://doi.org/10.1007/s00253-003-1404-6

Rawlings DE (ed) (1997) Biomining: theory, microbes and industrial processes. Springer/Landes Bioscience, Georgetown

Rawlings DE (2002) Heavy metal mining using microbes. Annu Rev Microbiol 56:65–91. https://doi.org/10.1146/annurev.micro.56.012302.161052

Rawlings DE (2005) Characteristics and adaptability of iron- and sulfur-oxidizing microorganisms used for the recovery of metals from minerals and their concentrates. Microb Cell Factories 4: 13. https://doi.org/10.1186/1475-2859-4-13

Rawlings DE, Johnson DB (eds) (2007a) Biomining. Springer, Heidelberg

Rawlings DE, Johnson DB (2007b) The microbiology of biomining: development and optimization of mineral-oxidizing microbial consortia. Microbiol-UK 153:315–324. https://doi.org/10.1099/mic.0.2006/001206-0

Rawlings DE, Silver S (1995) Mining with microbes. Biotechnology 13:773–778. https://doi.org/10.1038/nbt0895-773

Rawlings DE, Dew D, du Plessis C (2003) Biomineralization of metal-containing ores and concentrates. Trends Biotechnol 21:38–44. https://doi.org/10.1016/s0167-7799(02)00004-5

Riekkola-Vanhanen M (2013) Talvivaara mining company – from a project to a mine. Miner Eng 48:2–9. https://doi.org/10.1016/j.mineng.2013.04.018

Watling HR (2006) The bioleaching of sulphide minerals with emphasis on copper sulphides – a review. Hydrometallurgy 84:81–108. https://doi.org/10.1016/j.hydromet.2006.05.001

Wu A, Yin S, Wang H, Qin W, Qiu G (2008) Technological assessment of a mining-waste dump at the Dexing copper mine, China, for possible conversion to an *in situ* bioleaching operation. Bioresour Technol 100:1931–1936. https://doi.org/10.1016/j.biortech.2008.10.021

Chapter 2
Design, Construction, and Modelling of Bioheaps

David Grant Dixon and Hector Mark Lizama

Abstract Bioheaps have become an important tool for the recovery of metals from low-grade ores. This has primarily involved bioleaching of copper and more recently other transition metals such as nickel, but mineral heaps have also been used to biooxidise refractory gold ores as the initial stage for recovering this precious metal. In the first part of this chapter, the design, construction, and operation of bioheaps is considered. The *modus operandi* of bioheaps, the different configurations of leach pads (static and dynamic) and the central roles of integrated irrigation and aeration systems are described. The second part of the chapter gives a detailed and comprehensive mathematical model of the heap bioleaching process and how this has been refined over time. A case study of simulated heap bioleaching at a copper mine (Quebrada Blanca) in northern Chile, using the HeapSim2D model, where key objectives were to simulate chalcocite leaching and identify the key parameters that control heap temperature, is outlined.

Keywords Heap bioleaching · Copper sulfides · Heap leach modelling

2.1 Introduction

Bioheaps are bioreactors, albeit very large ones, and comprise biocatalysts (microorganisms), substrates, additives, reaction processes, transport processes, and reactor geometry. They are easily the most massive reactors ever constructed, and are measured in areas of square kilometers and volumes of millions of cubic meters (Fig. 2.1). Bioheaps have been utilised since the 1980s, and dump leaching (passive

D. G. Dixon (✉)
Department of Materials Engineering, University of British Columbia, Vancouver, BC, Canada
e-mail: dixon@mail.ubc.ca

H. M. Lizama
Technical Services Trail, Teck Metals Ltd., Trail, BC, Canada
e-mail: hector.lizama@teck.com

D. B. Johnson et al. (eds.), *Biomining Technologies*,
https://doi.org/10.1007/978-3-031-05382-5_2

Fig. 2.1 Life cycle of a heap. Construction of the leach pad includes installation of the primary liner, the overliner, and the PLS collection system (**a**). Ore from the mine is crushed and then agglomerated in an agglomerating drum (**b**). The heap is built by stacking agglomerate using different types of stacker assemblies (**c** and **d**). The heap can be a static pad configuration, where new lifts are stacked on top of old ones (**e**). In a dynamic pad configuration, leached residues are removed to make way for fresh agglomerate. The residues can simply be removed by haul truck (**f**) or by bucket wheel excavator (**g**)

bioleaching of low-grade mine waste dumps) has been practised for much longer. Hence, there is a large repository of knowledge regarding their design and operation.

A bioheap is an engineered pile of ore material, within which iron- and/or sulfur-oxidising microorganisms are cultivated to catalyse the oxidation of sulfide minerals. This chapter focuses on the use of bioheaps to solubilise base metals from their sulfide ores rather than for the pretreatment (biooxidation) of gold-bearing sulfides.

2.2 The *Modus Operandi* of Bioheaps

The most common bioheaps currently in operation recover copper from secondary copper sulfides such as chalcocite (Cu_2S). Secondary sulfide ores also often contain various amounts of copper oxides, which most often take the form of basic salts. Oxide minerals, such as brochantite ($CuSO_4{\cdot}3Cu(OH)_2$) readily dissolve in acid. In contrast, the oxidation of sulfide minerals in a bioheap is accomplished with ferric ions acting as the primary oxidant, supplied by the dissolution of iron-bearing minerals (often predominantly pyrite) within the ore itself. Addition of iron to the heap system is usually not required, as only a small concentration in solution is necessary to facilitate mineral oxidation, and this is continuously recycled between its two oxidation states.

Chalcocite is oxidised in two steps. In the first step, roughly 40% of the copper is liberated rapidly, leaving behind a copper sulfide known as secondary covellite (*blaubleibender*):

$$5\,Cu_2S + 4\,Fe_2(SO_4)_3 \rightarrow 4\,CuSO_4 + 8\,FeSO_4 + Cu_6S_5 \qquad (2.1)$$

The rest of the copper is recovered more slowly from the *blaubleibender* covellite to leave behind elemental sulfur:

$$Cu_6S_5 + 6\,Fe_2(SO_4)_3 \rightarrow 6\,CuSO_4 + 12\,FeSO_4 + 5\,S \qquad (2.2)$$

Ferrous iron and reduced sulfur are oxidised to ferric iron and sulfate by iron- and sulfur-oxidising microorganisms using oxygen as the electron acceptor. This is the essence of heap bioleaching:

$$4\,FeSO_4 + O_2 + 2\,H_2SO_4 \rightarrow 2\,Fe_2(SO_4)_3 + 2\,H_2O \qquad (2.3)$$

$$2\,S + 3\,O_2 + 2\,H_2O \rightarrow 2\,H_2SO_4 \qquad (2.4)$$

Reactions (2.3) and (2.4) only occur at significant rates in the presence of suitable microorganisms. Microorganisms do not attack the sulfide minerals directly, as even those attached to mineral surfaces mediate mineral dissolution by their regeneration of ferric iron and production of sulfuric acid. In order for these reactions to proceed within a bioheap, both water and oxygen are required. Sulfur biooxidation often does not satisfy the net demand for acid, especially if the heap contains significant quantities of acid-soluble gangue minerals, and addition of extraneous (sulfuric) acid is then required.

When the ore bed is fresh and the copper recovery rate is high, the effluent draining the heap (pregnant leach solution; PLS) will have a high copper concentration and is sent directly to the recovery plant. When the heap is older, and the rate of copper extraction has diminished, the effluent solution (intermediate leach solution; ILS) is recycled or sent to newer heaps to increase the copper tenor to an economical level prior to recovery. The most commonly used process to produce copper from PLS is solvent extraction and electrowinning (SX-EW; Kordosky 2002).

2.3 Design and Construction of Bioheaps for Processing Mineral Ores

Irrigated biomining operations have two major engineering designs: heaps and dumps. Heaps tend to have large surface footprints and are usually about 10 m in height, although multiple lifts (layers of ore) are commonplace. The height of ore dumps, on the other hand, can range from 15 to 100 m. Higher grade ore is generally destined for heaps while relatively low grade run-of-mine ore (often containing about 0.2% of target metal) is generally leached in dumps. Dumps are often highly profitable because of their low costs, and while leaching in dumps can involve periods of many years, they are a reliable revenue stream. Dump leaching has been described in a number of reviews elsewhere (e.g., Figueroa et al. 2015); and this chapter focuses on engineered bioheaps. In addition to heaps and dumps, "valley fills", as their name indicates, are bioheaps that are built within the confines of a valley and can be similar to either a heap or dump leach operation (Reyes et al. 2015).

Ore contained in heaps is crushed to a relatively uniform particle size and heaps are constructed to have ore beds that are as homogeneous as possible. In contrast, material used to construct dumps ranges in size from very large boulders to fine dust, and is handled as little as possible to reduce costs.

2.3.1 Leach Pad Configuration

There are two types of leach pad configuration used in bioheaps: static and dynamic (Fig. 2.1). In a dynamic pad, the leached residues are removed from the leach pad at the end of the leach cycle, and fresh ore is placed to start a new leach cycle. In a static pad, the leached residues are left in place, and a fresh lift added on top. A static pad can have many lifts but only the top one is actively leached. In dynamic pads, lifts are placed and then removed.

2.3.2 Leach Pad Disposition

The leach pad has a number of functions. It has to support the weight of the heap, promote drainage of PLS from the heap but prevent it from escaping to the ground underneath, and to hold and protect the PLS drain lines that form part of the PLS collection system. The most important component of the leach pad is the primary liner. This is a geomembrane that acts as an impermeable barrier to catch and contain the percolating leachates. Ground preparation is essential for a stable foundation; any deformation will put stress on the primary liner. An underliner, a layer of compact regolith with minimum hydraulic conductivity to control potential seepage, and an overliner of free draining sand and gravel, are used to protect the primary liner from damage. Some leach pads have double or composite liners (two geomembranes with a layer of sand and gravel between them) for additional protection. The ore is placed on top of the overliner, to generate ore beds with bulk densities ranging from 1.5 to 1.9 t m^{-3}.

The leachate collection system is incorporated into the overliner, and consists of a grid of evenly spaced perforated pipe networks that feed main pipes that drain the heap. All of the leaching solution applied to the heap must be able to drain through the overliner, and may have to travel hundreds of meters horizontally before it can exit the heap. Any holdup will flood the overliner and the ore bed above it. This is referred to as the phreatic head, analogous to the water table in soil. A very high phreatic head can result in spectacularly catastrophic liquefaction failure (sidewall blowout; Lupo 2010). Drainage must be facilitated; for this reason, a leach pad is always sloped, and the angle of slope is also a function of the geotechnical stability of the ore and influenced by particle size and other physical properties.

Some static pad heaps have liners placed in between lifts. These inter-lift liners can be geomembranes with an underliner and overliner or simply a compacted layer of solid material. The purpose of the inter-lift liner is to prevent solutions irrigating more than one lift at a time ("communication between lifts"). Inter-lift liners are an added expense because they require each lift to have its own PLS collection system. However, they are important in preventing excessive phreatic head. Static pad heaps without inter-lift liners need to install additional safeguards such as downhole pumps or wick drains.

2.3.3 *Ore Bed Construction*

Ore beds are constructed to be as homogeneous as possible, to help ensure that all available sulfide mineral surfaces are exposed to the leaching solution. This is facilitated by crushing the ore to reduce the particle size. Typically, ore particles in a bioheap are about 25 mm in diameter. Agglomeration is often used to reduce solution flow problems due to the presence of fines. Crushed ore is passed through a rotating drum fitted with internal acid sprays. The tumbling action and acid addition cause the fines to stick to the coarser particles, forming stable aggregates of more uniform size. Acid application during agglomeration also contributes to uniform adjustment of the ore pH suitable for bioleaching. Agglomerate is then stacked by conveyors (or using trucks) to the required lift height, the quality of the agglomerate bed dictating whether the heap will meet the target recovery. The agglomerate needs to have abundant void spaces to allow it to be permeable to both leaching solution and air.

2.3.4 *Irrigation*

Once the heap is built, the irrigation system is placed on its surface. This consists of a grid of evenly spaced irrigation lines fitted with evenly spaced drip emitters. The drip emitter spacing is determined by the physical properties of the agglomerate. Tighter spacing increases the uniformity of agglomerate wetting ("sweep efficiency"). Emitters require minimum flow rates to work effectively, and when the leaching solution is divided among increasing numbers of emitters some will fail to deliver solution, resulting in irrigation being uneven. Precipitates or scaling can clog or block emitter flow and anti-scalants may be added to prevent this.

Typical irrigation rates range from 1 to 12 L h^{-1} m^{-2} of heap surface. The irrigation rate used is limited at the low end by the lift height and at the high end by the properties of the (agglomerated) ore bed. Acid and other reagents are consumed and depleted as the leaching solution percolates through the ore bed; taller lifts with more ore to percolate through have greater depletion. Increasing flow rates ensure adequate acid delivery to the bottom of the heap, but also put more stress on the stability of the agglomerated ore.

2.3.5 *Aeration*

Many heaps have aeration systems installed within the overliner (Schlitt 2006). Air supplies the bioleaching microorganisms with both carbon dioxide, necessary for their growth, and oxygen to allow their oxidation of iron and sulfur, and humid air is a major redistributor of reaction heat in sulfide heaps. Maintaining a flow of air significantly greater than the flow of leaching solution may add 20 °C to the heap temperature, resulting in faster leach kinetics (Dixon 2000), though excessive aeration can lead to localized drying of the ore within a heap, thereby negatively impacting mineral leaching. A typical forced aeration system consists of a blower, a main header, and evenly spaced perforated lines placed under the ore bed. Air pressure inside the air lines and header falls rapidly with distance apart, limiting their useful length, and so numerous blowers are usually required to supply the entire heap. To ensure even distribution within the heap, the air must be fed from the bottom, contributing to the phreatic head if the overliner and leachate collection system are not well designed. The air pressure supplied by the blowers is generally equivalent to between 0.8 and 0.9 m of water, and a phreatic head higher than this cannot be overcome by the air blower.

2.4 Modeling Bioheaps

Significant progress has been made towards understanding the fundamentals of heap bioleaching in the last 30 years, including the development of many mathematical models (reviewed in Dixon 2003; Petersen and Dixon 2007; Marsden and Botz 2017). Modelling efforts fall into two basic categories: forecasting models and fundamental models.

Forecasting models are largely unconcerned with the scientific details of heap leaching, but rather with being able to predict the rate at which the value metals will be extracted. Such models can be simple mass balancing exercises, written into custom spreadsheets or built into various widely used process modelling software programs such as Metsim or SysCAD (Marsden and Botz 2017). Recently, a new class of forecasting models has been developed which takes advantage of "machine learning". These models (e.g., Saldaña et al. 2019) are "trained" with data from actual operations, and then can predict the outcome from similar operations. While they can be very useful for tracking or predicting the performance of existing operations, they are of very limited use for actually understanding what factors are driving the performance of heaps, or of designing heaps around new leaching technology.

Fundamental models attempt to represent the relevant physics and chemistry of heap leaching in mathematical form. Recent examples of such models include HeapSim (Petersen and Dixon 2007) and models based on CFD software published by Bennett et al. (2012) and others. Of all the models recently published, only HeapSim has attempted to deal explicitly with the problem of heap hydrology at the "micro" level of the single drip emitter, and even this attempt was based on outmoded assumptions of fluid flow and solute transport (the so-called "Turner structure"; Turner 1958). Other models developed since then have applied the classic equations of hydrology to mathematical models of heap leaching, but have adopted the "macro" approach of modelling heap hydrology, with boundary conditions taken as average fluxes over very large areas of the heap. As a result, none have recognised the primary importance of the single drip emitter in determining the basic hydrological characteristics of a heap leach. It was this shortcoming that motivated the development of HeapSim2D.

2.4.1 *HeapSim2D*

The rationale of HeapSim2D (and HeapSim before it) was to develop a systematic approach to heap leach modelling that could be adapted to any heap leaching scenario. The basic code of HeapSim2D consists of five coupled models: (a) solution flow, (b) solute transport (for as many solutes as required), (c) gas flow, (d) gas species transport (O_2 and CO_2), and (e) heat flow. In addition, there is a *Reaction Network Model* which includes the kinetic rate laws for every important chemical reaction in the heap (leaching reactions, gangue acid consumption, gas–liquid mixing, ferrous oxidation, bacterial growth, etc.) and a stoichiometric matrix which combines the various rate laws into net consumption and production terms for each solid, solute and gas species to be used as source terms within their various transport models.

2.4.1.1 Solution Flow

A realistic model of solution flow is arguably the single most important aspect of any heap leach model. Furthermore, modelling efforts at the University of British Columbia quickly uncovered the primary importance of drip emitter spacing on the development of flow patterns within heaps and columns. In the original version of HeapSim, this flow was represented by the so-called "Turner Structure". In this model, flowing solution is confined to vertical flow channels in plug flow, while most of the solution in the heap is stagnant, trapped within the pores of mineral agglomerates and within the larger interstitial spaces between flow channels. This stagnant solution only exchanges solutes with the flowing solution by molecular diffusion. This simplistic model was somewhat successful in simulating flow and solute transport through leaching columns. However, the parameters of this model are difficult to relate to the actual physical attributes of the packed bed. Also, this model ignores the well-developed field of vadose-zone hydrology, which has contributed accepted models for simulating flow and solute transport in unsaturated soils.

A heap surface is assumed to be irrigated by a set of drip emitters spaced at regular intervals, $2R$, each discharging leach solution at an equal rate of U L m^{-2} h^{-1}. Due to the symmetry of the drip emitter layout, the surface of the heap can be subdivided into identical, nearly cylindrical elements of diameter $2R$ and depth Z, with the drip emitter placed at the origin of the vertical axis. To account for the heap surface properly, the drip radius is defined as:

$$R = \sqrt{\frac{XY}{\pi}}$$

where X and Y are the drip emitter spacings in the x and y directions, respectively (Fig. 2.2).

The solution flow pattern for the heap can be conveniently described by analyzing the flow in this single representative, cylindrical volume. Because of the axial symmetry around the vertical axis, the infiltration process can be viewed as axisymmetric flow. This dramatically simplifies the numerical modelling effort.

Assuming the density of water to be virtually constant (it only varies by $\pm 2\%$ over the entire temperature range of stability), then water volume represents water mass and the 2D axisymmetric equation of water volume continuity (EOC) is:

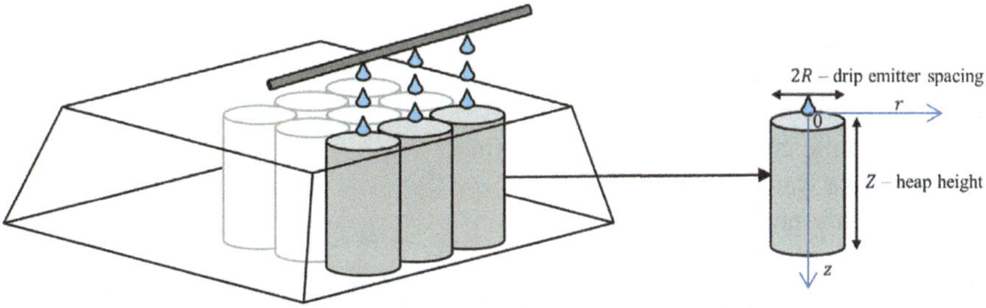

Fig. 2.2 Schematic diagram of the heap as a collection of identical cylinders each with its own drip emitter, as represented in HeapSim2D

$$-\nabla \cdot \mathbf{v}_{\mathrm{w}} = -\frac{\partial v_{\mathrm{w},z}}{\partial z} - \frac{1}{r}\frac{\partial (r v_{\mathrm{w},r})}{\partial r} = \frac{\partial \theta}{\partial t} - \frac{M_{\mathrm{w}}}{\rho_{\mathrm{w}}} s_{\mathrm{w}} \tag{2.5}$$

where

$\mathbf{v}_w$	water volume flux (a vector) (m^3 water m^{-2} s^{-1})
z	depth (m)
r	radius (m)
θ	volumetric water content (m^3 water m^{-3})
t	time (s)
s_{w}	net rate of water production (kmol m^{-3} s^{-1})
ρ_{w}	water density (taken as 1000 kg m^{-3} water)
M_{w}	water molecular mass (18.015 kg $kmol^{-1}$)

Expressing Darcy's law in a 2D axisymmetric configuration, and assuming no pressure gradients within the air phase, the water volume flux may be written in an isotropic, unsaturated porous medium as:

$$\mathbf{v}_{\mathrm{w}} = \frac{k k_{\mathrm{rw}}}{\mu_{\mathrm{w}}}(\rho_{\mathrm{w}} g \nabla z + \nabla p_{\mathrm{c}}) = K_{\mathrm{w}} k_{\mathrm{rw}} (\nabla z + \nabla h_{\mathrm{c}}) \tag{2.6}$$

where

$$v_{\mathrm{w},z} = \frac{k k_{\mathrm{rw}}}{\mu_{\mathrm{w}}}\left(\rho_{\mathrm{w}} g + \frac{\partial p_{\mathrm{c}}}{\partial z}\right) = K_{\mathrm{w}} k_{\mathrm{rw}}\left(1 + \frac{\partial h_{\mathrm{c}}}{\partial z}\right) \tag{2.7}$$

and

$$v_{\mathrm{w},r} = \frac{k k_{\mathrm{rw}}}{\mu_{\mathrm{w}}}\frac{\partial p_{\mathrm{c}}}{\partial r} = K_{\mathrm{w}} k_{\mathrm{rw}} \frac{\partial h_{\mathrm{c}}}{\partial r} \tag{2.8}$$

where

k intrinsic permeability (m^2)
μ_w water viscosity (kg m^{-1} s^{-1})
g gravitational acceleration (9.81 m s^{-2})
K_w hydraulic conductivity $=k\rho_w g/\mu_w$ (m s^{-1})
k_{rw} relative water permeability (–)
p_c capillary pressure (Pa)
h_c capillary head $=p_c/\rho_w g$ (m)

In order to solve these equations, constitutive relations are required for k_{rw} and h_c. The most popular function for describing the soil (or ore) water retention curve (relating the water content θ and "matric potential" or capillary head h_c) is the van Genuchten equation:

$$S_e = \frac{1}{\left[1 + (h_c/h_{c,0})^n\right]^m} \tag{2.9}$$

where the "effective saturation" S_e is defined thus:

$$S_e = \frac{\theta - \theta_r}{\theta_s - \theta_r} \quad \text{or} \quad \theta = \theta_r + (\theta_s - \theta_r)S_e \tag{2.10}$$

and where

n an exponent related to the pore size distribution (–)
m an exponent with values over the range $0 < m < 1$ (–)
$h_{c,0}$ a capillary head scaling factor, often referred to as the "air entry" head (m)
θ_s saturated volumetric water content (also the effective macro-porosity) (m^3 water m^{-3})
θ_r residual (fully drained) volumetric water content (m^3 water m^{-3})

Using Mualem's capillary tube bundle model, van Genuchten derived the van Genuchten-Mualem (VGM) formula which defines the relative hydraulic permeability k_{rw} as a function of capillary head h_c:

$$k_{rw} = \frac{\left\{1 - (h_c/h_{c,0})^{n-1}\left[1 + (h_c/h_{c,0})^n\right]^{-m}\right\}^2}{\left[1 + (h_c/h_{c,0})^n\right]^{\frac{m}{2}}} \tag{2.11}$$

Combining Eqs. (2.10) and (2.11) and simplifying by taking $n = (1-m)^{-1}$ yields the most widely used VGM formula defining the relative water permeability k_{rw} as a function of water saturation, S_e:

$$k_{\mathrm{rw}} = \begin{cases} S_{\mathrm{e}}^{\frac{1}{2}}(1-y)^2 & S_{\mathrm{e}} < 1 \\ 1 & S_{\mathrm{e}} \geq 1 \end{cases} \quad \text{where} \quad y = \left(1 - S_{\mathrm{e}}^{\frac{1}{m}}\right)^m \tag{2.12}$$

Equation (2.9) may be rearranged to express capillary head as a function of S_{e}:

$$h_{\mathrm{c}} = h_{\mathrm{c},0}\left(\frac{y}{S_{\mathrm{e}}}\right)^{\frac{1}{mn}} \tag{2.13}$$

Thus, according to the VGM formalism, the relative water permeability and capillary head may be specified as unique functions of effective saturation, S_{e} (the fraction of macroporosity filled with flowing water) as expressed in Eqs. (2.12) and (2.13), respectively.

Essentially, there are two forces driving the distribution of water within the ore bed beneath a drip emitter: gravity and capillary pressure. Gravity pulls the water downward. Capillary pressure drives the water into fine pore spaces, and can thus pull the water sideways ("wicking action"). To illustrate this point, the water distribution within a column of ore 4 m tall and 1 m in diameter under a single drip emitter has been simulated. The applied solution flux is taken as 10 L m^{-2} h^{-1}. All other hydrological parameters are taken as measured for a sample of copper heap ore from the Mantoverde mine in Chile (Afewu and Dixon 2008). Results of the simulation are shown in Fig. 2.3 for five different values of the capillary head parameter, $h_{\mathrm{c},0}$: 0, 0.01, 0.02, 0.04, and 0.08 m.

A value of $h_{\mathrm{c},0} = 0$ implies no capillary force, and thus no wicking action. Under this condition, the water flows straight downward from the drip emitter to the bottom of the column with zero spreading. This "rivulet flow" is what might be observed in a column of uniform glass beads or coarse pebbles with no fines. Obviously, a heap leach cannot be operated under such a condition, as the ore not directly along the drip axis would never be wetted, let alone leached. However, it bears noting that the ore in these simulations never exceeds an effective saturation of about 50%. Thus, even with all of the water confined to a narrow flow channel along the drip axis, the ore is in no danger of becoming saturated. In the other simulations, with non-zero values of $h_{\mathrm{c},0}$, the maximum saturation directly below the drip emitter is only about 40%.

With even a very small capillary head parameter of $h_{\mathrm{c},0} = 0.01$ m, the degree of water spreading is already significant. However, water does not reach the 0.5-m radial boundary of the cylinder until $h_{\mathrm{c},0} = 0.04$ m (incidentally, $h_{\mathrm{c},0}$ was determined for the Mantoverde ore used in this study to be roughly 0.05 m). At a value of $h_{\mathrm{c},0} = 0.08$ m, the water is more or less evenly spread across the entire diameter of the column from about 2 m below the drip emitter. Hence, $h_{\mathrm{c},0}$ can be seen as the "master parameter" controlling the degree of water spreading in the ore, so knowing this parameter is essential if one wishes to predict the effects of drip emitter spacing.

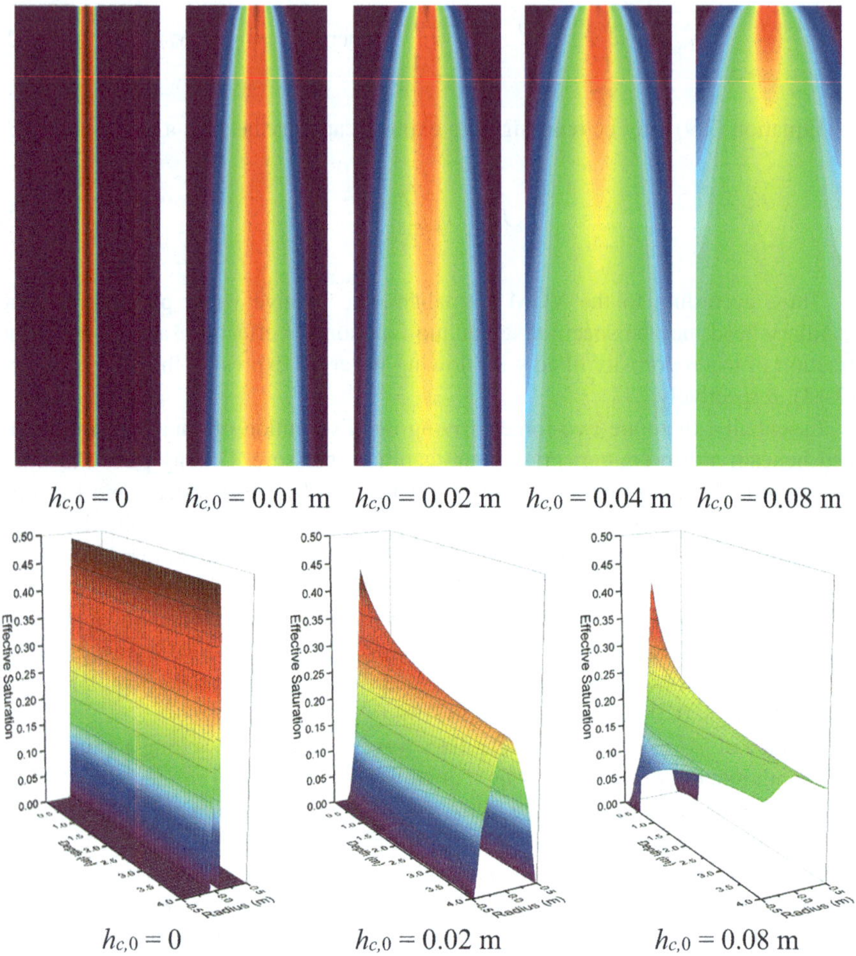

Fig. 2.3 The predicted effect of the capillary head parameter on the degree of solution spreading in a 4×1 m column under a single drip emitter at an applied flowrate of 10 L m^{-2} h^{-1}. The colours depict different levels of saturation

2.4.1.2 Solute Transport

Solute transport in unsaturated flow depends on the combined phenomena of advection and hydrodynamic dispersion, both of which depend on the prevailing flow field. Equation (2.14) represents the conservation of mass of a solute (or microorganism), including terms for solute motion, accumulation, and production or consumption by chemical (or biochemical) reactions within the solution:

$$-\nabla \cdot \mathbf{n}_{\mathrm{A}} = -\frac{\partial n_{\mathrm{A},z}}{\partial z} - \frac{1}{r}\frac{\partial (r n_{\mathrm{A},r})}{\partial r} = \frac{\partial}{\partial t}(\theta c_{\mathrm{A}}) - s_{\mathrm{A}} \tag{2.14}$$

where

$\mathbf{n}_A$ molar flux of solute A (a vector) (kmol m^{-2} s^{-1})
c_A molar concentration of solute A (kmol m^{-3} water)
s_A net molar production rate of solute A (kmol m^{-3} s^{-1})

Molar flux is given by Fick's law corrected for bulk flow, which, assuming dilute solution, maybe written thus:

$$\mathbf{n}_{\mathrm{A}} = \mathbf{v}_{\mathbf{w}} c_{\mathrm{A}} - \mathbf{D}_{\mathrm{A}} \cdot \nabla c_{\mathrm{A}} \tag{2.15}$$

where

$\mathbf{D}_{\mathrm{A}}$ dispersivity of solute A (a symmetric tensor) (m^3 water m^{-1} s^{-1})

Dispersion is a flow-driven process which tends to push solutes beyond where they would be carried by bulk flow alone. It behaves like molecular diffusion, but is typically several orders of magnitude more intense. The dispersivity in columns has been measured and found to be a quadratic function of solution flux, though dispersion has only a minor effect on solute transport. The variable flux of solution, as predicted by the solution flow model, carries solutes to different parts of the ore bed at different rates in a manner that looks much like diffusive flux but is actually advective flux at different flowrates. This is probably the most important outcome from this part of the model. Also, given that dispersion is not greatly important, molecular diffusion is insignificant as a source of solute motion and is therefore typically ignored in hydrological transport problems. In HeapSim2D it is not ignored, but rather assigned a common value for all solutes to simplify the mathematics of the problem.

2.4.1.3 Gas Flow

The 2D equation of gas mass continuity is:

$$-\nabla \cdot (\rho_{\mathrm{g}} \mathbf{v}_{\mathbf{g}}) = -\frac{\partial}{\partial z}\rho_{\mathrm{g}} v_{\mathrm{g,z}} - \frac{1}{r}\frac{\partial}{\partial r} r \rho_{\mathrm{g}} v_{\mathrm{g,r}} = \frac{\partial}{\partial t}(\theta_{\mathrm{s}} - \theta)\rho_{\mathrm{g}} - M_{\mathrm{g}} s_{\mathrm{g}} \tag{2.16}$$

where

$\mathbf{v}_{\mathbf{g}}$ gas volume flux (a vector) (m^3 gas m^{-2} s^{-1})
ρ_g gas mass density (kg m^{-3} gas)
$(\theta_s - \theta)$ volume occupied by gas (m^3 gas m^{-3})
s_g net rate of total gas production (kmol m^{-3} s^{-1})
M_g gas molecular mass (kg $kmol^{-1}$)

Gas mass flux is given by Darcy's law, which may be written in isotropic, unsaturated water–air media thus:

$$\rho_g \mathbf{v}_g = \frac{k k_{rg} \rho_g}{\mu_g} \left(\rho_g g \nabla z - \nabla p_g \right) = K_g k_{rg} \left(\rho_g \nabla z - \frac{1}{g} \nabla p_g \right) \tag{2.17}$$

$$\rho_g v_{g,z} = K_g k_{rg} \left(\rho_g - \frac{1}{g} \frac{\partial p_g}{\partial z} \right) \qquad \rho_g v_{g,r} = -K_g k_{rg} \frac{1}{g} \frac{\partial p_g}{\partial r} \tag{2.18}$$

where

k_{rg} relative gas permeability (–)
μ_g gas viscosity (kg/m/s)
K_g gas conductivity $= k \rho_g g / \mu_g$ (m s^{-1})
p_g gas pressure (Pa)

The ideal gas law gives:

$$p_g = \frac{n_g R T}{v_g} = \frac{\rho_g R T}{M_g} \quad \text{or} \quad \partial p_g = \frac{R T}{M_g} \partial \rho_g \tag{2.19}$$

where

n_g gas molar flux (kmol m^{-2} s^{-1})
R universal gas constant (8314.5 J kmol^{-1} K^{-1})
T temperature (K)

This allows Darcy's law to be restated completely in terms of gas density.

The relative gas permeability may be specified as a unique function of effective saturation:

$$k_{rg} = (1 - S_e)^2 \left(1 - S_e^{1+2/\lambda} \right) \tag{2.20}$$

where

λ the Brooks-Corey (BC) parameter (Brooks and Corey 1964)

Thus, the gas permeability decreases as the bed becomes more saturated, i.e., the gas must flow around the water, an effect which is quite dramatic even at relatively low saturation levels. The BC factor was estimated for the Mantoverde ore mentioned above to be about $\lambda = 3$. However, the gas permeability is not terribly sensitive to this value. The BC model implies that gas permeability decreases sharply as saturation increases from the fully drained condition. Hence, air pumped into the bottom of a heap will largely avoid the zones directly beneath drip emitters.

The output of the gas flow model is the gas density field, and these vary by only about ± 1 g m^{-3} across the entire depth of the heap (relative to an atmospheric air density of about 1200 g m^{-3}), corresponding to a pressure drop of only about

100 Pa, implying that the driving force for moving air across a heap is extremely small. Virtually all of the resistance to airflow into a heap resides within the air distribution network, and possibly the phreatic head of standing water within the heap overliner. Hence, the most important design question for heap aeration becomes how large to make the holes in the air distribution pipes.

2.4.1.4 Gas Species Transport

The 2D equation of solute continuity for a single gaseous species B is given thus:

$$-\nabla \cdot \mathbf{n}_{\mathrm{B}} = -\frac{\partial n_{\mathrm{B,z}}}{\partial z} - \frac{1}{r}\frac{\partial (r n_{\mathrm{B,r}})}{\partial r} = \frac{\partial}{\partial t}[(\theta_{\mathrm{s}} - \theta)\, c_{\mathrm{B}}] - s_{\mathrm{B}} \tag{2.21}$$

Molar flux is given by Fick's law, which may be written thus:

$$\mathbf{n}_{\mathrm{B}} = \mathbf{v}_{\mathbf{g}}\, c_{\mathrm{B}} - D_{\mathrm{B}} \nabla c_{\mathrm{B}} \tag{2.22a}$$

$$n_{\mathrm{B,z}} = v_{\mathrm{g,z}} c_{\mathrm{B}} - D_{\mathrm{B}} \frac{\partial c_{\mathrm{B}}}{\partial z} \qquad n_{\mathrm{B,r}} = v_{\mathrm{g,r}} c_{\mathrm{B}} - D_{\mathrm{B}} \frac{\partial c_{\mathrm{B}}}{\partial r} \tag{2.22b}$$

The gas diffusivity may be represented as:

$$D_{\mathrm{B}} = \varepsilon_{\mathrm{D,g}} D_{\mathrm{B,0}} \tag{2.23}$$

where $\varepsilon_{\mathrm{D,g}}$ is typically a power-law function of $(\theta_s - \theta)$ with a positive exponent, such that the effective gas diffusivity is approximately 10–20% of the true diffusivity.

For those gas species that are sparingly soluble in the aqueous phase (including O_2 and CO_2) the source term s_B in Eq. (2.21) will constitute a rate of gas–liquid mixing. For example, for O_2:

$$\begin{aligned} -s_{\mathrm{O_2(g)}} = r_{\mathrm{aO_2}} &= k_{\mathrm{l}}\, a_{\mathrm{v}} \left[K_{\mathrm{O_2}}(T) \cdot p_{\mathrm{O_2}} - c_{\mathrm{O_2(aq)}}\right] \\ &= k_{\mathrm{l}}\, a_{\mathrm{v}} \left[K_{\mathrm{O_2}}(T) \cdot RT \cdot c_{\mathrm{O_2(g)}} - c_{\mathrm{O_2(aq)}}\right] \end{aligned} \tag{2.24}$$

where

r_{a} gas absorption rate (kmol m^{-3} s^{-1})
k_{l} liquid-side gas–liquid mass transfer coefficient (m s^{-1})
a_{v} volume-specific gas–liquid interfacial area (m^2 m^{-3})
K Henrian gas solubility coefficient (kmol m^{-3} water Pa^{-1})

This same term appears with the opposite sign in Eq. (2.14) for those dissolved gas species, along with the appropriate reaction rate terms. Gas–liquid mixing rates can be important for heap leaching, especially at elevated temperatures, and it has been hypothesised that gas–liquid mixing rates dictate the absolute speed limit for heap leaching kinetics in hot heaps (Petersen and Dixon 2003).

2.4.1.5 Heat Flow

The final model element is heat flow, or more precisely, enthalpy transport. This model is essentially the 2D expansion of a 1D model presented by Dixon (2000). Interested readers are directed to that paper for a thorough discussion of heat flow in heaps.

The 2D equation of enthalpy continuity is:

$$-\nabla\cdot\mathbf{q} = -\frac{\partial q_z}{\partial z} - \frac{1}{r}\frac{\partial(rq_r)}{\partial r} = \frac{\partial}{\partial t}\left[\rho_b h_s + \theta\rho_w h_w + (\theta_s - \theta)\rho_g h_g\right] - Q \quad (2.25)$$

where

$\mathbf{q}$ enthalpy flux (a vector) (W m^{-2})
h specific enthalpy (of water w, gas g, and solids s) (J kg^{-1})
Q net specific production rate of heat (W m^{-3})

Enthalpy flux is given by Fourier's law, which may be written thus:

$$\mathbf{q} = \rho_w \mathbf{v_w} h_w + \rho_g \mathbf{v_g} h_g - \left[\theta k_w + (\theta_s - \theta)k_g + (1 - \theta_s)k_s\right]\nabla T \quad (2.26)$$

where

T temperature (K)
k thermal conductivity (W m^{-1} K^{-1})

One key feature of this model is that every point within the heap is assumed to be at 100% relative humidity. Since the vapour pressure of water is a strong function of temperature, the evaporation and condensation of water occur in equilibrium with the local temperature, and this causes significant movement of heat in the gas phase in the form of latent heat of vaporization. In warmer heaps at high aeration rates, this can become the dominant mode of the movement of heat through the heap. The heat generated by the oxidation of sulfide minerals increases the temperature of the solution as it flows downward, thus generating a vertical temperature gradient. Blowing air at a significant rate from the heap base carries water vapour from the warmer heap base upward, and if done with enough intensity, this can completely redistribute the heat, causing the upper half of the heap to become the warmest part. At very high aeration rates, excess heat can be blown out of the heap as water vapour and/or lead to localised drying within the heap.

The heat generation in a bioheap is actually easy to estimate. A survey of the heat of reaction of the oxidation of sulfide minerals and ferrous ions by dissolved oxygen has revealed that the heat of reaction is very nearly constant relative to the moles of oxygen reduced. The general heat of oxidation is about 400 MJ $kmol^{-1}$ O_2 consumed, or 100 MJ $kmol^{-1}$ of electrons transferred. This value is valid to within about ±3% for nearly every oxidation reaction of interest within bioheaps. It is important to note that the ferric oxidation of sulfide minerals involves very little exchange of heat. Only the reduction of oxygen generates significant heat. Hence, an accounting of all relevant ions in the heap effluent can close the "electron balance" and thus be converted directly into the overall heat generation (Petersen and Dixon 2002).

Another important aspect of the heat flow model is the heap surface boundary condition. HeapSim2D accounts for all environmental factors on the transfer of heat to and from the heap surface, including diurnal temperature cycles, solar irradiation, grey body radiation to the night sky, evapotranspiration, and convection due to wind. It can also simulate the effects of installing a heat barrier (such as plastic film) on the heap surface.

2.4.1.6 Reaction Network Modelling

In the reaction network model, all of the source terms s_A in Eq. (2.14) must be specified. Given the number of chemical reactions involved, modelling even a relatively simple scenario can be fairly complicated. Each solute, gas species, and reactive solid must be represented by a mole balance, and each mole balance is a combination of chemical reaction rates. For example, Eqs. (2.1) and (2.2) describe the 2-stage (bio)leaching of chalcocite. Each reaction involved in mixed copper oxide and sulfide heap leaching is represented with a rate equation, and reaction rates are assembled into mole balances to generate the component rates required for the source terms of Eq. (2.14). Finally, reaction heat is estimated from the rate of oxygen absorption (although it may be defined more precisely from each reaction individually if desired):

$$\text{Heat (estimated)} \quad Q = 4 \times 10^8 \cdot r_{\mathrm{aO_2}} \tag{2.27}$$

Each of the rate terms r in the mole balances of the different mineral dissolution/precipitation reactions represents a kinetic rate law, which must be determined carefully for each reaction. In the case of the leaching reactions, these rates must be measured for each specific case.

The leaching rate of any mineral generally depends on three factors: particle topology, thermal factors, and chemical factors. Particle topology includes the degree of mineral liberation—the degree to which the mineral of interest is liberated from the surrounding gangue and exposed to solution. It also includes the mode of particle leaching—how the size and shape of particles change as they are leached, and whether they are exposed or covered by a porous product layer. The thermal effect is almost always represented by the Arrhenius rate law. Chemical factors include the effect of oxidants and reductants, acids, dissolved gases, and catalysts (mostly chemical reagents). These three factors are multiplicative. Hence, the leaching rate law for any mineral can be written as:

$$\frac{dX}{dt} = k(T)f(C)g(1 - X) \tag{2.28}$$

where

X	fraction of mineral reacted (conversion)
t	time
$k(T)$	a rate constant which is a function of temperature
$f(C)$	a "chemical function" of solution composition
$g(1 - X)$	a "topological function" of fraction unreacted

Any determination of leaching kinetics usually begins with the topological function. This often takes some very familiar forms, which go by names such as "shrinking sphere" or "shrinking core". While these "stock" functions are often appropriate for tank leaching of fine particles, they are generally inappropriate for the leaching of large, imperfectly liberated mineral particles in heaps and columns. For this, the "general" model is typically more appropriate:

$$\frac{dX}{dt} = \frac{(1 - X)^{\varphi}}{\tau} \tag{2.29}$$

where

φ	an empirical exponent
τ	a "timescale" constant with units of time

This model has been shown capable of modeling the leaching of a distribution of particle sizes (Dixon and Hendrix 1993) and can often be fit to column leaching kinetics. If φ is less than 1, then τ represents the time for complete leaching. If $\varphi = 2/3$, then this model defaults to the shrinking sphere model. Like that model, the general model usually applies to situations that are under chemical reaction control. Hence, the timescale generally takes the following form:

$$\tau = \frac{D_0^m}{k(T)f(C)} \tag{2.30}$$

where

D_0	particle size
m	an empirical constant which is typically close to 1

The thermal function is almost always taken as the Arrhenius rate law, which can be written as:

$$k(T) = k(T_0) \exp\left[-\frac{E_a}{R}\left(\frac{1}{T} - \frac{1}{T_0}\right)\right] \tag{2.31}$$

where

E_a the Arrhenius activation energy (MJ kmol^{-1})
T_0 a reference temperature (K)

The temperature T_0 is typically where most of the experiments are conducted.

The chemical function is the most interesting of these. In ferric iron leaching situations, the chemical function almost always falls into one of three types as predicted by electrochemical theory:

$$\text{Type I} \quad f(C) = \left[Fe^{3+}\right]^{1/2}$$

$$\text{Type II} \quad f(C) = \left(\frac{\left[Fe^{3+}\right]}{\left[Fe^{2+}\right]}\right)^{1/2}$$

$$\text{Type III} \quad f(C) = \left[Fe^{3+}\right]$$

Oxidation of sulfide minerals is electrochemical in nature, such that the overall reaction is a combination of anodic (oxidation) and cathodic (reduction) reactions on the mineral surface. In ferric leaching, the reduction reaction is the reduction of ferric to ferrous iron, and the oxidation reaction is the breakdown of the mineral itself. The mineral acts as a semiconductor through which the two electrochemical half-reactions exchange electrons. Type I implies that both the anodic and cathodic reactions are slow, and therefore, rate-limiting. Type II implies that the anodic reaction (mineral breakdown) is rate-limiting but that the cathodic reaction (ferric reduction) is virtually instantaneous. Type III implies that the anodic reaction is rapid such that the mass transfer of ferric to the mineral surface is the rate-limiting step.

Pure instances of these three types are rarely observed, and "hybrid" situations between two of the three types are observed. For example, if the reaction is essentially Type II, but ferric iron reduction is not instantaneous, then Hybrid Type I–II kinetics are observed:

$$f(C) = \left(\frac{\left[Fe^{3+}\right]}{K + \left[Fe^{2+}\right]}\right)^{1/2} \tag{2.32}$$

This model can default to Type I or Type II depending on the value of K. Also, the exponent is rarely exactly ½, but must be determined empirically. Alternatively, there could be a rapid anodic reaction largely under mass transfer control, but ferric iron reduction is also relatively slow. In this case, Hybrid Type I–III kinetics might be operating:

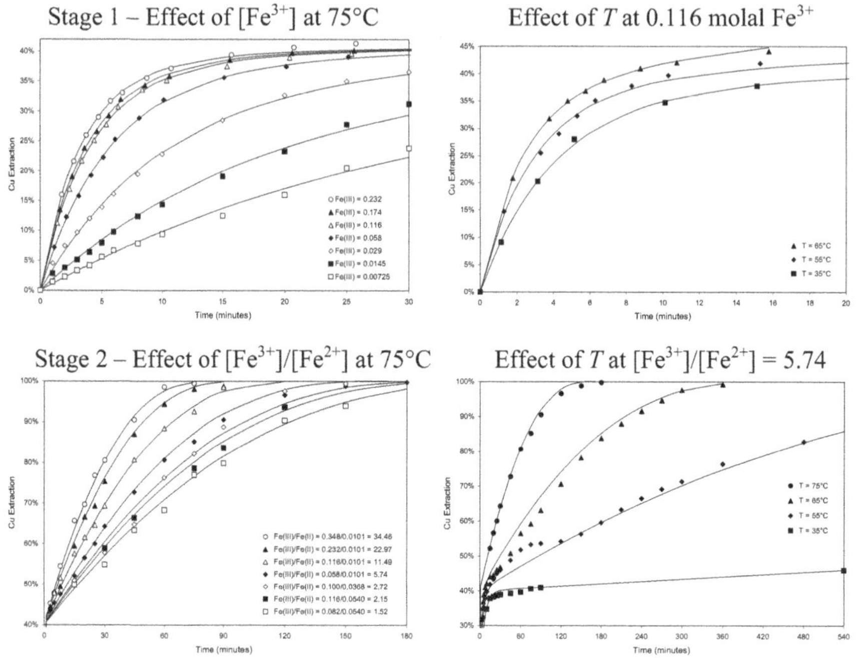

Fig. 2.4 Chalcocite oxidation by ferric sulfate solutions, depicting stage 1 (top) and stage 2 (bottom)

$$f(C) = \sqrt{K^2 + [Fe^{3+}]} - K \tag{2.33}$$

This model can default to Type I or Type III depending on the value of K. It also bears noting that this hybrid is closely simulated by the Type I model, but with an exponent somewhere between 0.5 and 1.0, depending on the value of K.

To demonstrate these principles, the kinetics of chalcocite dissolution can be considered. Chalcocite oxidation proceeds in two stages, as described previously (Eqs. (2.1) (stage 1) and (2.2) (stage 2)). Stage 1 is very rapid, dependent on particle size, and releases roughly 40% of the copper from the mineral, leaving behind an altered phase, *blaubleibender* covellite. This reaction is only limited by the supply of ferric iron and, in a stirred-tank leaching system, it will typically go to completion in the first reactor vessel, if not in the repulp slurry mix tank. Stage 2 is much slower, largely independent of particle size, and produces elemental sulfur.

Figure 2.4 shows the results of leaching Stage 1 chalcocite, first at 75 °C and over a range of ferric concentrations, and then at 0.116 molal ferric concentrations over a range of temperatures. The kinetics are strongly dependent on the ferric concentration, but only weakly dependent on temperature. This, and the extreme rapidity of the leach, implies mass transfer control. The rate law which was derived to fit this is:

$$\frac{dX_{Cu_2S}}{dt} = \frac{(1 - X_{Cu_2S})^{1.153}}{\tau_{Cu_2S}} \quad \text{where}$$

$$\frac{1}{\tau_{Cu_2S}(\text{min})} = \frac{200.4}{D_O(\mu m)} \exp\left(-\frac{18030}{8.314}\left(\frac{1}{T} - \frac{1}{308}\right)\right)\left(\sqrt{(0.090)^2 + c_{Fe^{3+}}} - 0.090\right) \tag{2.34}$$

Even though the reaction appears to be strictly controlled by the diffusion rate of ferric iron, the Type I–III hybrid was used. A value of $K = 0.090$ implies mixed control of mass transfer and ferric reduction kinetics, suggesting that the chalcocite surface (of stage 1) is not particularly favourable for ferric iron reduction. If the Type I–III hybrid is replaced by a simple power-law function with respect to ferric iron concentration (like Type I), then the exponent would be equal to about 0.68. As this is closer to 0.5 than 1.0, it seems fair to conclude that this leaching reaction is closer to Type I than Type III, and that the reduction kinetics of ferric iron is more rate-limiting than its mass transfer.

Figure 2.4 also shows the results from stage 2 leaching of chalcocite, at 75 °C and over a range of solution potentials (represented by various ferric/ferrous concentration ratios), and also at a ferric-ferrous ratio of 5.74 over a range of temperatures. The kinetics are modestly dependent on the ferric/ferrous ratio, but strongly dependent on temperature, implying chemical reaction control. The rate law which was derived to fit these data (and which was used to generate the solid curves in the plots) is:

$$\frac{dX_{Cu_{1.2}S}}{dt(\text{min})} = \frac{(1 - X_{Cu_{1.2}S})^{0.598}}{\tau_{Cu_{1.2}S}} \quad \text{where}$$

$$\frac{1}{\tau_{Cu_{1.2}S}(\text{min})} = 0.00921 \exp\left(-\frac{99230}{8.314}\left(\frac{1}{T} - \frac{1}{348}\right)\right)\left(\frac{c_{Fe^{3+}}}{0.00482 + c_{Fe^{2+}}}\right)^{0.376} \tag{2.35}$$

A Type I–II hybrid provides the best fit to the data implying that, while the anodic (mineral dissolution) reaction is slow (favouring Type II), ferric iron reduction kinetics are also relatively slow (favouring Type I), suggesting that *blaubleibender* covellite is also not a very favourable surface for ferric reduction.

The kinetics of chalcocite oxidation shown above were measured for finely ground pure mineral particles in a stirred tank. These results are only partially valid for heap leaching. The thermal and chemical functions should remain the same, but the topological function will change. Hence, the rate constants (200.4 and 0.00921 in Eqs. (2.34) and (2.35), respectively) and the topological exponents (1.153 and 0.598 in those equations) need to be determined independently from column tests, while all the other parameters should transfer directly.

2.4.1.7 Microbial Kinetics

Any model of heap bioleaching requires a model of microbial growth and activity. While there is considerable controversy over the best way to represent growth kinetics in a heap leaching context, at a minimum, the growth rate expression needs to represent the effects of microbial population, temperature, and important limiting and inhibiting factors (Ojumu et al. 2006).

In HeapSim2D, the growth rate of each microbial strain is described by Michaelis–Menten (or Monod) kinetics, modified with a collection of terms, as follows:

$$\frac{\mathrm{d}Y}{\mathrm{d}t} = k_{\mathrm{g}} Y \{ f_{\mathrm{g}}(T) [\Pi (1 + k_{\mathrm{e}}) - k_{\mathrm{e}}] - f_{\mathrm{d}}(T) \} \tag{2.36}$$

where

$$\Pi = \prod \left(\frac{C_{\mathrm{L}}}{K_{\mathrm{L}} + C_{\mathrm{L}}} \right) \cdot \prod \left(\frac{K_{\mathrm{I}}}{K_{\mathrm{I}} + C_{\mathrm{I}}} \right)$$

and where

Y	the population of microbial cells
k_{g}	the growth rate constant
$f_{\mathrm{g}}(T)$	a growth rate temperature function (e.g., the Ratkowski function)
$f_{\mathrm{d}}(T)$	a death rate temperature function
k_{e}	endogenous decay rate constant
K_{L}, C_{L}	Monod constants and concentrations of growth-limiting factors
K_{I}, C_{I}	Monod constants and concentrations of growth-inhibiting factors

Typically, rather than modelling actual microbial species, we assume that there is one dominant iron oxidiser and one dominant sulfur oxidiser within each relevant range of temperatures—mesophilic, moderately thermophilic, and extremely thermophilic. Hence, there may be as many as six "virtual" microbial species accounted for, depending on the situation.

With the microbial concentrations, the rates of ferrous and sulfur oxidations can be calculated assuming microbial yields as follows:

$$\text{Ferrous oxidation}: 4r_{\mathrm{Fe}} = \sum_{i=1}^{I} Y_{\mathrm{Fe},i} f_{\mathrm{g,Fe},i}(T) \Pi_{\mathrm{Fe},i} \left(\frac{k_{\mathrm{g,Fe},i}}{y_{\mathrm{g,Fe},i}} + k_{\mathrm{m,Fe},i} \right) \tag{2.37}$$

where

$$\Pi_{\mathrm{Fe},i} = \frac{c_{\mathrm{O}_2}}{K_{\mathrm{O,Fe},i} + c_{\mathrm{O}_2}} \frac{c_{\mathrm{Fe}^{2+}}}{K_{\mathrm{Fe},i} + c_{\mathrm{Fe}^{2+}}} \frac{K_{\mathrm{Y,Fe},i}}{K_{\mathrm{Y,Fe},i} + Y_{\mathrm{Fe},i}} \left[1 - \exp\left(-\frac{c_{\mathrm{H_2SO_4}}}{K_{\mathrm{H,Fe},i}}\right)\right]$$

$$\text{Sulfur Oxidation} : 2r_{\mathrm{S}} = \sum_{j=1}^{J} Y_{\mathrm{S},j} f_{\mathrm{g,S},j}(T) \Pi_{\mathrm{S},j} \left(\frac{k_{\mathrm{g,S},j}}{y_{\mathrm{g,S},j}} + k_{\mathrm{m,S},j}\right) \tag{2.38}$$

where

$$\Pi_{\mathrm{S},j} = \frac{c_{\mathrm{O}_2}}{K_{\mathrm{O,S},j} + c_{\mathrm{O}_2}} \frac{c_{\mathrm{S}}}{K_{\mathrm{S},j} + c_{\mathrm{S}}} \frac{K_{\mathrm{Y,S},j}}{K_{\mathrm{Y,S},j} + Y_{\mathrm{S},j}}$$

where

y_g microbial yield
k_{m} maintenance rate constant
i,j indices of iron- and sulfur-oxidising strains, respectively

and where the Monod terms for ferrous oxidation include oxygen and ferrous limitations, and population and acid inhibitions, and the Monod terms for sulfur oxidation include oxygen and sulfur limitations and population inhibition. Measuring all of these various Monod constants is a significant challenge, and is not often possible within the context of a heap leach modelling study. Also, it is often important to include terms for dissolved metals and other solutes that can seriously inhibit microbial growth if allowed to build up in heap leaching solutions. Microbial kinetics in heaps requires much further study.

2.4.2 *Modelling Case Study*

HeapSim2D has been used to simulate chalcocite heap bioleaching at the Quebrada Blanca (QB) mine located in northern Chile (Liu and Hashemzadeh 2016). The objective of the study was to simulate chalcocite leaching and identify the key parameters that control heap temperature. The model was calibrated by matching heap performance indicators such as copper recovery, copper and iron concentration, acid consumption, pH, and E_h to site data from four of the 23 heaps. After that, the model was validated by predicting site data from two other heaps.

The results of one validation simulation are shown in Fig. 2.5. As shown, all of the performance indicators were closely simulated by HeapSim2D. On this basis, it was assumed that the collection of model parameters found to achieve this validation represented the heap with reasonable accuracy, and could therefore be used to make predictions about heap performance when certain operating parameters were changed. HeapSim2D includes an "event manager" which allows the user to program changes in flowrates, aeration rates, and solution temperatures and compositions at predetermined times during the simulation, which is how the model is able to

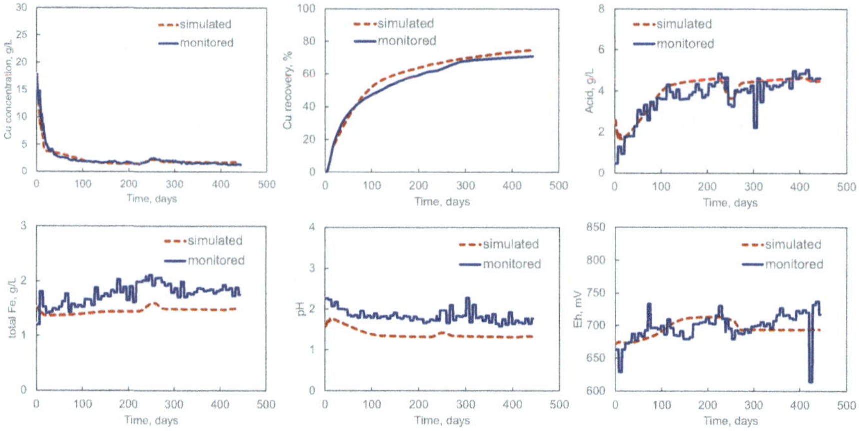

Fig. 2.5 Results of a validation simulation of a heap at Quebrada Blanca using HeapSim2D

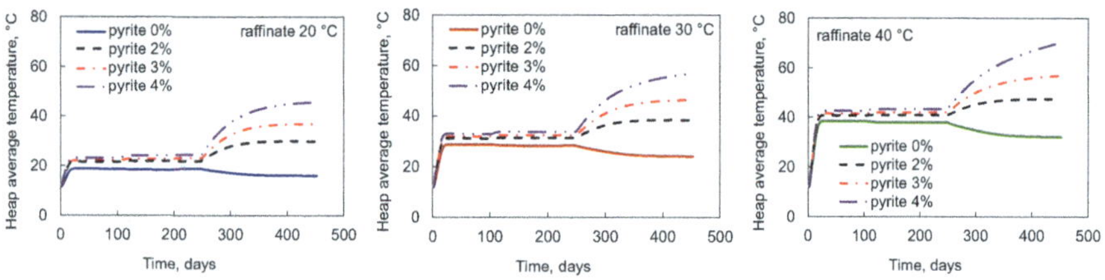

Fig. 2.6 Simulations of average heap temperature showing the effects of raffinate temperature and the addition of finely divided pyrite

capture sudden changes in the monitored data. In this particular simulation, the heap is 8 m tall, with a drip emitter spacing of 55 cm. The heap was aerated at a rate of 0.23 $m^3\ m^{-2}\ s^{-1}$ for the first 250 days, after which aeration ceased. The heap was irrigated at a rate of 12 $L\ m^{-2}\ h^{-1}$ for the first 100 days, 9 $L\ m^{-2}\ h^{-1}$ from day 101 to day 245, and 1.8 $L\ m^{-2}\ h^{-1}$ from day 246 to day 450, at which point the simulation ended.

With the proper parameters identified, HeapSim2D was used to predict the temperature profile in the heap assuming that finely divided pyrite flotation tails were blended with agglomerated ore. Heap temperature was sensitive to the raffinate temperature and the extent of reaction of the pyrite. When the raffinate flow rate was rapid (9 $L\ m^{-2}\ h^{-1}$), heap temperature was controlled by the raffinate temperature; when the raffinate flow rate was relatively slow (1.8 $L\ m^{-2}\ h^{-1}$), heat removal by the PLS was insignificant and the heat generated by pyrite reaction accumulated and led to a sharp increase in the heap temperature.

Typical simulated temperature profiles are shown in Fig. 2.6. When the flowrate was high (during the first 245 days), the heap reached the raffinate temperature, and pyrite oxidation had little effect. Once the flowrate was decreased (after day 245), the temperature increased in proportion with the mass fraction of pyrite added. While this analysis is based on many assumptions about the oxidation rate of the pyrite that would have to be confirmed in practice, it suggests that adding fine pyrite to a bioheap could be a very effective way to increase heap temperature, which could be instrumental for recovering copper from primary sulfides.

Another promising use of HeapSim2D is the simulation of catalytic heap bioleaching. A new technology is currently being developed for recovering copper from low-grade chalcopyrite ores using soluble catalysts with thiocarbonyl functional groups (Dixon et al. 2019). HeapSim2D is being used to simulate how these organic catalyst compounds are distributed within the heap, and to model their adsorption and consumption characteristics, as well as their effect on local leaching kinetics. This will provide critical guidance for determining those operating parameters that give optimal economic performance of the heap.

References

Afewu KI, Dixon DG (2008) Calibrating a 3D axisymmetric water and solute transport model for heap leaching. In: Hydrometallurgy 2008, Proceedings of the Sixth International Symposium, SME, pp 1098–1109

Bennett CR, McBride D, Cross M, Gebhardt JE (2012) A comprehensive model for copper sulphide heap leaching, part 1: basic formulation and validation through column test simulation. Hydrometallurgy 127–128:150–161. https://doi.org/10.1016/j.hydromet.2012.08.004

Brooks RH, Corey AT (1964) Hydraulic properties of porous media. Hydrology papers (3), Colorado State University

Dixon DG (2000) Analysis of heat conservation in copper sulfide heap leaching. Hydrometallurgy 58:27–41. https://doi.org/10.1016/S0304-386X(00)00119-5

Dixon DG (2003) Heap leach modeling – the current state of the art. In: Hydrometallurgy 2003, Proceedings of the 5th international symposium, vol 1, TMS, pp 289–314

Dixon DG, Hendrix JL (1993) Theoretical basis for variable order assumption in the kinetics of leaching of discrete grains. AICHE J 39:904–907. https://doi.org/10.1002/aic.690390520

Dixon DG, Olvera-Olmedo O, Asselin E, Ghahremaninezhad A, Ren Z, Mora N (2019) Process for leaching metal sulfides with reagents having thiocarbonyl functional groups. Canadian Patent 2973612, issued 15 January 2019

Figueroa A, Silva P, Arriagada F, Peralta R (2015) Copper production model for San Francisco dump leaching operation. In: Proceedings of Hydroprocess 2015, Gecamin, Santiago, Chile, p 470–479

Kordosky GA (2002) Copper recovery using leach/solvent extraction/electrowinning technology: forty years of innovation, 2.2 million tonnes of copper annually. J South Afr Inst Min Metall 102:445–450. https://hdl.handle.net/10520/AJA0038223X_2713

Liu W, Hashemzadeh M (2016) Modelling copper recovery, total iron concentration and acid balance in chalcocite heap bioleaching. In: Proceedings of Heap Leach Mining Solutions 2016, InfoMine, Lima, Peru, pp 6–16

Lupo JF (2010) Liner system design for heap leach pads. Geotext Geomembr 28:163–173. https://doi.org/10.1016/j.geotexmem.2009.10.006

Marsden JO, Botz MM (2017) Heap leach modeling – a review of approaches to metal production forecasting. Miner Metall Proc 34:53–64. https://doi.org/10.19150/mmp.7505

Ojumu TV, Petersen J, Searby GE, Hansford GS (2006) A review of rate equations proposed for microbial ferrous-iron oxidation with a view to application to heap bioleaching. Hydrometallurgy 83:21–28. https://doi.org/10.1016/j.hydromet.2006.03.033

Petersen J, Dixon DG (2002) Thermophilic heap leaching of a chalcopyrite concentrate. Miner Eng 15:777–785. https://doi.org/10.1016/S0892-6875(02)00092.4

Petersen J, Dixon DG (2003) A modelling study of the dynamics of sulphide heap leach processes with a view to improved and novel applications. In: Proceedings of the 22nd International Mineral Processing Congress, IMPC XXII, vol 3, SAIMM, pp 1231–1239

Petersen J, Dixon DG (2007) Modeling and optimization of heap bioleaching processes. In: Rawlings DE, Johnson DB (eds) Biomining. Springer, Heidelberg, pp 153–176

Reyes A, Garma P, Parra D (2015) 3-D slope stability analysis of valley-fill heap leach pads. In: Proceedings of Heap Leach Solutions 2015, InfoMine, Reno, pp 161–178

Saldaña M, González J, Jeldres RI, Villegas Á, Castillo J, Quezada G, Toro N (2019) A stochastic model approach for copper heap leaching through Bayesian networks. Metals 9:1–15. https://doi.org/10.3390/met9111198

Schlitt WJ (2006) History of forced aeration in copper sulfide leaching. Miner Metall Proc 23:57–66. https://doi.org/10.1007/bf03403337

Turner GA (1958) The flow-structure in packed beds: a theoretical investigation utilizing frequency response. Chem Eng Sci 7:156–165. https://doi.org/10.1016/0009-2509(58)80022-6

Chapter 3
Engineering Designs and Challenges of Stirred Tank Systems

David William Dew, Gary Vernon Rorke, and Anne-Gwénaëlle Guezennec

Abstract Stirred tank reactor (STR) design is critical to the successful performance of a commercial tank bioleaching plant treating mineral sulfides. The reactor duty for a given feed composition and tonnage feedrate is defined by the bioleaching kinetics, oxygen demand, and heat generation, while the gas dispersion and oxygen transfer requirement drive STR design. The chapter briefly discusses key elements of design, including reactor configuration and geometry, importance of limiting fluid shear, agitator selection, and materials of construction. A model correlation for oxygen mass transfer as a function of agitator power and gas flow rate per reactor volume is presented and compared to commercial-scale process performance results for a thermophile STR and commercial-scale mesophilic bioleach reactors. The final section of the chapter presents details of a commercial thermophile bioleach plant and concludes with a description of a new development in the field of STRs—a low-energy floating agitator system for pond bioleaching.

Keywords Commercial stirred tank bioleaching · Reactor design · High-temperature bioleaching · Thermophiles · Oxygen mass transfer · Floating agitators · Pond bioleaching

D. W. Dew
University of Exeter, Cornwall, UK
e-mail: d.w.dew@exeter.ac.uk

G. V. Rorke
Prudentia Process Consulting, Spring Hill, QLD, Australia
e-mail: grorke@prudentiapc.com.au

A.-G. Guezennec (✉)
BRGM, Orléans, France
e-mail: a.guezennec@brgm.fr

D. B. Johnson et al. (eds.), *Biomining Technologies*,
https://doi.org/10.1007/978-3-031-05382-5_3

3.1 Introduction

This chapter discusses the engineering designs of stirred tank bioleach reactors for leaching sulfide minerals such as concentrates, or other similar sulfide ores or waste materials. The bioreactor design for the BIOX® process is described elsewhere (Chap. 4).

Essential elements of design are presented with a focus on gas dispersion and oxygen transfer including:

- Influence of temperature on bioleach performance and bioreactor design.
- Reactor configuration.
- Oxygen transfer and agitator power requirements.
- Comments on materials of construction.

3.2 The Bioleach Process: An Engineering Design Perspective

3.2.1 *Sulfide Mineral Oxidation*

The bioleaching of sulfide minerals with acidophilic bacteria and archaea is driven by Fe(III) as the oxidant and the regeneration of Fe(III) by microbial oxidation of Fe (II) using oxygen as the electron acceptor. In parallel, reduced sulfur species, and reaction products such as elemental sulfur, are oxidised to sulfate by microbial action, which generates sulfuric acid. Typical bioleaching cultures will consist of a mixed community of prokaryotes that oxidise either iron or sulfur, strains capable of oxidising both, and others (heterotrophs) that oxidise neither (see Chaps. 5 and 7). The microbial oxidation of Fe(II) is several orders of magnitude faster than chemical oxidation at low pH, allowing sustained high rates of mineral oxidation in the bioleaching process at atmospheric pressure.

The oxygen demand, heat of reaction, and acid demand, calculated according to the reaction stoichiometry for a range of sulfide minerals and for potassium jarosite precipitation, common to both mesophile and thermophile bioleaching operations, are summarised in Table 3.1. Heats of reaction were calculated using published heats of formation according to the overall reactions shown in the table (reference NIST[1] and NASA[2] data; Johnson and Steele 1981; Bard et al. 1985; Barton 1969).

The following points should be noted from the data presented in Table 3.1:

[1] https://webbook.nist.gov/chemistry/

[2] https://www1.grc.nasa.gov/research-and-engineering/ceaweb/topicshome/

Table 3.1 Heats of reaction, oxygen, and acid demand

Reaction	Heat of reaction MJ mol^{-1} mineral	MJ kg^{-1} mineral	MJ kg^{-1} sulfide	Oxygen (O_2) demand kg kg^{-1} mineral	Acid (H_2SO_4) demand kg kg^{-1} mineral
$FeS + 2.25\ O_2 + 0.5\ H_2SO_4 \rightarrow 0.5\ Fe_2(SO_4)_3 + 0.5\ H_2O$	−0.999	−11.364	−31.155	0.819	0.557
$FeS + 0.75\ O_2 + 1.5\ H_2SO_4 \rightarrow 0.5\ Fe_2(SO_4)_3 + 1.5\ H_2O + 2S^0$	−0.376	−4.274	−11.719	0.273	1.672
$FeAsS + 3.5\ O_2 + 0.5\ H_2SO_4 + H_2O \rightarrow H_3AsO_4 + 0.5\ Fe_2(SO_4)_3$	−1.472	−9.037	−45.890	0.688	0.301
$FeS_2 + 3.75\ O_2 + 0.5\ H_2O \rightarrow 0.5\ Fe_2(SO_4)_3 + 0.5\ H_2SO_4$	−1.553	−12.940	−24.209	1.000	−0.408
$CuFeS_2 + 4.25\ O_2 + 0.5\ H_2SO_4 \rightarrow CuSO_4 + 0.5\ Fe_2(SO_4)_3 + 0.5\ H_2O$	−1.771	−9.651	−27.618	0.741	0.267
$Ni_{4.5}Fe_{4.5}S_8 + 3.25\ H_2SO_4 + 17.625\ O_2 \rightarrow 4.5\ NiSO_4 + 2.25\ Fe_2(SO_4)_3 + 3.25\ H_2O$	−7.856	−10.176	−30.622	0.731	0.413
$1.5\ Fe_2(SO_4)_3 + 0.5\ K_2SO_4 + 6\ H_2O \rightarrow KFe_3(SO_4)_2(OH)_6 + 3\ H_2SO_4$	0.156	0.311	–	–	−0.587
$FeSO_4 + 0.25\ O_2 + 0.5\ H_2SO_4 \rightarrow 0.5\ Fe_2(SO_4)_3 + 0.5\ H_2O$	−0.103	−0.675	–	0.053[a]	0.323
$S^0 + 1.5\ O_2 + H_2O \rightarrow H_2SO_4$	−0.623	−19.437	–	1.497[b]	−3.1

[a] kg kg^{-1} $FeSO_4$
[b] kg kg^{-1} S^o

(a) Complete sulfide mineral oxidation is strongly exothermic, equivalent to −12 to −31 MJ kg^{-1} sulfide and − 46 MJ kg^{-1} sulfide in the case of complete oxidation of arsenopyrite.
(b) Oxygen demand for complete sulfide mineral oxidation is within the range 0.7–1.0 kg O_2 kg^{-1} mineral. A typical oxygen demand for complete oxidation of mixed iron sulfide or base metal sulfide concentrates is within the range 1.9–2.3 kg O_2 kg^{-1} sulfide.
(c) Biooxidation of pyrite generates 0.41 kg H_2SO_4 kg^{-1} mineral oxidised. The precipitation of K-jarosite results in generation of acid, 0.59 kg H_2SO_4 kg^{-1}.
(d) Partial oxidation of sulfide to sulfur reduces the oxygen demand substantially, for example, partial oxidation of pyrrhotite where the oxygen demand is reduced to 0.273 kg O_2 kg^{-1} FeS compared to 0.819 kg O_2 kg^{-1} FeS for complete oxidation to sulfate. This results in a corresponding decrease in the heat of reaction, from −11.36 MJ kg^{-1} FeS for complete oxidation to −4.27 MJ kg^{-1} FeS for partial oxidation of sulfide to elemental sulfur.

3.2.2 Rate of Mineral Leaching and Temperature

The rate of mineral leaching increases with temperature, but in the case of bioleaching, stirred tank bioreactors need to operate within a temperature range optimal for the bioleaching performance of the selected microbial consortium, though the component strains will likely have different temperature optima.

The optimal temperature of a microbial consortium is usually a compromise between the optimal temperatures of its members, and these can evolve with time by adaptation of the strains and changes in the community composition. For example, the consortium used in the KCC (Kasese Cobalt Company Ltd.) bioleaching plant was mainly composed of strains of *Leptospirillum ferriphilum* (optimal temperature: expected to be in the range 35–40 °C) and *Acidithiobacillus caldus* (optimal temperature: 45 °C). It contained a minor proportion of *Sulfobacillus* spp. and *Ferroplasma* spp. The first study carried out by Battaglia (1994) showed that the consortium grew optimally between 35 and 37 °C (the optimal temperature of *L. ferriphilum*, the dominant microorganism of the consortium) and the consortium's activity was completely inhibited at 40 °C. Further experiments were conducted by D'Hugues (1996) to adapt the consortium to higher temperatures (up to 48 °C): the cobalt dissolution rate obtained with the adapted culture remained rather constant between 35 and 46 °C but the cell concentration decreased significantly when the temperature was increased. Above 46 °C, rates of mineral dissolution and cell numbers both decreased drastically.

From these results, 42 °C was selected as the operating temperature. Given the high sulfide content of Kasese tailings, this temperature enabled a significant reduction in costs associated with the cooling system compared to an operating temperature of 35 °C, while maintaining a high extraction rate of cobalt and limiting the risk of microorganism washout. The KCC example shows that the choice of the

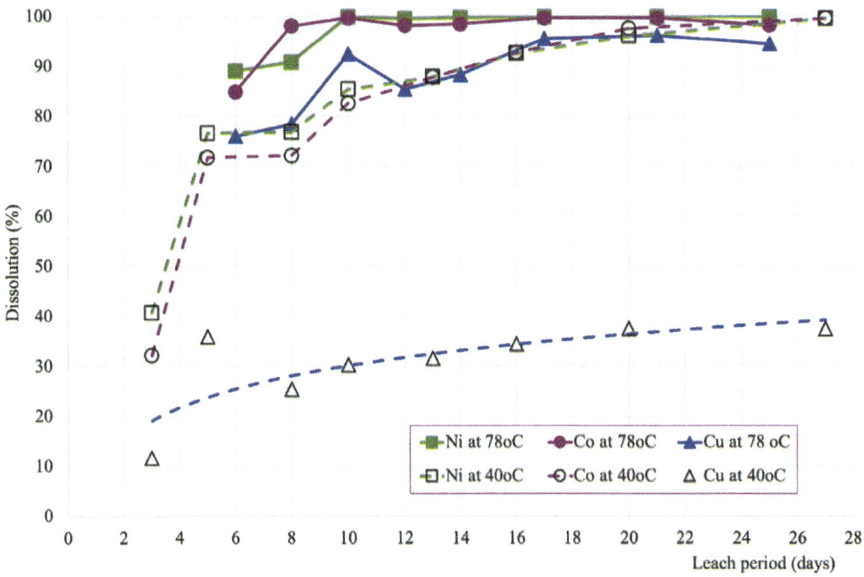

Fig. 3.1 Batch laboratory-scale STR bioleaching of Ni/Cu/Co mineral concentrate at 78 and 40 °C (after Holmes 2000)

operating temperature is guided by different constraints and is usually a compromise between the optimal temperatures of the selected microorganisms, the reaction kinetics, metal yields, and operating costs. Beyond these considerations, the fact that increasing the operating temperature might also promote higher jarosite precipitation needs to be considered, and this can significantly influence the downstream processing of the bioleached pulp and also limit ferric iron (oxidant) concentrations in solution, particularly for thermophile bioleaching at temperatures above 70 °C where hydronium jarosite will precipitate.

Commercial plants typically operate in the same range of conditions, with mesophilic bacterial cultures at temperatures in the range of 40–45 °C. Bioleaching with thermophile cultures has been used in the MesoTherm™ process (Chap. 4).

BHP Billiton demonstrated the benefits of bioleaching at higher temperatures (60–70 °C) using the type strain of *Sulfolobus* (now *Sulfuracidifex*) *metallicus* and more extremely thermophilic microbial strains operating at 75–80 °C (Dew et al. 2000), for treatment of Ni/Co/Cu concentrates and chalcopyrite concentrates. Mineral leaching in high-temperature bioreactors was reviewed by Norris et al. (2013), detailing the performance of a variety of thermoacidophilic archaea and describing a new culture (ICHT1) capable of operating at temperatures of 78 °C. The ICHT1 culture was used in the BioCOP™ commercial demonstration plant described later in this chapter. The improved performance of bioleaching at higher temperatures is illustrated in Fig. 3.1, showing results for batch laboratory-scale STR bioleaching tests with a Ni/Co/Cu sulfide concentrate (pyrrhotite 20%, pentlandite 52%, chalcopyrite 12%, pyrite 2.6%, gangue 12%) at 78 and 40 °C. The batch mesophile tests were carried out at 15% solids and the thermophile tests at 10% solids. The results show that the copper dissolution from chalcopyrite was limited to 40% at a bioleach temperature of 40 °C, consistent with the reported passivation of chalcopyrite in acidic ferric sulfate solutions at temperatures below 50 °C. The recovery of Ni, Co, and Cu dissolution improved significantly at the higher bioleach temperature of 78 °C, achieving final metal dissolutions of ~99% for Ni and Co and 95% for Cu. The results demonstrated that chalcopyrite passivation could be overcome at the higher leaching temperature.

3.2.3 *Dissolved Sulfate Salts and Metals Inhibiting Microbial Growth and Oxidation*

Various research groups have reported that elevated concentrations of sulfate can retard growth and iron oxidation by acidophilic prokaryotes. The effect of elevated sulfate concentrations on bioleaching of pyrite was reported by Basson et al. (2013). In tank bioleaching, the solids concentration is limited compared to non-biological processes due to the adverse effect of high solids loading which increases oxidant demand and limits microbial activity, primarily as a result of increased concentration of dissolved metal sulfate salts and cell damage due to shear forces in a slurry system. Commercial bioleach plants using mesophilic bacterial cultures at temperatures of 45 °C or less, typically operate at a solids content of 20% by mass. In the case of bioleaching of base metal concentrates with thermophilic archaea, solids concentrations are typically limited to 10–15% by mass.

3.3 General Process Design Elements

From a process design perspective, the reactor duty is determined from the following key parameters:

(a) The optimum operating temperature range dependent on the microbial culture selected.
(b) The sulfide and gangue mineral composition of the feed material.
(c) The rate of solids feed.
(d) The rate of sulfide mineral oxidation.
(e) The acid or alkaline (limestone/lime) demand for pH control of the process.
(f) The oxygen demand to sustain maximum leach rates.
(g) The heat generation from sulfide mineral oxidation and requirement for heat exchange, in order to control the temperature of the process within an optimum range to sustain microbial growth rates.

The key parameters are quantified by carrying out continuous pilot-scale tests. The results define the feed composition (chemical and mineralogical analyses results of representative feed samples), optimum solids concentration in the feed, rate of sulfide mineral oxidation as a function of reactor residence time, oxygen demand, and acid (sulfuric) or alkaline (limestone or lime) demand for pH control. Different temperatures and microbial cultures may be tested to determine the best conditions for maximum metal recovery and solids throughput. The following sections provide insights into the main components of industrial bioleach operations design.

3.4 Reactor Design

Whatever the process temperature, the tanks and agitators usually represent the main capital cost of the bioleaching plant. Therefore, the reactor design is critical to limit the cost while maximising process efficiency and revenues.

3.4.1 Reactor Configuration

The first parameter to be considered is the overall retention time which dictates the total volume of the reactors. Considering the bioleaching process as an autocatalytic process the ideal reactor configuration to minimise the overall leach time is a large volume STR (primary reactor) followed by a series of STRs (secondary reactors), with the primary reactor volume equal to the total volume of the secondary reactors. The large primary reactor is necessary to maintain maximum microbial growth rates and sustain high rates of mineral oxidation.

A study of oxidation data from BIOX® plants (Rorke 1997) has shown that a primary reactor volume equal to total secondary reactor volume is close to ideal and facilitates a tank configuration with equal volume tanks, simplifying design and reducing construction cost. A classical configuration consists of a primary stage configured with three equal-size reactors operating in parallel, in order to give the required retention time for microbial growth, followed by three similar-sized reactors operating in series to complete mineral oxidation. It is convenient to design bioleach plants with high tonnages in modules in order to avoid the need to build excessively large tanks, the largest bioleach tanks built to date have a capacity of 1500 m^3 (van Niekerk 2009).

3.4.2 Reactor Geometry

In mineral bioleach systems, it is common practice to build cylindrical reactors with height to diameter ratios of 1:1. Motivations for this geometry include the following:

- Significant bubble coalescence (small bubbles joining into larger bubbles) is usually observed after the high-shear zone around the agitator impeller, which drastically decreases gas surface area and increases bubble rising speed. Both of these factors make an extended residence time in a tall reactor unproductive.
- For the same tank volume, cylindrical tanks with a height/diameter ratio of ~1 allow the use of larger impeller diameters operating at lower tip speeds, which generates less shear, reducing damage to microbial cells.
- The tank height and impeller rotation speed are reduced for a given reactor volume compared to tall tanks of smaller diameter, minimising the impeller shaft bending moment and allowing the use of suspended impellers. This design eliminates the need for a foot bearing to support the base of the impeller shaft. In the case of mineral bioleaching, the slurry is abrasive, warm, and highly acidic and in this environment, a foot bearing is unlikely to last and is not a practical option.

3.4.3 Agitator Selection and Impeller Type

The main functions of the bioleach agitator and impeller system are gas dispersion and solids mixing. Important selection criteria are: energy efficiency, gas dispersion capability, mechanical reliability, and process performance at commercial scale. Their design requires specialist knowledge of a recognised agitator supplier experienced in the field of gas dispersion and oxygen transfer in slurry systems. Some examples are given below for illustration purposes.

Development of the BIOX® process in the early 1990s showed that high solidity axial flow impellers, such as the Lightnin® A315 impeller (SPXFLOW, North Carolina), provided greater energy efficiency and improved process performance compared to traditional high powered turbines such as the Rushton turbine (Harvey et al. 1999). Axial flow and high solidity impellers were also shown to be suitable for application in thermophile bioleaching where microbial strains are particularly sensitive to shear damage. The Kasese bioleach plant commissioned in 1998 used the BROGIM® mixing concept designed by Robin Industries (now Milton Roy Mixing) to achieve high oxygen transfer performances. It consists of two different impellers mounted on the same rotating shaft: an air dispersing flat blade radial turbine at the bottom, and low power-consuming propellers at the upper part aimed at maintaining homogeneity of all the phases in the tank (Bouquet and Morin 2006). More recently, AFX Mixing and Pumping Technologies Inc. have developed a combined lower downward-pumping axial flow impeller (P4) and an upper upward-pumping impeller (P3) which has shown improved energy efficiency for tank bioleaching applications; details are described in Chap. 12 for the Mondo Minerals bioleach plant.

3.5 Gas Supply Design

The supply of oxygen and carbon dioxide in a bioleach operation must be sufficient to meet the demand to sustain sulfide mineral oxidation and support the growth of the microbial community. It includes three main components: the injection of the gas, its dispersion into the pulp, and its transfer to the liquid phase, which are controlled by the gas flow, the geometry of the impeller, and the agitation power. The design of the gas supply and agitator system is a key issue for commercial application of bioleaching, representing a major part of capital and operating costs.

3.5.1 *Oxygen Supply and Control*

The oxygen demand in mesophile operations is satisfied by air sparged into the bottom of each reactor below the agitator. In thermophile bioleaching operations, although it is possible to use air in cases where the oxygen demand and hence air requirement is low, at higher duties, typical for bioleaching of base metal concentrates, the use of air is problematic and delivery of oxygen is required. The off-gas in a thermophile process carries as much as 50% by mass water vapour. In a high duty application such as the BioCOP™ demonstration plant (Sect. 3.7.2), if air is used, then the large volume of inert nitrogen passing through the reactor would cause huge water evaporation losses and resulting cooling. Large volumes of additional water would be required to keep the pulp diluted and both effects would prevent autothermal operation.

The supply of oxygen gas must be controlled to ensure the dissolved oxygen concentration is in the optimum range of 1.5–2.5 ppm. The dissolved oxygen level in solution must be above 1 ppm in order to sustain microbial growth during bioleaching. At dissolved oxygen concentrations above 5 ppm when sparging oxygen, thermophile cell growth is inhibited although mesophile cells can tolerate higher concentrations (between 13 and 17 ppm: such dissolved oxygen concentrations are possible as a result of high hydrostatic pressure at the bottom of large industrial tanks; Guezennec et al. 2017).

3.5.2 *Supply of Carbon Dioxide*

Carbon dioxide must be supplied to the bioreactor to sustain autotrophic microbial cell growth. In thermophile operations, CO_2 supply is essential since oxygen is injected instead of air. Even in mesophile operations, CO_2 provided by air injected into the tank is often not enough, especially under high-sulfide loading conditions (Guezennec et al. 2018). If there are no acid-soluble carbonate minerals present in the concentrate feed, or if limestone addition for pH control is not required, then CO_2

must be supplemented in a separate gas feed to the bioreactor to meet the required demand, typically 1–2% by volume of the inlet gas stream.

3.5.3 *Gas Transfer*

Oxygen transfer is critical to bioleach performance in a STR, given the high oxygen demand of sulfide biooxidation reactions and the low solubility of oxygen in aqueous solutions. The transport of oxygen from air bubbles to the cells can be represented by a number of steps and resistances, as detailed by Garcia-Ochoa and Gomez (2009). Taking into account that oxygen is only slightly soluble in water (and solubility decreases as temperature increases) it is commonly accepted that the greatest resistance for mass transfer is on the liquid side of the interface and the gas phase resistance can be neglected. The oxygen transfer rate (OTR) is usually described as the product of the driving force by the mass transfer coefficient (k_L) and the surface area of the gas bubbles (a). The driving force is the gradient between the concentration of the oxygen at the interface (C^*)—i.e., the oxygen saturation concentration at the liquid interface in equilibrium with the gas phase—and that in the bulk liquid (C_L).

During steady-state operations, the oxygen transfer rate must match the oxygen uptake rate (OUR), which leads to the common expression as follows:

$$\text{OTR} = \text{OUR} = k_L a(C^* - C_L) \tag{3.1}$$

In bioleaching operations, the biooxidation rates depend on the OTR which might become a limiting parameter if it is not sufficient. The oxygen uptake rate (OUR) can be calculated from measurement of process oxygen utilisation efficiency (U_{O2}) and the oxygen supply rate. It is determined as follows:

$$\text{OUR} = U_{O_2} \frac{O_{2-\text{in}} \cdot Q_{\text{gas}-\text{in}} \cdot \rho_{O_2}}{100.V} \left(\text{kg m}^{-3}\,\text{h}^{-1}\right) \tag{3.2}$$

where:

$O_{2\text{-in}}$ is the volume % oxygen in the inlet gas (dry basis)

ρ_{O2} is standard density of O_2 (kg m^{-3})

$Q_{\text{gas-in}}$ is volumetric gas flowrate in L h^{-1} at Standard Temperature and Pressure (STP)

V is the reactor operating volume (L)

Oxygen utilisation may be calculated from the conservation of N_2 flux between the inlet and the outlet using operating measurements according to the following equation:

$$U_{O_2} = 1 - \frac{O_{2\text{-off}}}{O_{2\text{-in}}} \cdot \frac{100 - O_{2\text{-in}} - CO_{2\text{-in}}}{100 - O_{2\text{-off}} - CO_{2\text{-off}}} \tag{3.3}$$

$CO_{2\text{-in}}$ is the volume % of carbon dioxide in the inlet gas (dry basis)
$O_{2\text{-off}}$ is the volume % of oxygen in the outlet gas (dry basis)
$CO_{2\text{-off}}$ is the volume % of carbon dioxide in the outlet gas (dry basis)
Standard Temperature and Pressure (STP) is defined as: 0 °C, 101.325 kPa.

The operating volume (V) may be standardised for all reactor designs as the total active volume of the reactor, i.e., including gas hold-up. This simplification could cause minor errors because of fluctuations in gas hold-up with gas delivery, but in the case of typical bioleach reactors, the error may be considered insignificant.

Oxygen utilisation efficiency with thermophiles, when using almost pure oxygen (+95%), was reported to be in the region of 80% (i.e., ~20% of the incoming oxygen escapes to the atmosphere). In the case of mesophiles, operating on air, the oxygen utilisation efficiency has been estimated to be ~40%.

3.5.4 *OTR Upscaling*

3.5.4.1 Volumetric Mass Transfer Coefficient ($k_L a$)

Gas to liquid (G/L) mass transfer requirements are usually determined at pilot scale and a correlation of $k_L a$ with superficial gas velocity is then established to scale up the process. However, extensive evaluation in large-scale pilot and commercial prototype plants at both mesophile (~40 °C) and thermophile temperatures (70–80 °C) and comparison with commercial BIOX® operations (Rorke 2005), indicated that correlations using superficial gas velocity were not reliable as a universal approach for reactor design and scale-up. Further analysis of data showed that a correlation based on the ratio of gassed volume to tank volume, instead of superficial gas velocity, provided a correlation that fitted all the reactor data analysed, independent of scale and operating temperature. The correlation, derived from measurements on operating pilot and commercial-scale bioleaching plants is shown in Eq. (3.4). It applies to bioreactors where single down-pumping axial-flow impellers are used in a reactor with a fixed geometry of height to diameter of approximately one to one. A correction for temperature was made using an exponential term and tested against measured data from both mesophile and thermophile bioleach operations (Rorke 2005).

$$k_L a = e^{a(T-75)+b} \mathrm{W} \left(\frac{1000P}{V}\right)^y \left(\frac{Q}{V}\right)^z \tag{3.4}$$

where:

e is the natural exponent
k_La is the overall oxygen transfer coefficient based on a log mean concentration driving force (as discussed in the next section)
$a = 0.0229$
$b = 0.219$
$w = 1.593$
$y = 0.509$
$z = 0.926$
T is the reactor temperature in °C
P is the agitator drawn power at motor in kW (called agitator power in the next sections)
V is the total tank volume to overflow launder in m^3
Q is the inlet gas flow in $m^3\ s^{-1}$ on a dry basis at the bulk slurry temperature and pressure mid-point between agitator and sparge ring

$$Q = \frac{Q_{\text{gas-in}} \times 101.325 \times (T + 275.13)}{3600 \times 275.13 \times P_{\text{CSat}}} \tag{3.5}$$

P_{CSat} is the absolute (static) pressure between a point between the sparge ring and agitator in kPa (see Eq. (3.11))

$$P = F_g P_U \tag{3.6}$$

where:

P_U is un-gassed agitator power in kW and F_g, the gassed to un-gassed power ratio, experimentally derived value.

The formula for un-gassed agitator power for a single agitator is:

$$P_U = N_P \cdot F_{\text{pr}} \cdot \rho_{\text{slurry}} \cdot \left(\frac{N}{60}\right)^3 D^5 \tag{3.7}$$

where:

N_P is agitator power number defined by the agitator supplier. Value is 0.75 for an A315 impeller
F_{pr} is a proximity factor depending on reactor geometry and on impeller position (the factor value is 1.08 for the reactors described in Sect. 3.7)
ρ_{slurry} is slurry density (kg L^{-1})
N is agitator speed (rpm, revolutions per minute)
D is agitator diameter (m)

The power drawn by an axial-flow Lightnin® A315 agitator, or equivalent design, increases with increasing gas rate until the point of flooding, which may be defined as the point when the impeller pumping action is overcome by the gas flow and effective gas dispersion and solids suspension is prevented. The value historically

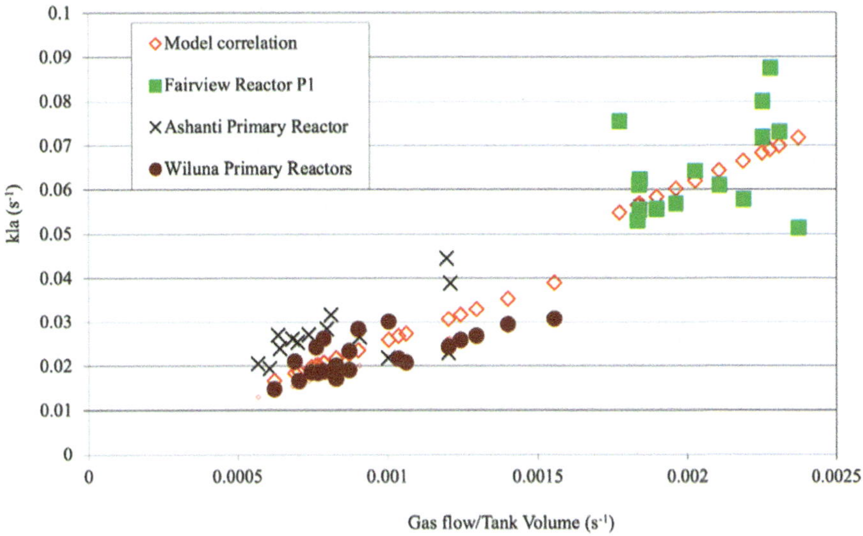

Fig. 3.2 Fit of model correlation (Eq. 3.4) to measured k_La data from commercial BIOX® plants as a function of gas flow to tank volume

assigned to F_g for bioleach reactors operating with air has been 1.13. In the case of high-temperature bioreactors (thermophile bioleach applications) using oxygen, where the gas rates are low because of the purity of the oxygen and high utilisations, it is likely that this factor should be closer to 1. Results from the 300 m^3 Industrial Test Reactor, described in Sect. 3.7.1 (Rorke 2005), showed excellent agreement (within 5%) between drawn and calculated power using Eqs. (3.6) and (3.7), given above. The fit of Eq. (3.4) to commercial BIOX® plant test data is demonstrated in Fig. 3.2. The calculated k_La values show a good correlation to plant results considering the variation of airflow measurements and fluctuation in measured k_La values on a commercial scale plant.

3.5.4.2 Driving Force

In an air system the value of C^* in Eq. (3.1), assuming a well-mixed STR, may be determined by the oxygen content of the inlet or exit gas, since the difference between the oxygen content of the gas exiting reactor (residual gas) and that of the gas entering the reactor is relatively small (usually below 9% v/v). In the case of sparging with high purity oxygen gas the difference between the inlet and residual, or exit gas oxygen content, can be as much as 40% (v/v). This makes the selection as to whether C^* is based on the inlet or exit gas compositions critical. In this case, a better approach is to estimate the effective saturated dissolved oxygen by using the log mean value of the inlet and outlet gas concentrations (dry basis; % v/v).

$$\frac{\text{OUR}}{3600} = \frac{k_{\text{L}}a \cdot \left(\left(C^*_{\text{Inlet}} - C\right) - \left(C^*_{\text{Offgas}} - C\right)\right)}{\ln\left(\frac{C^*_{\text{Inlet}} - C}{C^*_{\text{Offgas}} - C}\right)} \quad (\text{ppm s}^{-1}) \tag{3.8}$$

where:

C is the dissolved oxygen in the bulk liquid (ppm); same as C_L in Eq. (3.1)
C^*_{Inlet} is the maximum saturated dissolved oxygen concentration (ppm) based on the partial pressure (on a dry basis) of the oxygen in the inlet gas (ppm)
C^*_{Offgas} is the maximum saturated dissolved oxygen concentration (ppm) based on the partial pressure (on a dry basis) of the oxygen in the gas exiting the reactor (ppm)

Fluid flow studies in STRs have shown that outside of the agitator zone oxygen mass transfer rates are very low (Bakker 1992). In light of this, the saturated dissolved oxygen value (C^*) for both inlet and off gas are calculated under the temperature and pressure (P_{CSat}) conditions at the midpoint between agitator and sparge ring. In a standard tank, the reactor diameter is the same as the operating height, the agitator diameter to tank diameter ratio (D/T) is typically within the range of 0.35–0.45, the agitator is positioned at 80% of its diameter from the reactor floor and the sparge ring is located at midpoint between the agitator and reactor floor. Taking all this into consideration the height at which to determine the pressure can be estimated by:

$$z' = \frac{0.6 \cdot Z \cdot D}{d} \tag{3.9}$$

In the case of a standard tank with equal height-to-diameter ratio Eq. (3.9) simplifies to:

$$z' = 0.6 \cdot D \tag{3.10}$$

where:

D is the agitator diameter (m)
d is the tank diameter (m)
Z is the active slurry height (to overflow) (m)

The total pressure at point midway between sparger and agitator (z') can then be calculated by:

$$P_{\text{Csat}} = P_{\text{atm}} + \rho_{\text{Slurry}} \cdot g \cdot (Z - z') \cdot (1 - \varepsilon) \tag{3.11}$$

where:

P_{Csat} is the absolute pressure in zone between agitator and sparge ring (kPa)
P_{atm} is the local atmospheric pressure (kPa)

ρ_{slurry} is the slurry density (kg L^{-1})
g is the gravitational constant (9.81 m s^{-2})
ε is the fractional gas holdup

From this, it is possible to calculate the dissolved oxygen at a known temperature and partial oxygen pressure.

$$C_i^* = \left(\frac{1000}{760 \cdot (760 - \rho \text{PH}_2\text{O})}\right) \cdot \left(\frac{1000 \cdot P_{\text{CSat}}}{133.32} - \rho \text{PH}_2\text{O}\right) \cdot \left(\frac{32 \cdot \alpha}{22.41}\right) \cdot (760 - \rho \text{PH}_2\text{O}) \cdot \text{O}_{2i} \quad (3.12)$$

where:

C_i^* and O_{2i} are the saturated dissolved oxygen and gas fraction (v/v), respectively, of stream i
ρPH_2O is the vapour pressure of water in mm Hg at bulk slurry temperature
α is the Bunsen coefficient for oxygen solubility[3]

3.6 Other Design Criteria

3.6.1 Additional Agitator Performance Criteria

3.6.1.1 Shear Rate Limitation

It is generally accepted that there is an absolute limit to the amount of shear a microorganism will tolerate. The thermophilic microorganisms currently used do not possess rigid cell walls, and are thus more susceptible to shear damage than their mesophile counterparts with rigid cell walls. The shear rate at which this occurs is commonly equated to the tip speed of the agitator impeller.

$$\text{Agitator tip speed} = \pi \cdot D \cdot \frac{N}{60} \ \left(\text{m s}^{-1}\right) \quad (3.13)$$

The limiting tip speed has been found to increase with scale. Studies in the development of thermophile bioleaching reactors showed that in a 5 m^3 pilot-scale reactor the maximum tip speed was limited to 5 m s^{-1}. In the 1260 m^3 primary reactor of the BioCOP™ prototype plant, the agitator successfully operated at a tip speed of 7 m s^{-1}.

[3] https://srdata.nist.gov/solubility/IUPAC/SDS-7/SDS-7.pdf

3.6.1.2 Point of Flooding

The agitator impeller size must be adequate for gas dispersion and operate well away from the point of flooding. The point of flooding may be defined as the point when the impeller is not capable of deflecting the rising gas flow and the flow pattern in the reactor is dominated by the rising gas flow rather than by the effect of the impeller. Each type of agitator has a flooding point unique to the impeller type and size. Agitators designed for low-power solid suspensions, such as hydrofoils, flood easily (low capability for gas dispersion). Axial-flow impellers with high solidity ratios (ratio of the portion of area physically occupied by the blades to the area of the horizontal plane in which the agitator rotates), such as the Lightnin® A315, can disperse a large volume of gas before flooding occurs. The gas handling capability may be defined by the impeller aeration number. The aeration number (N_a) is defined by:

$$N_a = \frac{60 \cdot Q}{N \cdot N_q \cdot D^5} \tag{3.14}$$

where:

Q is the volumetric gas flow at the impeller (taken as midpoint between the impeller and gas sparger for simplicity of design) in $m^3\ s^{-1}$
N is agitator rotational speed in revolutions per minute
D is agitator diameter (m)
N_q is the flow number for the agitator (0.73 for a Lightnin® A315)

In the case of the LIGHTNIN® A315 impeller, the point of flooding may be estimated to occur when the aeration number is equal to or greater than 0.75. This is called the flooding aeration number (N_{af}). When designing semi-industrial or industrial bioleaching operations, a safety margin is applied; for example, the aeration number is set equal or less than 75% of the flooding aeration number.

3.6.1.3 Solids Suspension

It is essential for a successful continuous bioleach operation that solid material does not build up in the reactor vessels ("sanding out"). In order to ensure this does not happen there must be adequate agitation to promote solids mixing. Adequate agitation requires that the agitator is sufficiently sized and the tank is correctly baffled. Undersized agitators for solids suspensions are usually not an issue in bioreactors because of the enormous power required for oxygen transfer and operation away from the point of impeller flooding. The ability of an agitator to suspend solids may be simply estimated by the slurry flux across the horizontal area of the tank:

$$\text{Slurry flux} = \frac{\text{pumping rate}}{\text{tank area}} = \frac{60.4.N_q.N.D^3}{\pi.d^2} \left(\text{m min}^{-1}\right) \tag{3.15}$$

A simple rule of thumb that applies to solids suspensions treating milled concentrates or ores is that the minimum slurry flux for solid suspensions should be 10 m min^{-1}. This is not an absolute rule as particle sizes vary between projects typically from a d_{80} of 12 μm to 75 μm and as a result solids settling rates are likely to be different. This rule was developed for coarser material and so it is a safe assumption for most bioleach applications. As mentioned previously, solid suspensions seldom need to be considered because of the amount of energy required for oxygen transfer and it is sufficient to check particle size and settling rates against slurry rates of flux under design operating conditions.

3.6.1.4 Gas Holdup

Gas holdup is the volume of the reactor occupied by the gas that is sparged into the reactor to provide oxygen and is a function of the residual unused volume of gas as opposed to simply the gas volume delivered to the vessel. The volume of unused gas is dependent on the oxygen demand and oxygen utilisation.

Using data from measurements of gas holdup in commercial-scale bioreactors (100–500 m^3 volume) with equal diameter to height ratio using axial-flow impellers, Rorke (2005) showed the fraction of gas holdup can be estimated using the following empirical equation:

$$\varepsilon = \frac{Q_N \cdot Z}{36 \cdot (2 \cdot Q_N + 8.5 \cdot V)} \tag{3.16}$$

where:

ε = Fractional gas holdup
Q_N = Gas flow delivered to the reactor in m^3 h^{-1} at STP for mesophile operations or residual gas flow exiting the reactor for thermophile operations
Z = the height of active reactor volume (m)
V = active reactor volume (m^3)

Fractional gas holdup for typical aerated commercial bioleach reactors (500–1000 m^3 volume) is 0.06 to 0.15.

3.6.2 *Reactor Cooling Circuit*

Bioleaching reactor cooling is most efficiently achieved using internal baffle coils connected to either a closed cooling circuit or an open circuit. In a closed circuit, heat from the hot bioleach cooling water return is transferred to another fluid (raffinate, water, or even air) via heat exchangers. The more traditional cooling circuit is an open circuit using evaporative cooling. In this case, the hot cooling water return from the bioleach reports to an evaporative cooling tower. This constant evaporation leads to build-up of salts, and a portion of the water needs to be constantly purged (blowdown) to keep salt levels in check. Anti-scalant and biocide are also required to minimise scaling and buildup of algae and other microorganisms. Leaks in cooling coils have occurred on operating plants in the past, and have resulted in the cooling water entering the bioleach reactors causing process inhibition or loss of the bioleach microbial community.

3.6.3 Control of Leach Solution Chemistry

The control of pH is an essential aspect of bioleaching to maintain the required Fe (III) concentrations necessary to sustain mineral leach rates and maintain microbial growth rates. The slurry pH is ideally in the range of pH 1.2–2.

Microbial growth rates may be retarded by high metal sulfate concentrations, As (III) concentration, and low levels (mg L^{-1} concentration) of residual solvent extraction reagents and flotation reagents. The effect of dissolved metals, sulfate, and organic reagents on microbial activity must be evaluated during pilot-scale studies and the process design must consider mitigating action required to prevent negative effects on the bioleach process performance.

3.6.4 Bioleach Solid–Liquid Separation

The requirements for solid–liquid separation and solids washing are important parts of the bioleach process, but are very dependent on the specific concentrate being treated and the metals to be recovered. Counter-current decantation (CCD) thickeners are typically used where solids separation and washing are required. Vacuum- or pressure-filtration may be used in combination with or as an alternative to the use of a CCD circuit. Filter cake production may also be required, if, for example, the residue is going to be further treated in a separate process.

3.6.5 Materials of Construction

An important consideration for thermophile bioleaching operations is the materials of construction. Materials of construction suitable for mesophile bioleaching temperatures (40–45 °C), such as 316 L stainless steel, were shown not to hold up in environments of 60–80 °C applied in thermophile bioleaching. In thermophile bioleaching higher-grade stainless steels are required for all tank internals, such as SAF2205 or 904 L. In the case of the BioCOP™ demonstration plant, the tanks themselves were constructed from concrete with a refractory tile lining according to the Stebbins design; at the time of construction this option was found to be more cost-effective considering the cost of high-grade stainless steel and saving due to less agitator supporting steel structures required to support the agitator gearbox and impeller shaft (Harvey et al. 1999; Rorke 2005). Standard corrosion tests using process solutions under operating conditions are advised to confirm the suitability of a selected stainless steel for application in a bioleach STR.

3.7 Examples of Commercial-Scale Designs and Performance

3.7.1 The Industrial Test Reactor (ITR) High-Temperature Bioreactor Plant

3.7.1.1 ITR Plant Description

Development research by Billiton in the period 1995–2000 demonstrated that thermophile bioleaching was a viable treatment route for a variety of base metal sulfide concentrates. In order to provide a sound basis for thermophile bioreactor design and scale-up an Industrial Test Reactor (ITR) facility was constructed at the Pering Mine, Reivilo, South Africa. The plant was commissioned in October 2000.

The ITR consisted of a 50-m^3 and a larger 300 m^3 bioreactor. Both reactors were fully instrumented to monitor bioleach performance. The reactors were operated under industrial-scale conditions, in order to test the design and evaluate materials of construction. Of particular importance were the control of the oxygen supply and measurement of oxygen transfer rates to test design correlations (Batty and Rorke 2006).

The ITR plant consisted of two sections. The first portion of the plant was the concentrate preparation and regrind circuit. The second major section of the plant included the thermophile bioleach reactors and their support units. The 300 m^3 and 50 m^3 reactors both had a height-to-diameter ratio of 1:1. The reactors were constructed by Stebbins using their proprietary technology for acid-resistant tile-lined tanks. Ancillary units to support the bioleach operation included: bulk liquid oxygen and carbon dioxide storage facility; two oil-free air compressors with a cooler and air dryer system; gas mixing station, for blending oxygen, carbon dioxide, and air (Rorke 2005).

3.7.1.2 Dissolved Oxygen and Gas Mixing Process Control

Dissolved oxygen was measured continuously by a dissolved oxygen probe immersed in the reactor. This information was then used in a cascaded control algorithm to control the oxygen gas flow to the reactor and maintain the required dissolved oxygen concentration in the bioreactor slurry.

The process required a specific gas composition in terms of oxygen, nitrogen, and carbon dioxide content. Bulk liquid oxygen, bulk liquid CO_2 as well as compressed air were available for gas supply. Oxygen was diluted with a controlled air dosage. A simple control feedback loop from an oxygen gas analyser to a control valve on the compressed air supply line controlled the oxygen content of the bioreactor feed gas. A second control loop from a gas analyser to a control valve on the CO_2 supply line controlled the CO_2 content of the feed gas.

3.7.1.3 Oxygen Utilisation and Agitator Performance

The data produced by the ITR facility was used to develop and test a new correlation (final general version of the correlation is given in Eq. (3.4)) for oxygen transfer using test results from the 50 m^3 reactor during bioleach operation. The correlation was then tested against operating results from the 300 m^3 reactor. The standard operating conditions were 78 °C and not less than 90% O_2 (v/v). Figure 3.3 shows the measured k_La data from the 50 m^3 reactor at different impeller power inputs and in the 300 m^3 reactor at a fixed impeller power input, as a function of gas volume to reactor volume. The model correlation showed a good fit to the test data, confirming the validity of the correlation for high-temperature bioleaching with the given bioreactor configuration and agitator design.

3.7.2 The BioCOP™ Demonstration Plant

The BioCOP™ technology using bioleaching was developed by BHP Billiton to treat copper concentrates not amenable to smelting. A joint venture company called Alliance Copper Limited was formed between BHP Billiton and Codelco to exploit the technology (Batty and Rorke 2006). Process engineering design for bioleaching at moderate temperatures using mesophiles is well understood based on the established BIOX® process. In early 2000, bioleaching with thermophile cultures at temperatures of greater than 70 °C had not been tested and a prototype commercial demonstration plant was essential to prove the process performance.

The demonstration plant was designed to produce 20 kt of copper per annum, which was sufficient to cover the plant operating costs and provide good scale-up data for a larger commercial plant. The plant consisted of regrind circuit, concentrate pre-leach, bioleach section, solid–liquid separation, and PLS storage. PLS produced

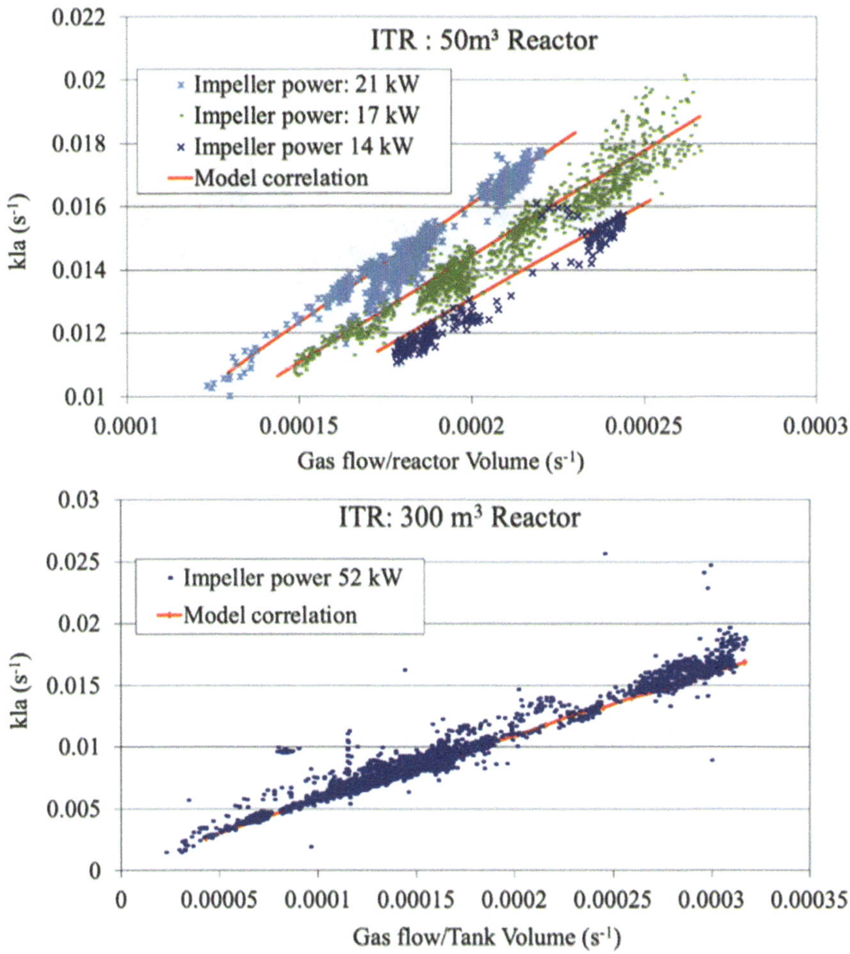

Fig. 3.3 Comparison of k_La model correlation (Eq. 3.4) to measured k_La performance data for the steady-state bioleach operation of the 50 m^3 and 300 m^3 ITR plant reactors at (78 °C)

was treated at the solvent extraction/electrowinning plant at Chuquicamata Mine to produce copper metal. The bioleach section consisted of six 1260 m^3 bioreactors. The bioreactors were based on the design of the reactors tested in the ITR facility (see Sect. 3.7.1) using Stebbins concrete and ceramic tiles construction, and used Lightnin® A315 agitators. All wetted metal parts of the bioreactor were made from SAF2205 duplex stainless steel. At the time of plant commissioning the 5-m diameter A315 agitator was the largest ever constructed.

The reactors operated at a temperature of 78 °C and each received an oxygen feed containing approximately 97% (v/v) oxygen. Each reactor was also equipped with an online dissolved oxygen probe, a reactor off-gas analyser and there were centralised feed gas analysers. Together these instruments allowed online measurement of

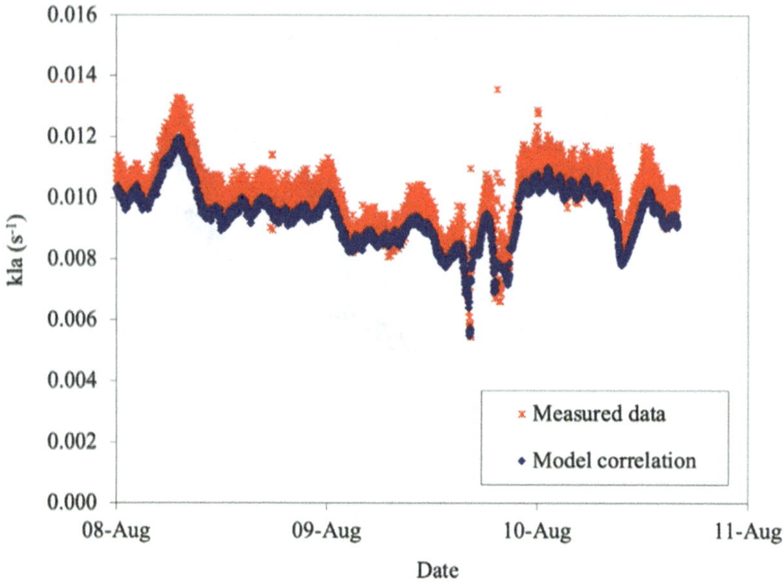

Fig. 3.4 Measured oxygen mass transfer coefficient (k_La) versus predicted values using correlation given by Eq. (3.4) for bioreactor P2 of the BioCOP™ commercial demonstration plant operating at an impeller power input of 237 kW

oxygen transfer data. The measured results compared to predicted values using the correlation given by Eq. (3.4) are shown in Fig. 3.4. On average the oxygen transfer coefficient was conservatively under predicted by 6%, which translates to an error of ~3% with respect to oxygen utilisation.

The BioCOP™ commercial demonstration plant construction and commissioning were completed in August 2003 and the first pregnant leach solution (PLS) was produced in September 2003. The design operating tonnage to produce 20 kt copper per year was achieved by November 2003.

3.8 New Developments in the Field of Bioleach STR

Although stirred tank bioleaching has proven to be highly robust and easy to implement, it carries considerably higher capital and operating costs than heap leaching and its use is usually restricted to high-value minerals or mineral concentrates. Several attempts have been made to design new bioreactor configurations in order to improve the process economics and to widen its application to lower value targets. The "low-duty bioleaching reactor" is a new concept developed as an intermediate process option to conventional heap or stirred tank leaching. It consists of using floating agitators to inject gases as well as to suspend solids in the solution. This technology enables bioleaching:

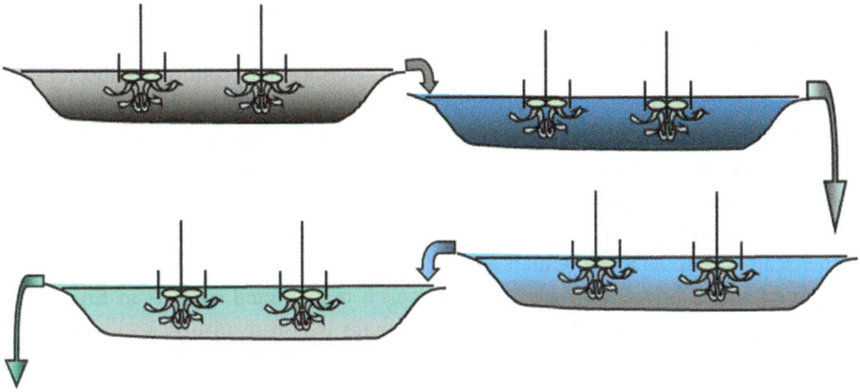

Fig. 3.5 Schematic of the low-duty bioreactor concept

- At higher solid load than in conventional stirred tank bioreactor
- In ponds instead of costly tanks (Fig. 3.5)

The concept has been tested at pilot scale with a small floating agitation device built on the model of TurboxAL agitators designed by MRM and Air Liquide (France) for water treatment applications (Guezennec et al. 2016). The dimensions of the industrial agitator were divided into four for the purpose of the study (750 mm diameter and 850 mm height). Different types of sulfide materials were tested: a copper concentrate, high sulfide tailings, and a low sulfide ore. For the first two case studies, the bioleaching tests were performed with the KCC consortium (Sect. 3.2) at 42 °C whereas, for the last one, the tests were performed with an indigenous consortium at 50 °C. At 20% (w/w) solids concentration, similar bioleaching rates were reached with the floating agitator as with a classical STR. At 30% (w/w), an adaptation regime (based on a progressive increase of the solids concentration) was applied to maintain the same level of performance compared to 20% solids concentration. The quality of the solids suspension was checked by sampling the pulp at different locations in the reactor. An overall uniform solids concentration was obtained from the bottom to the surface of the pulp.

One of the main challenges of this concept is the management of heat, since no cooling or heating system can be implemented in a pond. The heat generated by the reactions must be balanced by the heat losses at the pond surface and walls and by the process inputs. A numerical model was developed to assess the heat balance for different sulfide contents from 2.5 to 20% (Loubiere et al. 2021). The simulations demonstrated that it is possible to maintain a suitable temperature (between 40 and 48 °C) by controlling the fresh pulp inlet conditions (flow rate and temperature) and the aeration (flow rate and O_2 partial pressure).

This concept, patented by BRGM, Milton Roy Mixing, and Air Liquide, has been designed as a compromise option combining the main advantages of heap and STR bioleaching. In particular, the leaching kinetics and the recovery yields will be in the same order of magnitude as in an STR. Ponds and lagoons will replace the tanks, which significantly reduces the capital costs. The process is designed to operate at higher solids load than a typical STR (up to 30% in the low-duty bioreactor, vs. 15–20% in a STR), which will reduce the volume of the reactor. The low-duty bioreactor is thus expected to have lower CAPEX but the same level of revenue as a STR. It will also involve lower surface areas than in heap leaching processes. These advantages will enable the broadening of application to a wider range of materials. In particular, the low-duty bioreactor might be a good option for low-grade sulfidic materials (primary ores as well as tailings) that have too low value to be treated in an STR and which are not suitable for heap leaching, either because they are located in areas where available space is restricted such as in Europe, or because they contain minerals such as carbonate which disturb heap irrigation or because they are already finely ground and thus not suitable for heap construction such as with flotation tailings. This new concept of bioreactor also has the advantages of easy maintenance, continuous service, and flexible operation.

3.9 Concluding Remarks

Bioleach stirred tank reactor (STR) design is driven primarily by oxygen transfer to meet the oxygen demand of sulfide mineral biooxidation, but design must also ensure efficient solids mixing, heat transfer to control operating temperature and conditions that allow optimal performance of the adapted microbial consortium to maintain design output. Innovation of STR agitator systems has reduced energy consumption for oxygen mass transfer improving the economics of tank bioleaching. The new low-duty STR described and under development aims to improve economy of operation for the treatment of low-value materials.

References

Bakker A (1992) Hydrodynamics of stirred gas-liquid dispersions. Ph.D. Dissertation, February 1992, Delft University of Technology, Delft

Bard AJ, Parsons R, Jordan J (1985) Standard potentials in aqueous solution, 1st edn.

Barton PB (1969) Thermochemical study of the system Fe-As-S. Geochim Cosmochim Acta 33(7): 841–857. https://doi.org/10.1016/0016-7037(69)90031-3

Basson P, Gericke M, Grewar TL, Dew DW, Nicol MJ (2013) The effect of sulphate ions and temperature on the leaching of pyrite. III. Bioleaching. Hydrometallurgy 133:176–181. https://doi.org/10.1016/j.hydromet.2013.01.008

Battaglia-Brunet F (1994) Optimisation de la lixiviation bactérienne d'un concentré de pyrite cobaltifère. PhD Thesis. Aix-Marseille 1 University. Available at: https://www.theses.fr/1994AIX11039

Batty JD, Rorke GV (2006) Development and commercial demonstration of the BioCOP™ thermophile process. Hydrometallurgy 83:83–89. https://doi.org/10.1016/j.hydromet.2006.03.049

Bouquet F, Morin D (2006) BROGIM®: a new three-phase mixing system testwork and scale-up. Hydrometallurgy 83:97–105. https://doi.org/10.1016/j.hydromet.2006.03.040

D'Hugues P (1996). Biolixiviation d'une pyrite cobaltifère: optimisation des conditions de culture et bilans gazeux (These de doctorat). PhD Thesis. Aix-Marseille 1 University. Available at: https://www.theses.fr/1994AIX11062

Dew DW, Buuren CV, McEwan K, Bowker C (2000) Bioleaching of base metal sulphide concentrates: a comparison of high and low temperature bioleaching. J SAIMM 100:409–414

Garcia-Ochoa F, Gomez E (2009) Bioreactor scale-up and oxygen transfer rate in microbial processes: an overview. Biotechnol Adv 27:153–176. https://doi.org/10.1016/j.biotechadv.2008.10.006

Guezennec AG, Archane A, Ibarra D, Jacob J, D'Hugues P (2016) The use of oxygen instead of air in bioleaching operations at medium temperature. In: Proceedings of the 8th International Mineral Processing Congress: IMPC 2016, Canadian Institute of Mining Metallurgy and Petroleum (CIM), p 2438

Guezennec AG, Joulian C, Jacob J, Archane A, Ibarra D, Gregory R, Bodenan F, D'Hugues P (2017) Influence of dissolved oxygen on the bioleaching efficiency under oxygen enriched atmosphere. Minerals Eng 106:64–70. https://doi.org/10.1016/j.mineng.2016.10.016

Guezennec AG, Joulian C, Delort C, Bodénan F, Hedrich S, D'hugues P (2018) CO2 mass transfer in bioleaching reactors: CO2 enrichment applied to a complex copper concentrate. Hydrometallurgy 180:277–286. https://doi.org/10.1016/j.hydromet.2018.08.006

Harvey PI, Batty JD, Dew DW, Slabbert W, van Buuren C (1999) Engineering considerations in bioleach reactor design, BioMine 99, Perth August 1999

Holmes PR (2000) Thermophile Pilot Run, unpublished report Billiton PR00/113

Johnson GK, Steele WV (1981) The standard enthalpy of formation of chalcopyrite (CuFeS2) by fluorine bomb calorimetry. J Chem Thermodyn. 13:991–997. https://doi.org/10.1016/0021-9614(81)90076-8

Loubiere C, Olmos E, Menard Y, Guezennec AG (2021) Heat balance and CFD coupling strategy for the scale-up of an innovative bioleaching process. In: Proceedings of 6th BioProScale Symposium

Norris PR, Burton NP, Clark DA (2013) Mineral sulfide concentrate leaching in high temperature bioreactors. Minerals Eng 48:10–19. https://doi.org/10.1016/j.mineng.2013.01.001

Rorke GV (1997) Optimisation of BIOX® plant configuration by non-linear mapping: Preliminary study, Gencor unpublished report (1997)

Rorke GV (2005) Bioleach Bioreactor Design-Methodology, unpublished BHP Billiton Report 2005

Van Niekerk J (2009) Recent advances in BIOX® Technology, Hydrometallurgy Conference 2009, SAIMM, 2009

Chapter 4
Bioprocessing of Refractory Gold Ores: The BIOX, MesoTHERM, and ASTER Processes

Jan Albert van Niekerk, Craig Bradley van Buuren, and Johan Waldemar Olivier

Abstract The BIOX® process for the treatment of refractory gold concentrates has been in commercial operation for 35 years following the commissioning of the first commercial installation at the Fairview Gold Mine in 1986. The process has been proven to operate successfully over a wide range of concentrate feed grades and mineralogies, climatic conditions, and site altitudes.

In 2020, Metso Outotec launched its novel two-stage MesoTHERM BIOX technology, a hybrid biooxidation process specifically targeting certain oxidised refractory samples which typically consume high levels of cyanide during the leaching process. Traditionally, cyanide consumption represents a significant operating cost in most gold leach circuits with conventional BIOX plants included. MesoTHERM technology application treating refractory gold samples has seen cyanide consumption reduced by up to 50% compared to mesophile BIOX products.

In 2010, Metso Outotec introduced the ASTER™ process for improved water balances in commercial BIOX applications in both arid and tropical regions. Previously, a common challenge for BIOX plants was the inability to recycle thiocyanate-containing water due to the low tolerance of the organisms. ASTER, a biological thiocyanate destruction process, produces a non-toxic solution for reuse in BIOX. Four commercial ASTER plants are in operation with the fifth ASTER plant under construction in Zimbabwe.

Keywords BIOX · MesoTHERM · ASTER · Biooxidation · Refractory · Gold · Mesophile · Thermophile · Thiocyanate · Detoxification

J. A. van Niekerk (✉) · C. B. van Buuren · J. W. Olivier
Gold Technology Group, Metso Outotec, Centurion, Gauteng, South Africa
e-mail: jan.vanniekerk@mogroup.com; craig.vanbuuren@mogroup.com; Waldemar.olivier@mogroup.com

D. B. Johnson et al. (eds.), *Biomining Technologies*,
https://doi.org/10.1007/978-3-031-05382-5_4

4.1 Introduction

A hallmark of the BIOX technology has always been the continued development of the process and understanding of the operation of the BIOX bacterial culture and BIOX plants. Metso Outotec has continued the development of the understanding of the microbial cultures used in the BIOX process and the effect of changes in process parameters on the culture. Metso Outotec has also continued the development of the equipment used in the implementation of the BIOX process with the development of the OKTOP BIOX reactor and agitator.

Metso Outotec's MesoTHERM technology uses a combination of mesophilic and thermophilic microbial consortia as a hybrid, two-stage biooxidation process where mesophiles are used to realise the primary stage sulfide oxidation and thermophiles to bring about the latter stage, near-complete sulfide oxidation. This new BIOX process was evaluated in the early 2000s to investigate the potential of such a hybrid process in producing residues that consume less cyanide when leached. Traditionally, cyanide consumption represents a significant operating cost in most gold leach circuits and Metso Outotec was looking at positioning its conventional BIOX technology to compete more aggressively against peer refractory technologies. Early batch work using thermophiles showed that a two-stage MesoTHERM technology application to refractory concentrates yielded cyanide consumption reductions up to 50%, down from 20 kg t^{-1} to around 8 kg t^{-1} to 10 kg t^{-1} when compared to the leach consumptions from conventional mesophile BIOX processing.

Over the past two decades, development of the MesoTHERM technology spanned initial laboratory batch biooxidation tests to continuous piloting, engineering scale-up and large-scale culture robustness and demonstration testing at the Fairview Gold Mine, part of Pan African Resources' Barberton Mines Limited's metallurgical complex. This technology was officially launched in June 2020 and aims to support existing and future BIOX users to achieve further cost savings without any overall impact on metallurgical performance.

In 2010, Metso Outotec introduced the ASTER process for improved water balances in commercial BIOX applications by supporting tightening environmental legislation on water usage and availability in both arid and tropical regions. Previously, a common challenge for BIOX plants was the inability to recycle thiocyanate-containing water due to the low tolerance of the organisms. ASTER, a biological thiocyanate and cyanide destruction process, produces a non-toxic solution for reuse in BIOX. Four commercial ASTER plants are in operation with the fifth ASTER plant under construction in Zimbabwe.

The rights to the BIOX, MesoTHERM, and ASTER process are currently held by Metso Outotec (Finland) Oyj following the successful acquisition of Biomin Technologies S.A. in 2015.

4.2 The BIOX Process

4.2.1 *Current and Historic BIOX Plants*

A complete list of the current and historical BIOX plants is given in Table 4.1. A short description of the new BIOX plants is given below, with full details of the current and historical BIOX plants available in the literature.

4.2.1.1 Runruno BIOX Plant

The Runruno Gold Project is located on the island of Luzon, approximately 200 km north of Manila in the Philippines. The mine is a surface mine operation and uses the BIOX and carbon in leach (CIL) processes to recover gold, and the ASTER process for detoxification of the CIL residue prior to discharge to the Residue Storage Impoundment (RSI). The BIOX circuit is designed to treat 140,000 tonnes per annum (t y^{-1}) of concentrate with a design capacity of 404 tonnes per day (t day^{-1}) of concentrate at a design sulfide sulfur grade of 17%.

The Runruno BIOX Plant is the first application of the Metso Outotec BIOX Generation III design principles, focusing on ease of operation and maintainability. During development and implementation, the project team focused on identifying opportunities to optimise capital efficiency while not compromising on the operability or efficiency of the process or equipment.

Construction of the processing plant started in 2013 with the turnkey BIOX and ASTER inoculum build-up facilities set up in 2014. Operator training commenced early in 2015 in preparation for commissioning of the BIOX plant later in the year.

Table 4.1 Current and historical BIOX plants

Mine	Year commissioned	Capacity (t day^{-1} concentrate)	Reactor size (m^3)
Fairview, South Africa	1986	62	340
Sao Bento, Brazil	1990	150	550
Harbour Lights, Australia	1991	40	160
Wiluna, Australia	1993	158	480
Obuasi, Ghana	1994	1000	900
Coricancha, Peru	1998	60	262
Fosterville, Australia	2005	211	900
Suzdal, Kazakhstan	2005	520	650
Jinfeng, China	2007	790	1000
Bogoso, Ghana	2007	820	1500
Kokpatas, Uzbekistan	2009	2138	900
Runruno, Philippines	2016	404	1300
Cam & Motor, Zimbabwe	2021	100/*200*[a]	1200

[a]Future Phase 2 capacity

Typhoon Lando (Koppu), however, had a direct impact on the mine and plant site and resulted in significant delays in completion of the construction and obtaining the final permits required for commissioning. Ore commissioning was resumed in 2016 with the first gold from oxide and transitional material produced. The first BIOX product was pumped to the CIL section in November 2016.

4.2.1.2 Cam & Motor BIOX Plant

Gold mining activities at the Cam & Motor Gold Mine located near the city of Kadoma in Zimbabwe date back over 100 years utilising oxide ores and later producing a gold concentrate for treatment at a central roasting facility. Development of the new modern processing plant commenced with the installation of a 2000 t day^{-1} oxide treatment plant. Planning for the sulfide plant expansion commenced at the same time resulting in the installation of a flotation plant at the Cam & Motor Gold Mine in 2018.

BIOX was selected as the preferred technology for the oxidation of the refractory sulfide ore and construction of the BIOX plant commenced in 2019. The plant was designed for an ultimate capacity of 200 t day^{-1} concentrate at a sulfide-sulfur grade of 21.4% with the project implemented in two phases commencing with the installation of a 100 t day^{-1} Phase 1 BIOX plant. The Cam & Motor facility includes an ASTER plant for efficient water management.

The Phase 1 BIOX plant consists of three 1200 m^3 OKTOP BIOX reactors including dual OKTOP 3105 impellers for gas dispersion. The capacity will be expanded during Phase 2 with the installation of an additional three OKTOP BIOX reactors. Commissioning of the Phase 1 BIOX plant commenced in 2021 including inoculum build-up and operator training with first gold production in the second quarter of 2022.

4.2.2 Generation 3 and Generation 4 BIOX Design Development

The BIOX design has evolved over the years with the incorporation of the knowledge and experience from every project into the design of the latest BIOX plant. This, coupled with focused research and development programmes, led to the launch of the Generation 3 BIOX design philosophy in 2013. The main focus of the Gen 3 design was delivering improved process robustness and ease of operation. It comprises four main themes:

- Increased robustness of the process
- Process improvements
- Improved BIOX service offering
- Improved knowledge transfer

A structured process was followed to collate all the information from the recent BIOX commissioning programmes, BIOX plant audit campaigns, and operating experiences from the BIOX users. It was realised that there was a need to not only address the technical and process-related issues but also to implement a more comprehensive service package for BIOX clients.

The Generation 4 design aims to achieve reduction in the cost structure of a BIOX project by addressing the main capital and operating cost items. The main capital cost items include the BIOX tanks, agitators, and blowers while power and reagents are the two main operating cost items.

The development of the BIOX design programme and implementation philosophy is a continuous process and will continue beyond the successful roll-out of the Gen 3 and Gen 4 designs.

4.2.2.1 Generation 3 BIOX Design

The BIOX design and associated process development and implementation knowledge have developed continuously since commissioning of the first BIOX reactors at the Fairview mine in 1986. The first-generation BIOX plants commissioned up to 1998 were, with the exception of the Obuasi plant in Ghana, typified by smaller, lower-duty BIOX reactors. The second generation of BIOX plants commissioned between 2005 and 2010 was larger with higher duty BIOX reactors. Throughout this time, the lessons learned from each new and operating plant, combined with dedicated research and development programmes, were included in the following design. This led to the development and delivery of the Generation 3 BIOX design philosophy with the goal of increasing the robustness of the technology and increasing the service offering to the client. Four main focus areas of improvement were identified:

- *Increased Robustness*: The focus is to increase the overall robustness of the BIOX design by offering an improved BIOX design package, setting guidelines for BIOX agitator design and fabrication, and to ensure that the impact of ore and concentrate variability on the long-term performance of the circuit is well understood and addressed during each design phase.
- *Process Improvements*: The major potential problem areas experienced during the operation of a BIOX plant, including frothing in the BIOX circuit, BIOX sparge ring chokes, and colloidal gold losses experienced over the CCD circuit, were attended to.
- *Expanded BIOX service offering*: Improved guidance is provided to new operations on unit processes up and downstream from the BIOX circuit. In addition, a series of value adding products and services are also offered to assist with various needs of the clients.
- *Improved knowledge transfer*: The objective is to effectively transfer key BIOX knowledge to the correct target audience during each of the project phases.

The following section will give some examples of the implementation of the Gen 3 design principles during the execution of a new project, although not all aspects can be covered.

Improved Mechanical Reliability of BIOX Agitators A BIOX reactor mechanical specification was drafted to guide the design engineers and equipment suppliers as to what mechanical design considerations must be taken into consideration and what standards need to be met during the design of these units. The document addresses all mechanical design and basic fabrication requirements for both the internal and external reactor components. Safety factors and hydraulic service factors need to be sufficient to limit fatigue-related failures of shafts and gearboxes.

Design for Variability The Gen 3 design takes into account a more comprehensive evaluation of how the predicted variation in the feed ore may impact the plant design requirements. The plant design criteria are aligned with the sulfide-sulfur resource model and expected mining plan. Batch variability analysis and continuous test work are normally performed to map the metallurgical performance over as large as possible ore type distribution within the project resource. Cost-effective initiatives such as increasing the capacity of the BIOX feedstock tank and planning for ROM pad blending strategies can be used to limit the impact of ore variability.

Process Improvement Continuous process improvements are essential to address both problems experienced on the operating plants as well as opportunities identified for improving process performance or reducing costs. Improvements made are communicated to all BIOX users on a regular basis and if applicable, incorporated as standard practice in new BIOX plant designs.

Froth Suppression System Frothing of the BIOX reactors is not a new phenomenon and some of the plants have experienced severe frothing during and after commissioning. Frothing is normally caused by a combination of factors that include changes in the mineralogy of the concentrate, type and addition rates of flotation reagents, and bacterial activity. Though using anti-foaming agent is recommended for all BIOX clients, the Gen 2 BIOX process design package did not include a froth suppression system.

Improved BIOX Service Offering Inoculum maintenance and build-up is often an area of concern for the project team since it has a significant impact on the overall commissioning schedule. As part of value adding products, integrated inoculum build-up facilities can be supplied to potential clients. This can include the milling and flotation sections with all screening and liquid–solid separation equipment, the BIOX pilot plant with all associated services and equipment and the ASTER pilot plant also with all associated services and equipment. These pilot facilities can be installed in an existing or new laboratory facility or can be supplied as turnkey containerised facilities for stand-alone operation at the mine site.

Improved Knowledge Transfer One of the most important observations from the Generation 2 BIOX plant operations is the need for proper knowledge transfer to ensure that the appropriate information reaches the correct audience. In the BIOX

Generation 2 plants, training and knowledge transfer were targeted at the plant operators, metallurgists, and plant managers, each according to their required level of understanding. However, production pressure has a significant influence on the operational and management decisions taken. Not having a full understanding of the potential implication of decisions on the performance of the BIOX process may lead to an incorrect operational philosophy which in turn leads to poor BIOX performance. Significant improvements in performance have been seen when the correct BIOX operational procedures and standards are followed and short-term production pressures are allowed to take a second place in view of long-term process stability.

4.2.2.2 Generation 4 BIOX Design

Continued development of the process is critical to improve process efficiency and reduce cost. The Generation 4 BIOX design focused on addressing the main capital and operating cost drivers to affect a step change lower in the cost structure of the BIOX process.

Reduction of Cyanide Consumption Cyanide consumption during leaching of the biooxidation product is one of the main operating costs for most operating BIOX plants. The mechanisms for cyanide consumption were investigated in detail, leading to the development of the MesoTHERM process, which is discussed in Sect. 4.3.

Improved Aeration and Agitation System The biooxidation reactor train creates the core of the BIOX process performance. The BIOX process requires large amounts of oxygen for oxidation of sulfidic compounds and thus the main criteria for agitation are to provide sufficient oxygen mass transfer from air to solution. The power requirement for agitation and aeration to supply oxygen to the process constitutes a significant portion of the BIOX capital and operating costs.

Recent development work has shown that a dual impeller system will deliver a step change in efficiency compared to the standard single large hydrofoil impeller systems. Various impeller configurations were tested using a laboratory test reactor to identify the optimum configurations. The optimum configurations were then tested using the 21 m^3 BIOX test reactor at the Fairview BIOX plant. Water tests were done to determine the oxygen mass transfer coefficients ($k_L.a$) and gas handling capabilities of the different configurations under typical power input and gas flow rates expected in a commercial BIOX reactor. The 21 m^3 test reactor was then operated as a primary BIOX reactor using fresh concentrate feed from the Fairview BIOX plant, confirming the performance of the impellers under operating BIOX conditions.

4.2.3 *BIOX at Sub-zero Temperatures*

During the BIOX process sulfide minerals such as pyrite, arsenopyrite, and pyrrhotite are oxidised by a microbially assisted chemical leaching process. These oxidation reactions are highly exothermic and on average 30 MJ kg^{-1} sulfide oxidised is generated depending on the mineralogy of the concentrate. The BIOX microbial culture needs to be maintained at a temperature of around 40 °C and the reactors are cooled by circulating water through internal cooling coils. The amount of cooling required per reactor is dependent on the extent of sulfide oxidation and the magnitude of heat losses from the reactor, mainly in the form of convection and radiation.

The Suzdal BIOX plant in Kazakhstan was the first BIOX plant to operate in sub-zero temperatures. The original design capacity of the plant was 192 t day^{-1} of concentrate at 12% sulfide-sulfur. Detailed heat balance calculations performed for the Suzdal BIOX process indicated that all the reactors, except the last secondary reactor, would require cooling even at the minimum design ambient temperature of −50 °C. Convection heat losses from this reactor could be reduced by placing insulation material with a low thermal conductivity on the outside of the reactor. The following changes were made to the standard design to meet the challenges posed by the sub-zero temperatures at Suzdal:

1. The feedstock tank, counter-current decantation, neutralisation, water recovery, and reagent make-up sections were positioned inside the main building for control and maintenance purposes.
2. Covered walkways provide access to the top of the BIOX reactors and all the major process lines and control valves (air, water, and sulfuric acid) were placed inside the covered walkways for maintenance purposes.
3. Additional standby power was provided to run certain sections of the BIOX plant during extended power outages.

The Suzdal BIOX plant successfully ran through the 2005/6 winter and sulfide oxidation rates above design were achieved.

4.3 The MesoTHERM Process

Metso Outotec's MesoTHERM technology uses a combination of microbial consortia as a hybrid, two-stage biooxidation process where mesophiles are used to realise the primary stage sulfide oxidation and thermophiles to bring about the latter stage, near-complete sulfide oxidation.

While biooxidation may be described as a simple process, the actual oxidation of sulfide minerals is complex. Refractory gold and base metal-bearing minerals typically present sulfur in oxidation states of −2 and −1 in sulfide minerals (Sand et al. 1995). It is improbable that the oxidation path of sulfide to sulfate (S oxidation state, +6) is simple and direct during biooxidation as the microbial facilitated kinetics is slow and biooxidation likely occurs via single electron transfer steps with rate-limiting steps favouring the existence of metastable species of intermediate oxidation states.

This pathway focuses specifically on the oxidation of pyrite by iron-oxidising *Acidithiobacillus* spp. with thiosulfate as a transient intermediate. Acid-unstable thiosulfate is converted to polythionates and sulfites and soluble sulfates are the final stable species of biooxidation, which are removed by washing except the small fraction retained as jarosite. As may be inferred, the total reaction pathway associated with biooxidation is complex, resulting in the conversion of sulfides into potential cyanicides and finally to benign sulfates. These labile sulfur species, although unstable, may form to a significant extent and be part of the final product of the biooxidation stage and it was believed that the accumulation of the labile sulfur species as globules either inside or outside cells and/or on residual solids react with cyanide to increase cyanide consumption through the general reaction schemes shown as Eqs. (4.1–4.3) (Luthy and Bruce 1979), with thiocyanate as the principal remaining species.

$$S_xS^{2-} + CN^- \rightarrow \left[S_{(x-1)}\right]^{2-} + SCN^- \tag{4.1}$$

$$S_2O_3{}^{2-} + CN^- \rightarrow SO_3{}^{2-} + SCN^- \tag{4.2}$$

$$S^0 + CN^- \rightarrow SCN^- \tag{4.3}$$

The early batch biooxidation test work using the mesophile—thermophile combination showed that much lower cyanide consumptions were obtained when compared to the traditional mesophilic cultures, with effluent leach liquors containing lower thiocyanate levels. This pointed to a cleaner biooxidised residue being produced post the higher temperature thermophile biooxidation stage resulting in less reactive sulfide species. This early work showed that not only was the cyanide speciation different, but also the cyanide consumption was reduced by almost 50%.

These results appeared to support the notion that the higher thiocyanate formation and hence an increase in the cyanide consumption on the mesophile product was being promoted by higher labile sulfides carrying over to the leach with a lower level presenting from the MesoTHERM product. To ascertain whether nonbiological mechanical strategies could effect a reduction in the labile sulfide carry over to the leach further tests were conducted which saw highly oxidised mesophile residue undergo various conditioning treatments (Legodi et al. 2020). The types of pre-treatment conditions undertaken on the mesophile biooxidation product included:

- Extensive washing using acidified water followed by normal water.
- Heating of the biooxidation product to 80 °C followed by centrifuging and thereafter extensive washing using acidified followed by normal water.
- Extensive washing with acidified water followed by a 24 h period of caustic conditioning and thereafter normal water washing again.
- Attritioning of the biooxidation product (in a Denver cell at 1000 rpm for 4 h) followed by normal water washing.

These biooxidation residue pre-treatment steps to reduce the carryover of sulfur species to the cyanidation step were unsuccessful with most of the cyanide added converting to thiocyanate. Additional washing stages as well as fine grinding the solids did little to limit the thiocyanate formation with 84% and 87% cyanide reporting as thiocyanate, respectively. Heating to 80 °C and centrifuging and the low-temperature caustic conditioning only marginally reduced the fraction of the cyanide converted to thiocyanate viz. 67% and 68%, respectively. For a thermophile biooxidation residue, 19% of the cyanide was converted to thiocyanate.

4.3.1 MesoTHERM Biooxidation Cultures and Operating Conditions

The mesophile culture operating in the temperature range 38–42 °C was shown by qPCR analysis to include both iron and sulfur oxidising bacterial and archaeal species. The thermophilic culture operating in the range 60–68 °C indicated the presence of *Acidiplasma cupricumulans*, mostly dominant at moderate thermophilic temperatures, and species belonging to the *Metallosphaera* genus becoming more dominant at higher operating temperatures.

4.3.2 Pilot Scale (240 L and 1000 L) Development

During 2006 and later between 2015 and 2016, Metso Outotec commenced two continuous pilot programmes onsite at the Barberton Mines Fairview BIOX plant with the principal objectives to:

- Have access to actual fresh plant feed and biooxidation slurry.
- Determine the extent of gold dissolutions achievable in a two-stage mesophile—thermophile circuit.
- Determine the cyanide consumptions achievable in this biooxidation configuration.

In both the trials spanning the different thermophile reactor sizes and various trial periods, the piloting circuits were configured as thermophile reactors receiving a partially oxidised mesophile plant product, decanted to remove the acidic ferric sulfate (dominant species) solution with the solids reconstituted in water and serving as the sulfide substrate for the thermophile stage.

The thermophile stage was operated in the range 63 °C to 68 °C and process water supplemented with the standard BIOX nutrient suite was used to repulp the partially oxidised material which served as the feed. Low-pressure blower air from the plant was used to supply the oxygen demand and both campaigns operated well. Daily overflow samples were collected from the pilot plant and routinely leached to determine the MesoTHERM performance and were compared with the leach

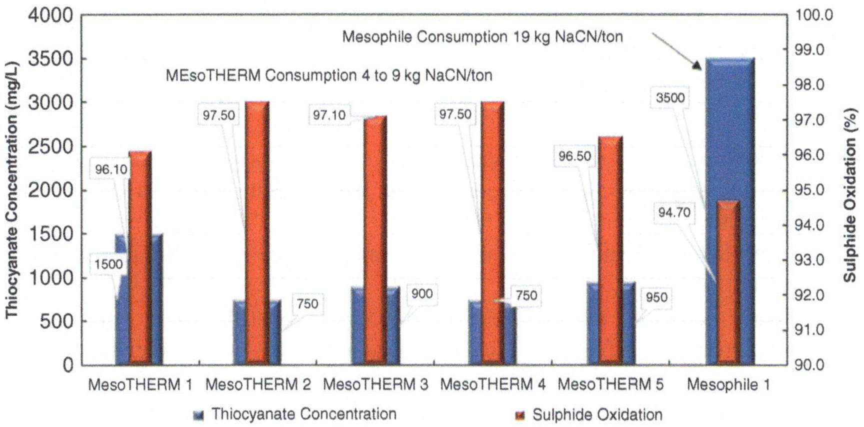

Fig. 4.1 MesoTHERM and mesophile leach liquor thiocyanate levels and corresponding sulfide oxidation

performance results derived from mesophile biooxidation residue. Figure 4.1 shows the summary results of selected leach solutions derived on the corresponding biooxidation residue batch leach and shows the respective residual thiocyanate concentration (in the cyanide leach effluent) and the corresponding extent of sulfide oxidations.

The figure shows that the leach liquors produced on the MesoTHERM residues contained lower levels of SCN and also highlights the difference in cyanide consumption obtained on the residues produced from the two biooxidation processes. The pilot trials were therefore successful in validating the early laboratory batch tests and demonstrated the feasibility of such a two-stage biooxidation circuit to maintain the overall process performance with a lower cyanide consumption.

4.3.3 Engineering Scale Up (21 m^3) and Large Demonstration (80 m^3)

The approach of utilising mesophiles for the initial extent of sulfide oxidation is quite novel as it lowers the sulfide loading and duty for the subsequent thermophile stage, thus resulting in a less onerous mass transfer requirement at the higher thermophile temperatures. This is quite important as at higher temperatures the solubility of oxygen is lowered and depending on the temperature, to achieve the required oxygen mass transfer, oxygen-enriched air may be required. Metso Outotec wanted to deliver a new yet simple biooxidation process that precluded the use of oxygen and, while the operation of the two pilot plants showed that the use of air was adequate to supply the oxygen requirement, further confirmatory work was required. The next stage of development was therefore planned to evaluate some engineering scale-up considerations in larger biooxidation reactors to ensure an adequate design basis could be established for the MesoTHERM technology. These scale-up and larger demonstration campaigns were conducted in 21 and 80 m^3 biooxidation reactors and the objectives were therefore to:

- Evaluate the robustness of the thermophile culture by exposing it to varying feed (pyrite and arsenopyrite) blends and operating the reactor in situ in the commercial circuit.
- Determine the tolerance of the culture to varying solids concentrations and total dissolved solutes.
- Evaluate the actual mass transfer rates achievable using air at temperatures ranging from 63 to 68 °C by measuring aqueous dissolved oxygen (DO) and off-gas concentrations.
- Operate the reactor at impeller tip speeds expected in commercial operation and monitor whether mechanical shear impacts microbial performance.
- Show that the same metallurgical performance can be achieved as before in the smaller scale piloting trials (cyanide consumption reduction, gold dissolutions, and sulfide oxidations).

The 21 m^3 thermophile reactor was integrated into the commercial circuit with an automated feed and aeration control system and was configured to receive fresh concentrate slurry or primary stage biooxidation product as feed. The reactor was fitted with a gas chimney to allow oxygen off-gas measurements to be undertaken and together with aqueous dissolved oxygen measurements recorded allowed for the mass transfer rates to be determined. The reactor operating philosophy was in much the same way as the earlier pilot trials receiving decanted partially oxidised slurry from the commercial primary reactors, diluted thereafter with process water and standard nutrients.

A summary of the MesoTHERM engineering scale performance results is shown in Fig. 4.2 which shows the comparative cyanide consumptions and gold dissolutions achieved on residues derived from the combined MesoTHERM circuit. The 21 m^3 data is plotted against the previously recorded pilot data and that typically obtained in a mesophile circuit which depicts the gold dissolution and cyanide consumption represented as horizontal lines.

Performance results from the 21 m^3 pilot trial compared favourably to the two earlier pilot trials and showed that the gold dissolutions were similar to the mesophile results and that cyanide consumptions were reduced to within 2.5–11 kg NaCN/ton on MesoTHERM biooxidation residues. Typical cyanide consumption achieved using only a mesophile biooxidation process on this material averaged 20 kg NaCN t^{-1} with virtually all cyanide being consumed in the leach.

The operation of the 21 m^3 trial, therefore, validated the earlier piloting and batch trials and allowed a larger in situ demonstration assessment of the thermophile technology by way of robustness, performance, and also allowed certain key operational parameters to be evaluated. A view of the larger 80 m^3 reactor is shown in Fig. 4.3 with some design and performance data derived summarised thereafter in Table 4.2.

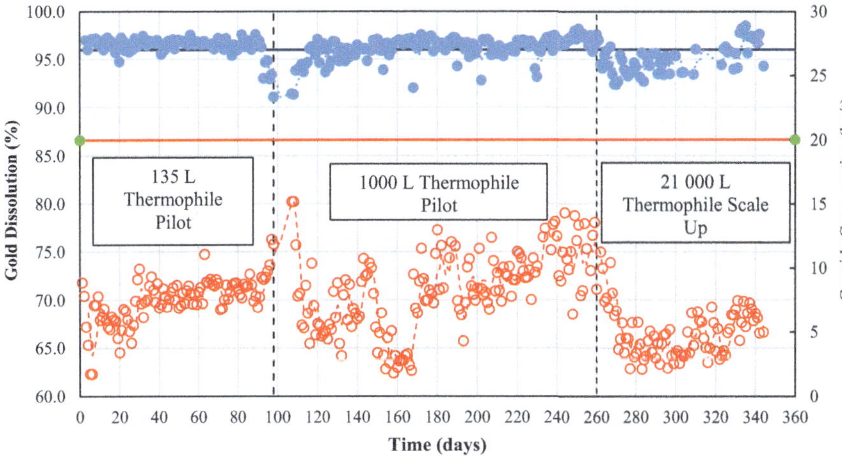

Fig. 4.2 Comparison of mesophile and MesoTHERM performance data in pilot trials and larger engineering scale (●, MesoTHERM, black line mesophile) gold extraction; (o MesoTherm, red line mesophile) cyanide consumption)

Fig. 4.3 Views of the engineering (21 m^3) and demonstration (80 m^3) scale thermophile reactors

Table 4.2 80 m^3 In situ thermophile operational measurements

Parameter	Range measured
Dissolved oxygen (mg L^{-1})	1.5–3.5
Sulfate concentration (g L^{-1})	30–90
Tip speed (m s^{-1})	≤ 8
Oxygen utilisation (%)	10–40
Gold dissolution (%)	90–97
Cyanide consumption (kg t^{-1})	5–10

4.3.4 MesoTHERM Circuit

The successful piloting, engineering scale-up assessments, and demonstration trialing of the MesoTHERM technology present a new generation BIOX process to specifically target ores that traditionally have realised high cyanide consumption. The MesoTHERM biooxidation technology leverages off the well-established and commercialised BIOX process and the novel design of splitting the oxidation duties and managing the metal and salt levels between the biooxidation stages allows a less onerous mass transfer and a more efficient thermophile operation. Figure 4.4 provides a schematic of the Metso Outotec MesoTHERM biooxidation circuit. In July 2020, Metso Outotec launched MesoTHERM as a process option for refractory gold extraction.

4.4 ASTER Process

4.4.1 Process Description

The ASTER (*A*ctivated *S*ludge *T*ailings *E*ffluent *R*emediation) process was developed to remove SCN^- and CN^- in process waters to very low levels. Environmental legislation associated with the land disposal of cyanidation tailings and water discharge is becoming increasingly stringent worldwide, enforcing the need to treat or recycle cyanide-contaminated water streams. The ASTER process enables the treatment of tailings effluent streams with SCN^- concentrations as high as ~4500 mg L^{-1} down to levels below 1 mg L^{-1}. This makes the treated tailings water suitable for recycling back to the BIOX process or the seasonal release to the environment (van Buuren et al. 2011).

An ASTER circuit comprises a series of aerated reactors containing microorganisms that hydrolyse the CN^- and SCN^- in tailings solutions or dilute slurries. Molasses and phosphorus are added as nutrients to sustain and promote microbial growth. The flow sheet, shown in Fig. 4.5, is typically configured as a number of primary reactors in parallel, feeding one or more secondary reactors in series. The circuit may also include a settler to accumulate sludge for recycling to the primary ASTER reactors and to clarify the detoxified ASTER solution. The presence of residual cations in some solutions will require a purge stream from the settler underflow to avoid heavy metal accumulation once adsorbed onto the sludge.

The typical hydraulic residence time of an ASTER circuit is between 6 and 12 h depending on factors such as the SCN^- and CN^- concentrations in the feed and operating temperature. Although initially developed to treat solutions, the process has been adapted to treat dilute slurries with feed solids concentrations of up to 6%. In the case of treating CIL tails, the slurry is first subjected to a cyanide detoxification stage using Air-SMBS and copper to bring the WAD CN^- present in the ASTER feed stream to below 10 mg L^{-1} to ensure optimal SCN^- degradation rates in the

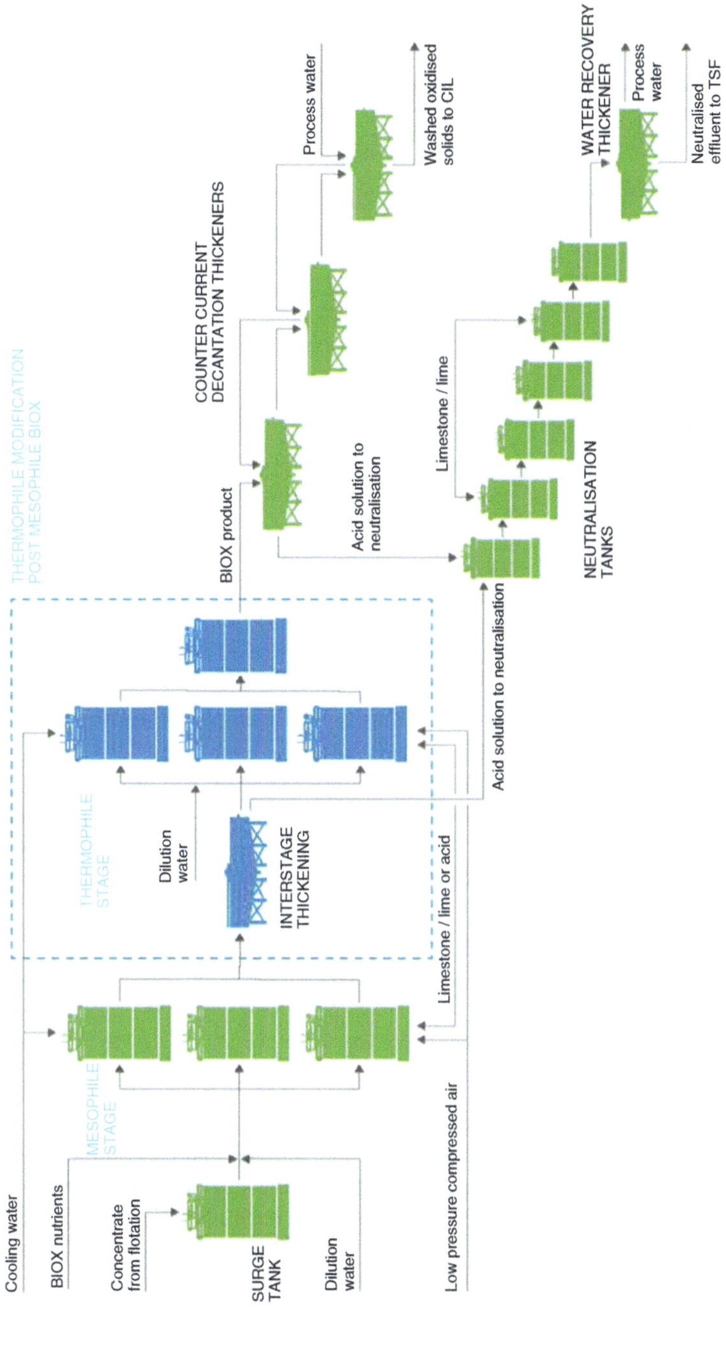

Fig. 4.4 MesoTHERM BIOX flowsheet for refractory gold

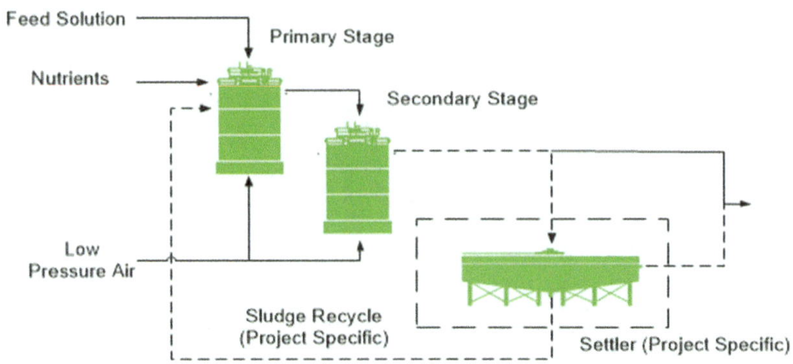

Fig. 4.5 Typical ASTER circuit configuration

Table 4.3 Typical ASTER circuit operating parameters

Parameter	Control range
Operating temperature	20–25 °C
Feed thiocyanate concentration	< 4500 mg L^{-1}
Feed free cyanide concentration	< 50 mg L^{-1}
Feed slurry density	0–6% solids
pH	7–9.5
Dissolved oxygen	> 2 mg L^{-1}
Organic nutrient: Molasses	0.15 kg m^{-3}
Inorganic nutrient: (fertiliser P, K)	0.13 kg m^{-3}

ASTER reactors. The operating parameters typically employed for the process are listed in Table 4.3.

Biological degradation of cyanide and thiocyanate in the ASTER process is achieved through a mixed consortium of bacteria, fungi, and algae known to enzymatically hydrolyse and oxidise cyanide (van Hille 2011). The proposed reaction pathways as well as the overall reactions are summarised in Eqs. (4.4–4.7):

$$SCN^- + H_2O \rightarrow HCNO + HS^- \quad (4.4)$$

$$HCNO + 2\,H_2O \rightarrow NH_4 + HCO_3^- \quad (4.5)$$

$$HS^- + 2\,O_2 \rightarrow SO_4^{2-} + H^+ \quad (4.6)$$

$$SCN^- + 3\,H_2O + 2O_2 \rightarrow HCO_3^- + NH_4 + SO_4^{2-} + H^+ \quad (4.7)$$

The primary reaction products are ammonia, sulfate, and carbon dioxide/bicarbonate. A total of four metabolic pathways for the degradation of cyanide and cyanide compounds have been described and have been identified in both prokaryotes (bacteria) and eukaryotes (fungi and algae; Dash et al. 2009): the hydrolytic, oxidative, reductive, and substitution/transfer pathways. The first three are responsible for the degradation of cyanide or thiocyanate to simpler molecules that can be assimilated, with the expenditure of metabolic energy, as carbon and nitrogen

Table 4.4 ASTER culture identification, role, and biological classification

ASTER dominant species	Role in SCN^- and / or CN^- degradation
Ralstonia eutropha	Degrades SCN^-
Bosea thiooxidans	Degrades SCN^-
Pseudomonas stutzeri isolate OC-10	Degrades CN^-
Microbacterium schleiferi strain 118	No direct role
Acinetobacter sp. ST-01	Degrades environmental pollutants
Or *Acinetobacter venetianus* strain L17	Degrades SCN^- and CN^-
Cuprivadus necator	

sources. The substitution/transfer pathway involves the direct assimilation of cyanide. The microbial community in the ASTER process, shown in Table 4.4, was isolated at the Pan African Resources—Barberton Mines during the 1990s and is constituted as a number of composites of organisms drawn from the Fairview tailings dam.

All major microbial species detected in the ASTER culture are classified as lowest risk with regard to potential pathogenicity according to the Technical Rules for Biological Agents (TRBA) classification and the US Centers for Disease Control and Prevention (CDC). As part of ongoing research and process monitoring, 16S rRNA gene amplicon sequencing was performed during 2019 on various Suzdal ASTER samples as well as in an ASTER laboratory liquid medium culture maintained by CeBER at the University of Cape Town. The resulting (non-chimeric) sequences were clustered and the respective operational taxonomic units (OTUs) were taxonomically assigned on the basis of representative sequences. For both the Suzdal ASTER reactor samples and the CeBER maintained ASTER culture, *Thiobacillus* OTU (OTU 15) was now dominant and is believed to be one of the key microorganisms currently involved in thiocyanate destruction.

4.4.2 Commercial ASTER Operations

There are currently four ASTER plants in operation globally with a fifth plant currently under construction. Basic details for the various ASTER plants are shown in Table 4.5 and images of two of the plants in Fig. 4.6.

4.4.2.1 Consort ASTER Plant

The ASTER plant at the Pan African Resources' Consort mine in South Africa was commissioned in 2010 and was designed to treat 320 m^3 day^{-1} of tailings solution containing SCN^- at an average feed concentration of 120 mg L^{-1} and a CN^- concentration between 10 and 30 mg L^{-1}. The circuit consists of four primary

Table 4.5 Commercial ASTER plants

Mine	Year commissioned	Capacity (m^3 day^{-1})	Feed SCN^- level (mg L^{-1})
Consort, South Africa	2010	320	150
Suzdal, Kazakhstan	2013	528	1200
Runruno, Philippines	2016	5000	350
Fosterville, Australia	2021	792	5000
Cam & Motor, Zimbabwe	2022	7292	300

Fig. 4.6 Views of Runruno (left) and Fosterville (right) ASTER plants

reactors in parallel and four secondary reactors in series for a total retention time of 12 h. Overflow from the final secondary reactor cascades into a clarifier, which returns thickened sludge to the primary reactors. The clarified overflow, containing a final SCN^- concentration of <0.5 mg L^{-1} is used as process water in the flotation plant producing concentrate for the Fairview BIOX plant. The introduction of the ASTER process has resulted in a significant improvement in the water balance across the operation, reducing the requirement for freshwater.

4.4.2.2 Suzdal ASTER Plant

The Suzdal ASTER plant located at the Suzdal Gold Mine in Kazakhstan was commissioned in 2013. The plant had an initial design feed rate capacity of 528 m^3 day^{-1} containing 1200 mg L^{-1} to 1500 mg L^{-1} SCN^-. Treatment rates have since increased to around 900 m^3 day^{-1} combined with SCN^- levels increasing to 2800 mg L^{-1}. The circuit consists of two aerated (non-agitated) tanks in series. A static settler is used to recycle thickened biomass to the primary reactor while simultaneously delivering a clear overflow solution used for process water.

4.4.2.3 Runruno ASTER Plant

The ASTER circuit at the Runruno BIOX Gold Mine in the Philippines was commissioned in 2016 (van Niekerk et al. 2017). The ASTER process is used to treat the CN^- and SCN^--containing CIL tailings slurry prior to discharge to the residue storage impoundment. For this application, the process was adapted to treat a dilute slurry containing 4.5–8% solids at 350 mg L^{-1} of SCN^- and ~30 mg L^{-1} of CN^- at a design treatment rate of 5000 m^3 day^{-1}. The circuit consist of four degradation reactors configured as three primary reactors fed in parallel and one secondary reactor treating the combined primary reactor overflow. No settler was included for this circuit. Extensive pilot test work was conducted in South Africa to validate and confirm the ability of the ASTER process to treat a dilute tailings slurry and to confirm the process design criteria. The success of the ASTER circuit is vital for the Runruno operation to manage the plant water balance, allowing a detoxified slurry to be deposited onto the Residue Storage Impoundment (RSI). Decant water from the RSI is circulated as process water to the process plant, including the BIOX circuit.

4.4.2.4 Fosterville ASTER Plant

The Fosterville ASTER plant marked the fourth commercialisation of the technology and was commissioned in the first quarter of 2021. The plant has a tailings solution treatment rate of 792 m^3 day^{-1} and is designed to treat up to 4000 mg SCN^- L^{-1}. The plant consists of six tanks with operating volumes of 180 m^3 each, followed by a static settler. The static settler allows for the recycling of thickened biomass to the primary ASTER tanks thereby increasing the biomass in these tanks, which enables a high degradation rate to be achieved. To achieve and maintain the high degradation rates, the six ASTER tanks can also have the feed configuration to the primary and secondary reactors modified to support variances in the SCN^- degradation load required. Detoxification of this high level of SCN^- to values <1 mg L^{-1} allows the detoxified solution to be recycled upstream and used as process water.

4.4.2.5 Cam and Motor ASTER Plant

Commissioning and ramp up of the Cam & Motor ASTER installation in Zimbabwe in the fourth quarter of 2022 will realise five successful landings for the technology. This ASTER plant has a tailings solution design treatment rate of 7292 m^3 day^{-1} and will detoxify SCN^- levels in the range 200 mg L^{-1} to 300 mg L^{-1}. The ASTER facility utilises some redundant tankage at the mine, thermal energy from the BIOX facility to maintain optimal tank temperatures, and also taps in to the BIOX aeration supply to target a SCN^- effluent level of <1 mg L^{-1}. Detoxified tailings solution is transferred to the process water pond and available for use upstream and downstream of the BIOX circuit as process water.

4.4.3 ASTER Circuit Design Considerations

This section describes the important parameters to take into consideration during the design and operation of an ASTER plant.

4.4.3.1 Feed Solution Properties

The ASTER process is suitable to treat both solution and dilute slurries over a wide range of feed SCN^- concentrations. Feed SCN^- concentrations as high as 5000 mg L^{-1} and feed solids concentrations up to 8% has been treated successfully. The degradation rate achieved is a function of the feed SCN^- concentration, biomass concentration, reactor retention time, and operating temperature. SCN^- degradation rates as high as 150 mg L^{-1} h^{-1} has been recorded for a well-adapted ASTER culture. The microorganisms are able to adapt, within limits, to changes in flow rates and substrate concentrations in the short term, providing robustness to the process.

4.4.3.2 Biomass Retention

It is important to maintain a sufficient microbial cell density in the ASTER reactors in the form of activated sludge. Biomass retention and accumulation, initially through the formation of sludge granules and through the recycling of the suspended sludge from the ASTER settler underflow is an important process requirement. For circuits with no settler included, circulation of inoculum from the secondary reactor to the primary reactors is effectively used during ASTER process upsets to restore the performance of the circuit.

4.4.3.3 *Pre-ASTER* Cyanide Destruction

It has been observed that CN^- concentrations in excess of 10 mg L^{-1} in the ASTER feed can have an inhibitory effect on the SCN^- degradation rate. When treating a leach tailings stream containing elevated cyanide concentrations an SO_2/air cyanide destruction step is included in the process flow sheet to reduce the CN^- concentration to below 10 mg L^{-1} in the ASTER feed. The addition of SMBS to the leach tails slurry has no inhibitory effect on the ASTER performance. The copper addition rate is also managed as it is a natural biocide and can result in ASTER culture inhibition at too high concentrations.

4.4.3.4 ASTER Process Operating Conditions

Published data (Patil and Paknikar 1999; Gokulakrishnan and Gummani 2006) indicate an optimum temperature of between 30 and 35 °C for the species identified in the ASTER culture. Test work and commercial scale operation have, however, indicated that operating temperatures in the range 20–30 °C have no detrimental effect on the performance of the ASTER circuit. The process can be operated at temperatures as low as 12 °C, though a significant reduction in the SCN^- degradation rate is observed as the temperature is lowered.

Maintaining a constant dissolved oxygen level (>3 mg L^{-1}) is crucial to ensure optimal performance of the ASTER circuit. It is important to ensure constant air supply via a reliable low-pressure blower in combination with an efficient sparge system. Optimal ASTER degradation rates are achieved in the pH range $7.0 \leq pH \leq 8.0$. The process is sensitive to pH changes with high operating pH levels (>9.5) more problematic than low pH (<7.0) levels. High feed pH values in excess of 9.5 negatively impact the ASTER performance even if the reactor pH is below 8.5.

Due to the short retention time in the ASTER circuit, automation of the circuit is encouraged to assist with process monitoring and control. This may include online pH and DO analyses and auto-titration for SCN^- and CN^- in the feed and product streams. Accurate mass and volumetric feed rate measurement and control is pivotal for the successful operation of an ASTER circuit.

4.4.3.5 ASTER Reaction Products

A significant challenge common to many chemical treatment processes relates to the management of reaction products, specifically the nitrogen-containing species. Ammonia and nitrate are significant contributors to the eutrophication potential of wastes and are covered by stringent discharge specifications in most countries. At low SCN^- and CN^- feed concentrations, the majority of the nitrogen-containing products are assimilated by the ASTER microorganisms to sustain their growth. At higher SCN^- feed concentrations ammonia is produced faster than it can be assimilated and accumulation to high concentrations (>700 mg L^{-1}) can occur preventing discharge to the environment without treatment. Similarly, at high SCN^- degradation rates the sulfate concentration in the effluent can exceed 4000 mg L^{-1}. This is significantly higher than the discharge specifications in most countries. These challenges could be alleviated where ASTER is implemented alongside a BIOX plant if the treated ASTER water can be recycled back to the bioleach reactors. For circuits where this is not possible, additional treatment may be required including denitrification.

4.4.4 ASTER Circuit Capital and Operating Costs

The mechanical equipment supply (MES) cost for an ASTER circuit is dominated by the cost of the ASTER reactors. The inside of the mild steel reactors is typically coated with a polyamine cured phenolic epoxy with the heat exchange coils and air spargers constructed from either 304 L or 316 L stainless steel. The settler and the low-pressure blowers are also significant cost items. The typical operating cost range for ASTER circuits is in the range of 0.4–1.0 US$ m^{-3}. Power consumption, often being the main operating cost, is determined by the aeration and agitation requirement to maintain the slurry in suspension and supply the required oxygen to the process. Slurry heating can also be a significant operating cost in colder climates if no waste heat is available.

References

Dash RR, Gaur A, Balomajumder C (2009) Cyanide in industrial wastewaters and its removal: a review on biotreatment. J Hazard Mater 163:1–11. https://doi.org/10.1016/j.jhazmat.2008.06.051

Gokulakrishnan S, Gummani SN (2006) Kinetics of cell growth and caffeine utilization by *Pseudomonas* sp. GSC 1182. Process Biochem 41:1417–1421. https://doi.org/10.1016/j.procbio.2005.12.018

Legodi W, Kweto B, Sandenbergh RF, Van Buuren C (2020) Strategies for reducing cyanide consumption during cyanidation of BIOX treated gold ores, Cape Town. In: International mineral processing congress, Cape Town, pp 4–5

Luthy RG, Bruce SG (1979) Kinetics of reaction of cyanide and reduced sulfur species in aqueous solution. Environ Sci Technol 13:1481–1487. https://doi.org/10.1021/jm00001a005

Patil YB, Paknikar KM (1999) Removal and recovery of metal cyanides using a combination of biosorption and biodegradation processes. Biotechnol Lett 21:913–919. https://doi.org/10.1023/A:1005550707798

Sand W, Gehrke T, Hallman R, Schippers A (1995) Sulfur chemistry, biofilm, and the (in) direct attack mechanism – a critical evaluation of bacterial leaching. Appl Microb Biotechnol 43:961–966. https://doi.org/10.1007/BF00166909

van Buuren C, Makhotla N, Olivier JW (2011) The ASTER process: technology development through to piloting, demonstration and commercialization. In: Proceedings of the ALTA 2011 Nickel-Cobalt-Copper, Uranium and Gold Conference, Perth 23–28 May 2011

van Hille RP (2011) Critical review of the ASTER technology for the treatment of cyanide and thiocyanate containing process water, Metso Outotec Internal Report

van Niekerk JA, Olivier JW Savelle M (2017) Commissioning of the BIOX and ASTER™ processes at the Runruno gold project, proceedings of the 7th world gold conference, Paper 9385, p 7–8

Chapter 5
Biomining Microorganisms: Diversity and *Modus Operandi*

Mark Dopson and Naoko Okibe

Abstract Consortia of biomining microorganisms catalyze metal sulfide dissolution for extraction and recovery of metals such as gold and copper by regenerating the ferric iron oxidant along with metabolizing the resultant elemental and reduced inorganic sulfur compounds. These microorganisms are from the Bacteria and Archaea domains with the Bacteria having generally lower growth temperatures while the Archaea comprise mostly moderate and extreme thermophilic species. All microorganisms used in current biomining operations are able to grow at acidic pH values along with metal tolerance systems that allow them to survive the multiple extreme conditions in leaching liquors. The bacterial and archaeal ferrous iron oxidation systems differ while their reduced inorganic sulfur compound metabolisms contain enzyme pathways that are similar but not identical between the two domains. In addition, dissolution of non-sulfidic ores such as oxyhydroxide minerals, and electronic wastes, can also be mediated via biogenic acidolysis and complexation or by microbial Fe-reducing activity. Recent advances in "omics" technologies have aided in identifying new acidophilic biomining species and future studies will continue to elucidate their *modus operandi* to aid in increasing rates of mineral dissolution.

Keywords Acidophile · Phylogeny · Iron oxidation · Sulfur oxidation · Autotrophy · pH

M. Dopson (✉)
Centre for Ecology and Evolution in Microbial Model Systems (EEMiS), Linnaeus University, Kalmar, Sweden
e-mail: mark.dopson@lnu.se

N. Okibe
Department of Earth Resources Engineering, Faculty of Engineering, Kyushu University, Fukuoka, Japan
e-mail: okibe@mine.kyushu-u.ac.jp

D. B. Johnson et al. (eds.), *Biomining Technologies*,
https://doi.org/10.1007/978-3-031-05382-5_5

5.1 Biomining Microorganisms

Biomining is an encompassing term for both "biooxidation" where the target metal of value is trapped within a metal sulfide mineral matrix such as microscopic particles of gold within arsenopyrite (FeAsS)/pyrite (FeS_2), and "bioleaching" where the target metal is an integral part of the sulfide mineral matrix, such as copper in chalcopyrite ($CuFeS_2$) and nickel in pentlandite ($(Ni,Fe)_9S_8$). Microorganisms that catalyze metal sulfide biomining are exclusively from the Archaea and Bacteria domains. Bioprocessing non-sulfidic ores and wastes may also involve eukaryotes, and is briefly discussed at the end of this chapter. The Eukarya also contain acidophilic species such as protozoa and the red alga *Galdieria sulphuraria* that have been identified in acid mine drainage (AMD) and a wide diversity of eukaryotes inhabit the acidic Rio Tinto in Spain. While these species can influence mineral dissolution in some situations (e.g., by predating on acidophilic prokaryotes), they do not have a direct role in catalyzing metal dissolution and are not discussed in this chapter.

Biomining microorganisms can be broadly separated by growth temperature optima with Bacteria typically growing at temperatures between 5 and 55 °C while biomining Archaea predominantly grow at higher temperatures, 37–80 °C although *Cuniculiplasma divulgatum* can grow as low as 10 °C. All characterized acidophiles have an optimum growth temperature above 15 °C though several bacterial (but not archaeal) species, such as *Acidithiobacillus ferrivorans*, can grow close to 0 °C. These temperature optima have been widely used to categorize acidophiles into psychrotolerant/eurypsychrophiles, mesophiles, thermotolerant/moderate thermophiles, and extreme thermophiles. However, these definitions are not fixed and have been variously defined between publications.

Microorganisms growing in typical biomining environments such as bioheaps, stirred-tank reactors, mine waters, and wastes are faced with a number of challenges that require them to thrive in multiple extreme conditions and they are termed "polyextremophiles" (reviewed in Harrison et al. 2013). An essential trait of sulfide mineral biomining organisms is the ability to grow at low pH: moderate acidophiles have been defined as having a pH optimum for growth of between 3.0 and 5.0, extreme acidophiles have an optimum between 1.0 and 3.0, and hyper-acidophiles are categorized with a pH optimum of <1.0. Acidophiles are widespread in the phylogenetic tree of life (Fig. 5.1) though it is unclear if these lineages have a common ancestor, if the ability to survive at low pH has independently evolved multiple times or genes coding for pH homeostasis mechanisms have been transferred between lineages by, for example, horizontal gene transfer. The most extreme acidophiles (e.g., *Picrophilus* and *Ferroplasma* spp.) can survive negative pH values and all acidophiles possess a range of proton homeostasis mechanisms (reviewed in Slonczewski et al. 2009). These mechanisms are not all present in every acidophile species, but include a reversed (compared to neutrophiles) inside positive membrane potential that varies with the external pH and thus inhibits proton influx to various

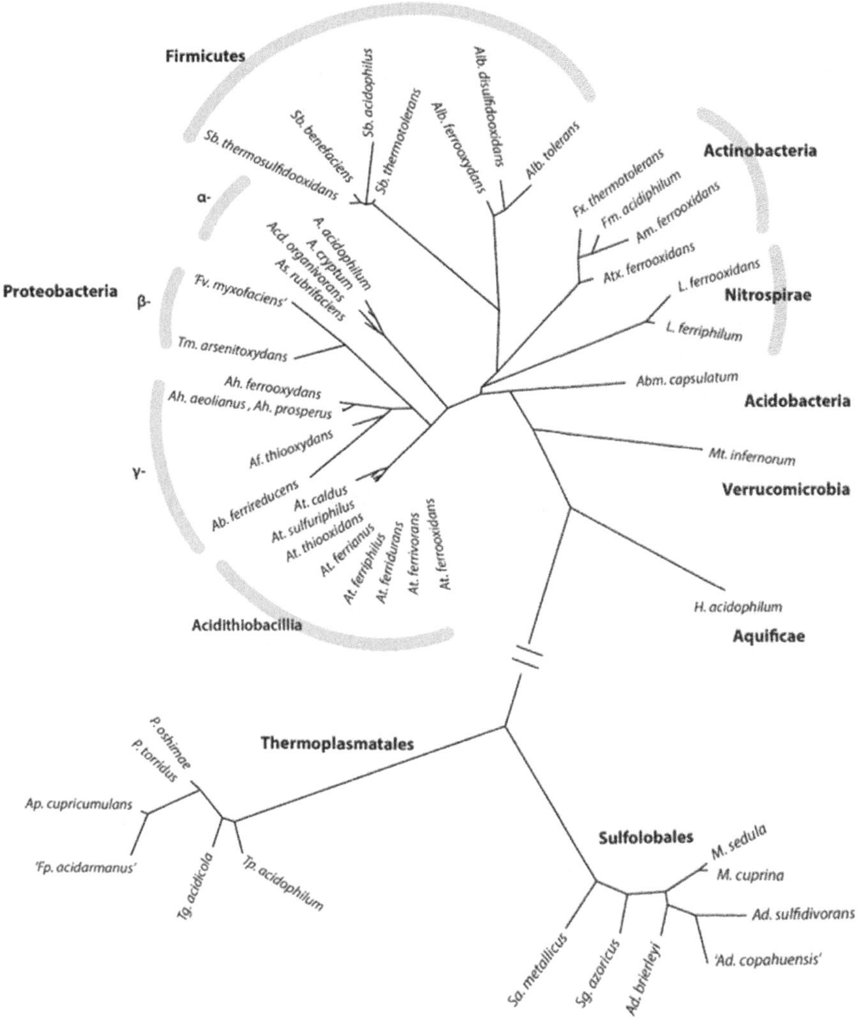

Fig. 5.1 Phylogenetic tree of selected biomining microorganisms. Unrooted phylogenetic tree of selected biomining microorganisms indicating bacterial phyla and archaeal orders. The tree was constructed using the MUSCLE (Multiple Sequence Comparison by Log-Expectation), PhyML algorithm, and TreeDyn visualization and annotation tool at www.phylogeny.fr/ (S. Santini (CNRS/AMU IGS UMR7256); PACA Bioinfo platform (IBISA)). The branch length between the Bacteria and Archaea has been cut for space considerations

degrees, proton exporters, membrane adaptations to inhibit proton influx, and proton consuming reactions.

Biomining microorganisms are also often challenged by elevated concentrations of soluble transition and other metals and metalloids that require multiple active or energy-consuming resistance mechanisms allied with passive chemical complexation of the free metal ions by sulfate (reviewed in Dopson and Holmes 2014; Dopson et al. 2014). The metal resistance mechanisms include efflux pumps to extrude metals from the cell, metal complexation systems, and enzymatic conversion of metalloids into less toxic forms. Acidic environments also pose challenges due to osmotic/salt stress (further discussed in Chap. 13). Chloride stress has received a large amount of attention due to the economic and environmental advantages of biomining using saline waters in countries including Australia and Chile. However, biomining environments more often present osmotic challenges due to high sulfate and other dissolved solutes. Finally, shear forces caused by abrasion of mineral particles in stirred-tank bioreactors with high pulp densities often more severely affect thermophilic Archaea than Bacteria. These polyextreme conditions can result in some acidophiles being unable to thrive in biomining environments such as many species being highly sensitive to chloride ions.

5.2 Microbial Role in Biomining Sulfide Mineral Ores

The role of acidophiles in the biomining of sulfide minerals (Fig. 5.2) is primarily to catalyze the abiotic oxidation of the metal sulfide bond. This occurs by iron-oxidizing microorganisms regenerating the chemical oxidant, ferric iron (Fe^{3+}) that

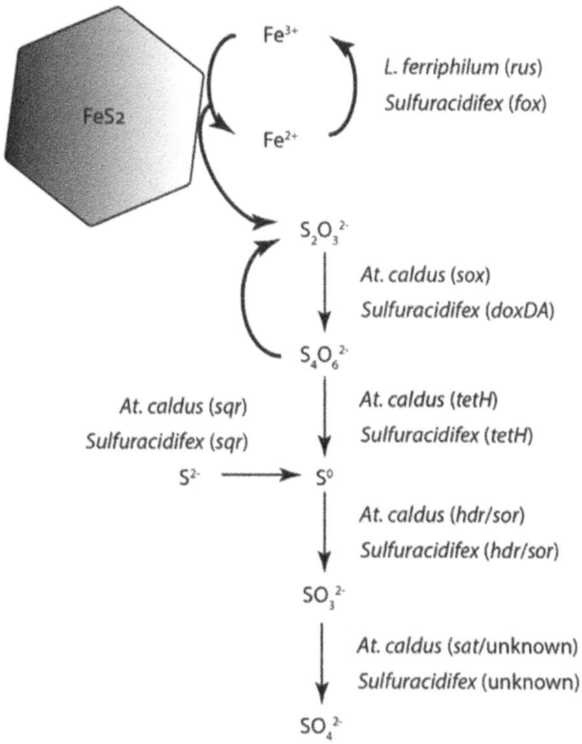

Fig. 5.2 Metal sulfide dissolution mechanisms. Interaction of abiotic attack on the mineral surface with biological catalysis of ferrous oxidation and a simplified RISC metabolism (not showing all products or reactions). Examples of common biomining species with relevant genes (in brackets) are given for the respective reactions. The *Sulfuracidifex* genus includes the ferrous iron and RISC-oxidizer *Sa. metallicus*. Figure adapted from Schippers and Sand (1999)

creates a catalytic cycle between abiotic mineral oxidation by ferric iron and biological oxidation of the ferrous iron (Fe^{2+}) product. Once the iron-oxidizing microorganisms become active, the biological ferrous oxidation is usually the more rapid reaction (depending on factors such as sufficient gas mass transfer, inhibition of microbial activity, and the reactivity of the mineral substrate) and ferric iron typically accumulates in leaching liquors. The initial sulfur product from oxidative attack on the mineral is either thiosulfate ($S_2O_3^{2-}$) produced from acid-insoluble metal sulfides (e.g., pyrite) or elemental sulfur (S^0) from acid-soluble minerals such as chalcopyrite ($CuFeS_2$; Schippers and Sand 1999). Sulfur-oxidizing acidophiles such as *Acidithiobacillus* and *Sulfobacillus* spp. catalyze the dissimilatory oxidation of elemental sulfur and reduced inorganic sulfur compounds (RISCs; sulfur oxyanions such as thiosulfate and tetrathionate) with the final product being sulfuric acid that provides the low pH conditions under which acidophiles grow optimally. RISCs and elemental sulfur are subsequently referred to collectively as "reduced S" in this chapter. An additional important role of sulfur-oxidizing acidophiles is the removal of potentially passivating layers of reduced S accumulating on the surface of e.g. chalcopyrite that hinders dissolution and results in poor copper yields. However, the exact nature of the passivating layer has been a topic of intense scrutiny and the role of polysulfides in surface passivation has been questioned. In addition to obligately autotrophic iron- and sulfur-oxidizers, facultative autotrophs and obligately heterotrophic microorganisms are also present in biomining environments. These taxa have been demonstrated to play a role in the community by consuming organic acids released by the acidophilic microorganisms such that they do not reach toxic concentrations that result in acidification of the cytoplasm and ultimately cell death. In addition, many heterotrophic species can also oxidize iron and/or reduced S and thereby contribute directly to sulfide mineral dissolution (reviewed in Rawlings and Johnson 2007).

Biomining mechanisms have been demarcated according to the interaction between the microbe and mineral surface (Rawlings 2002). Non-contact leaching is defined as the chemical oxidation of the metal sulfide bond by protons or ferric iron in solution that releases the products back into the leach liquor where they are further metabolized by iron- and sulfur-oxidizing members of the consortium. In contrast, contact leaching involves the same reactions but the microorganisms are attached to the mineral surface and the reactions occur in the extracellular polysaccharide layer whereby the oxidative reactants (such as ferric iron) can accumulate to orders of magnitude greater than in the bulk leaching liquor and thus, metal dissolution rates are accelerated. Finally, cooperative leaching is a union between microorganisms in contact with the mineral surface releasing reduced S, soluble metals, and mineral fragments into the leach liquor where they are available for further oxidative attack by ferric iron and protons via the non-contact method and microbial oxidation of reduced S to sulfate. The microorganisms carrying out these processes are in almost all cases consortia of species that each metabolize a part of the mineral dissolution rather than a single, generalist species performing both ferrous oxidation and complete conversion of reduced S to sulfuric acid.

5.3 Biomining Environments and Microbial Diversity

The different types of biomining environments include stirred-tank reactors, bioheaps, and mine waters that have differing physicochemical conditions, which select for various acidophiles (discussed further in Chap. 7). For instance, a stirred-tank reactor is a more controlled, homogenous environment with a fixed temperature and excess oxygen for aerobic energy conservation. Metal(loid)s and protons can accumulate to high concentrations in biomining tank reactors, especially if several are coupled in series. In addition, there is a selection pressure on the ability of the cells to grow and divide quickly under the imposed physicochemical conditions due to relatively fast throughput rates. These systems typically have a lower species diversity although the selected taxa can vary through the bioreactor series as the conditions alter with the degree of mineral dissolution. In contrast, bioheaps are much more heterogeneous with variations in metal sulfide contents, reactant concentrations, a decreasing temperature gradient from the bioheap core to the heap surface and gradients in oxygen concentration. In bioheaps, there is also a pressure not to be washed out that is countered by cell attachment to the mineral surfaces. This range in physicochemical conditions and selection pressures typically selects for a broader range of species with a suite of traits that enable them to exploit the various niches within the bioheap. Finally, AMD and mine waters are usually exposed to the environment with changes in physicochemical conditions (such as temperature over the course of a year) that also select for a typically high community diversity.

The design of microbial consortia for biomining can be via a "top down" or "bottom up" approach. The "top down" strategy is to inoculate a range of acidophiles with the expectation that the most efficient taxa will be selected while the "bottom up" strategy is whereby a consortium is chosen based upon known characteristics of the species making them suitable for the conditions within the biomining system (Rawlings and Johnson 2007). However, either strategy of inoculating a biomining tank or heap may ultimately prove futile as native populations present on the mineral may dominate the community over time. In addition, the community may develop by evolution (mutation/gene transfer) and/or adaptation (gene expression) as was observed in the Fairview biooxidation plant where genes for arsenic resistance were acquired by horizontal gene transfer.

In this chapter, acidophilic biomining microorganisms (Tables 5.1 and 5.2) are differentiated as Bacteria versus Archaea along with (1) iron-oxidizers, (2) reduced S-oxidizers, (3) iron- and reduced S-oxidizers, and (4) other species, and their diversity and *modus operandi* described.

Table 5.1 Selected traits of representative acidophilic bacterial genera and species were chosen based upon their relevance to biomining along with species with novel identifying characteristics

Genus and species	Selected traits
Iron-oxidizers	
"*Acidibacillus ferrooxidans*"	Facultative anaerobe; heterotroph; Fe^{2+} as electron donor; pH_{opt}1.8; T_{opt} ~ 30 °C
Acidiferrimicrobium australe	Facultative anaerobe; heterotroph; Fe^{2+} as electron donor; pH_{opt}3.0; T_{opt} 30 °C
Acidimicrobium ferrooxidans	Aerobe (reduces Fe^{3+} under oxygen limitation); autotroph/heterotroph; Fe^{2+} as electron donor; pH_{opt} ~ 2.0; T_{opt} 45–50 °C
Acidithrix ferrooxidans[a]	Facultative anaerobe; heterotroph; Fe^{2+} as electron donor; pH_{range} 2.0–4.4; T_{opt} 10°-30 °C
Ferrimicrobium acidiphilum[a]	Facultative anaerobe; heterotroph; Fe^{2+} as electron donor; pH_{opt} 2.0; T_{opt} 35 °C
Ferrithrix thermotolerans[a]	Facultative anaerobe; heterotroph; Fe^{2+} as electron donor; pH_{opt} 1.8; T_{opt} 43 °C
'*Ferrovum myxofaciens*'[a]	Aerobe; autotroph & diazotroph; Fe^{2+} as electron donor; pH_{opt} 3.3; T_{opt} 32 °C
Leptospirillum ferriphilum	Aerobe; autotroph; Fe^{2+} as electron donor; pH_{opt} 1.4–1.8; T_{opt} 30°–37 °C
Leptospirillum ferrooxidans	Aerobe; autotroph; Fe^{2+} as electron donor; pH_{range} 1.5–4.0
Iron- and sulfur-oxidisers	
'*Acidibacillus sulfuroxidans*'	Facultative anaerobe; heterotroph; Fe^{2+} and RISCs as electron donor; pH_{opt}2.9; T_{opt} ~ 43 °C
Acidiferrobacter thiooxydans[a]	Facultative anaerobe; autotroph and diazotroph; Fe^{2+} and RISCs as electron donor; pH_{opt} ~ 2.0; T_{opt} 38 °C; moderate osmophile
Acidihalobacter aeolianus	Aerobe; autotroph; Fe^{2+} and RISCs as electron donor; pH_{opt} 1.8; T_{opt} 36 °C; halotolerant
Acidihalobacter ferrooxydans	Aerobe; autotroph; Fe^{2+} and RISCs as electron donor; pH_{opt} 1.8; T_{opt} 36 °C; halotolerant
Acidihalobacter prosperus	Aerobe; autotroph; Fe^{2+} and RISCs as electron donor; pH_{opt} 2.0; T_{opt} 37 °C; halotolerant
Acidihalobacter yilgarnensis	Aerobic; autotroph; Fe^{2+} and RISCs as electron donor; pH_{opt}2.5; T_{opt} 30 °C; halotolerant
Acidithiobacillus ferrianus	Facultative anaerobe; autotroph; Fe^{2+}, S^0, and H_2 as electron donor; pH_{opt} ~ 2.0; T_{opt} ~ 30 °C
Acidithiobacillus ferridurans	Facultative anaerobe; autotroph; Fe^{2+}, RISCs, and H_2 as electron donor; pH_{opt} 2.1; T_{opt} 29 °C
Acidithiobacillus ferriphilus	Facultative anaerobe; autotroph; Fe^{2+} and RISCs as electron donor; pH_{opt} ~ 2.0; T_{opt} ~ 30 °C; some strains eurypsychrophilic
Acidithiobacillus ferrivorans	Facultative anaerobe; autotroph and diazotroph; Fe^{2+}, RISCs, and H_2 as electron donor; pH_{opt} 2.5; T_{opt} 28–33 °C eurypsychrophilic
Acidithiobacillus ferrooxidans	Facultative anaerobe; autotroph; Fe^{2+}, RISCs, and H_2 as electron donor; pH_{opt} 2.5; T_{opt} 30°–35 °C
Acidithiomicrobium P2	Aerobe; autotroph; Fe^{2+} and RISCs as electron donor; T_{opt} 50 °C
Alicyclobacillus ferrooxydans	Aerobe; heterotroph; Fe^{2+} and RISCs as electron donor; pH_{opt} 3.0; T_{opt} 28 °C

(continued)

Table 5.1 (continued)

Genus and species	Selected traits
Alicyclobacillus tolerans	Aerobe; heterotroph; Fe^{2+} and S^0 as electron donor; pH_{opt} 2.0–2.7; T_{opt} 37°–42 °C
Alicyclobacillus disulfidooxidans	Aerobe; heterotroph; Fe^{2+} and RISCs as electron donor; pH_{opt} 1.5–2.5; T_{opt} 35 °C
Sulfobacillus acidophilus	Aerobe; autotroph/heterotroph; Fe^{2+} and RISCs as electron donor; pH_{opt} ~ 2.0; T_{opt} 45°–50 °C
Sulfobacillus benefaciens	Facultative anaerobe; autotroph/heterotroph; Fe^{2+} and RISCs as electron donor; pH_{opt} 1.5; T_{opt} 39 °C
Sulfobacillus sibiricus	Aerobe; heterotroph; Fe^{2+} and RISCs as electron donor; T_{opt} 43 °C
Sulfobacillus thermosulfidooxidans	Facultative anaerobe; autotroph/heterotroph; Fe^{2+} and RISCs as electron donor; pH_{opt} 1.7–2.4; T_{opt} 50°–55 °C
Sulfobacillus thermotolerans	Aerobe; autotroph (weak)/heterotroph; Fe^{2+} & RISCs as electron donor; pH_{opt} 2.0; T_{opt} 40 °C
Elemental sulfur and inorganic sulfur compound oxidizers	
Acidicaldus organivorans	Facultative anaerobe; heterotroph; S^0 plus yeast extract as electron donor; pH_{opt} 2.5–3.0; T_{opt} 50°–55 °C
Acidiphilium acidophilum	Aerobe; autotroph/heterotroph; RISCs as electron donor; pH_{opt} 3.0–3.5; T_{opt} 25°–30 °C
Acidithiobacillus caldus	Aerobe; autotroph; RISCs as electron donor; pH_{opt} 2.0–2.5; T_{opt} 45 °C
Acidithiobacillus sulfuriphilus	Aerobe; autotroph; RISCs as electron donor; pH_{opt} 3.0; T_{opt} 25°–28 °C
Acidithiobacillus thiooxidans	Aerobe; autotroph; RISCs and H_2 as electron donor; pH_{opt} 2.0–2.5; T_{opt} 28°–30 °C
Hydrogenobaculum acidophilum	Aerobe; autotroph; H_2 and RISCs as electron donors; pH_{opt} 3.0–4.0; T_{opt} 65 °C
Other species	
Acidiphilium cryptum	Facultative anaerobe; heterotroph; pH_{range} 2.2–5.2; T_{opt} 35–41 °C
Aciditerrimonas ferrireducens[a]	Facultative anaerobe; autotroph and heterotroph; H_2 as electron donor; pH_{opt}3.0; T_{opt} 50 °C
Acidobacterium capsulatum	Facultative anaerobe; heterotroph; pH_{range} 3.0–6.0
Methylacidiphilium infernorum[a]	Aerobe; autotroph; CH_4 as electron donor; pH_{opt} 2.0–2.5; T_{opt} 60 °C

[a]Genus with a sole species

Abbreviations: RISCs, reduced inorganic sulfur compounds; pH_{opt} and T_{opt}, pH and temperature optimum, respectively

5.4 Bacteria

As discussed above, the majority of biomining species are iron- and/or sulfur-oxidizing Bacteria but some species can also grow using other inorganic and organic electron donors. For instance, some lithotrophic acidophiles (but not obligatory

Table 5.2 Selected traits of representative acidophilic archaeal genera and species were chosen based upon their relevance to biomining along with species with novel identifying characteristics

Genus and species	Selected traits
Iron-oxidizers	
Acidiplasma aeolicum	Facultative anaerobe; heterotroph; Fe^{2+} and RISCs as electron donor; pH_{opt} 1.4–1.6; T_{opt} 42–45 °C
Acidiplasma cupricumulans	Facultative anaerobe; heterotroph; Fe^{2+} as electron donor; pH_{opt} 1.0–1.2; T_{opt} 54 °C
"*Ferroplasma acidarmanus*"	Facultative anaerobe; heterotroph; Fe^{2+} as electron donor; pH_{opt} 1.2; T_{opt} 42 °C
Ferroplasma acidiphilum	Aerobe; heterotroph; Fe^{2+} as electron donor; pH_{opt} 1.7; T_{opt} 35 °C
Iron- and sulfur-oxidizers	
Acidianus brierleyi	Facultative anaerobe; facultative autotroph; Fe^{2+} and S^0 as electron donor; pH_{opt} 1.5–2.0; T_{opt} 70 °C
"Can. *Acidianus copahuensis*"	Facultative anaerobe; facultative autotroph; Fe^{2+}, RISCs, and H_2 as electron donor; pH_{opt} 2.5–3.0; T_{opt} 75 °C
Acidianus sulfidivorans	Facultative anaerobe; facultative autotroph; Fe^{2+} and S^0 as electron donor; pH_{opt} 0.8–1.4; T_{opt} 74 °C
Metallosphaera cuprina	Aerobe; facultative autotroph; Fe^{2+} and RISCs as electron donor; pH_{opt} 3.5; T_{opt} 65 °C
Metallosphaera sedula	Aerobe; facultative autotroph; Fe^{2+} and RISCs as electron donor; pH_{range} 1.0–4.5; T_{opt} ~ 75 °C
Sulfuracidifex metallicus	Aerobe; autotroph; Fe^{2+} and RISCs as electron donor; pH_{range} 1.0–4.5; T_{range} 50–75 °C
Sulfurococcus yellowstonii	Aerobe; facultative autotroph; Fe^{2+} and S^0 as electron donor; T_{range} 40–80 °C
Other species	
Cuniculiplasma divulgatum[a]	Facultative anaerobe; heterotrophic; pH_{opt} 1.0–1.2; T_{opt} 37–40 °C
Picrophilus oshimae	Aerobe; heterotroph; yeast extract as electron donor; pH_{opt} 0.7; T_{opt} 60 °C
Picrophilus torridus	Aerobe; heterotroph; yeast extract as electron donor; pH_{opt} 0.7; T_{opt} 60 °C
Stygiolobus azoricus[a]	Anaerobe; autotroph; H_2 as electron donor; pH_{opt} 2.5–3.0; T_{opt} ~ 80 °C
Thermoplasma acidophilum	Facultative anaerobe; heterotrophic; pH_{opt} 1.0–2.0; T_{opt} 55–59 °C
Thermogymnomonas acidicola[a]	Aerobe; heterotroph; pH_{opt} ~ 3.0; T_{opt} 60 °C

[a]Genus with a sole species
Abbreviations: RISCs, reduced inorganic sulfur compounds; pH_{opt} and T_{opt}, pH and temperature optimum, respectively

heterotrophic species) grow via dissimilatory oxidation of hydrogen (H_2) such as *Sulfobacillus acidophilus*, *Hydrogenobaculum acidophilum*, and some *Acidithiobacillus* spp., while *Methylacidiphilium infernorum* oxidizes methane (CH_4).

5.4.1 Iron-Oxidizers

Iron is the most abundant transition metal in the lithosphere and its bioavailability increases in extremely acidic environments (pH < 3) due to the enhanced chemical stability and solubility of both ferrous and ferric species. Under acidic pH conditions, ferric iron can complex with both sulfate and hydroxyl anions to form $Fe(SO_4)_2^-$, $Fe(SO_4)^+$, $Fe(OH)^{2+}$, and $Fe(OH)_2^+$. While the standard redox potential of the (uncomplexed) Fe^{3+}/Fe^{2+} couple is quoted as +0.77 V (pH 2), those of complexed ferric iron/ferrous iron are more negative (Johnson et al. 2012). Since the discovery of *(Acidi)thiobacillus ferrooxidans* in 1950, iron-oxidizing acidophiles have been deliberately used in biomining operations. The capacity to oxidize iron (reviewed in Bonnefoy and Holmes 2012) is found in phylogenetically diverse acidophilic Bacteria, according to the Genome Taxonomy Database (GTDB) classification (Parks et al. 2018), such as the phyla Proteobacteria (Acidithiobacillia, α, ß, and γ classes), Nitrospirae, Firmicutes, Actinobacteria, and Acidobacteria (Fig. 5.1 and Table 5.1). These taxa include iron-oxidizing facultative autotrophs and heterotrophic acidophiles that are typically important in consortia where organic carbon is added and/or present and iron oxidation is reported to be coupled to growth and organic carbon assimilation. This is demonstrated by typically increased growth yields from culture in the presence of iron. The phylogenetic diversity of iron-oxidizers includes autotrophs from the genera *Leptospirillum* and "*Ferrovum*" along with heterotrophs from the *Acidimicrobium*, *Ferrimicrobium*, and *Ferrithrix* bacterial genera (Table 5.1). In this section, those biomining species that are known only to use ferrous iron as an electron donor are described, and iron- and sulfur-oxidizers, and species known only to oxidize reduced S, in subsequent sections.

Leptospirillum ferriphilum is one of the most commonly identified Bacteria in biooxidation tanks. In general, *Leptospirillum* species proliferate in ferric iron-dominated systems due to specific iron-oxidation activities, greater affinity to ferrous ion and tolerance to ferric ion. For example, *L. ferriphilum* was identified as the prevailing iron-oxidizing species in 40 °C South African biooxidation tanks for gold recovery. *Leptospirillum* spp. have also been identified in both bioheap and leach liquors at the Dexing copper mine, China, as well as in bioleaching consortia in Chilean bioleaching operations. However, *Leptospirillum* spp. are not always the dominant acidophiles present, such as in the unaerated low-grade copper bioheap at the Zijinshan mine, China, where *Acidithiobacillus* spp. were found to have higher relative abundances. The high affinity of *L. ferriphilum* for ferrous iron that allows it to dominate in biooxidation tanks and efficiently catalyze arsenopyrite dissolution is not always a desirable trait, as this generates liquors of high redox potential (ORP; often E_H values > 900 mV). For instance, passivating layers that form on the mineral surface during chalcopyrite bioleaching are mitigated at low ORP values. However, the "strong" iron-oxidizer *L. ferriphilum* typically raises the ORP above the desired value and one strategy to increase chalcopyrite dissolution is to maintain a consortium dominated by "weak" iron-oxidizers (i.e., with lower affinities for ferrous iron) within the bioheap. The pragmatic challenge is to control the biomining consortium in the very large bioheaps used in industry that also contain a mineral-associated native consortium that may come to dominate over time.

Other ferrous iron-oxidizing species include the heterotrophic Actinobacteria *Ferrimicrobium acidiphilum* and *Ferrithrix thermotolerans* the type species of which were isolated from mine and geothermal sites, respectively. A low relative abundance of 16S rRNA gene sequences most similar to the *Ferrimicrobium* genus were identified from the Zijinshan copper bioheap along with at a pilot-scale bioheap treating a low-grade nickel ore in Talvivaara, Finland. 16S rRNA gene sequences related to the *Ferrithrix* genus have been identified at a chalcopyrite mine in Touro, Spain, and tailings from the copper mine in Malanjkhand, India. Further iron-oxidizing Actinobacteria include *Acidithrix ferrooxidans*, which is characterized by cell growth in long, entangled filaments, and originally from an acidic stream in Wales, and which has also been identified in the Huaxi AMD creek, China; *Acidithrix* strain C25 isolated from iron-rich pelagic aggregates ("iron snow") in a German acidic pit lake; and *Acidiferrimicrobium australe* isolated from metal-containing acidic water from Trongol coal mine, Chile. In addition, *Acidimicrobium ferrooxidans* has been shown in laboratory tests to catalyze the oxidative dissolution of both pyrite and chalcopyrite. It has also been investigated as a model organism for the acidophile response to chloride ions during pyrite bioleaching whereby it altered its membrane proteins and generated osmoprotectants. Strains of the iron-oxidizing Firmicute, "*Acidibacillus ferrooxidans*" have been isolated from a variety of sources from around the world, including weathered sulfidic regolith (the proposed type strain) and pH neutral drainage at a low-grade copper mine in Brazil. "*Acidibacillus*" 16S rRNA gene sequences have also been identified in DNA extracted from AMD emanating from the Lancaster mine, South Africa. Finally, the obligately autotrophic iron-oxidizer "*Ferrovum myxofaciens*" has been identified in several sulfide mineral-related environments including in AMD streamer-type biofilms at low temperatures. While "*Fv. myxofaciens*" can accelerate oxidative dissolution of pyrite in pure culture, it has yet to be demonstrated whether it has an active role in biomining operations.

5.4.2 *Iron- and Reduced Sulfur-Oxidizers*

A wide range of acidophilic Bacteria, including members of the *Acidiferrobacter*, *Acidithiobacillus*, *Sulfobacillus*, and *Alicyclobacillus* genera are capable of oxidizing ferrous iron and reduced S (Table 5.1). The most intensely studied iron- and sulfur-oxidizer is *At. ferrooxidans,* first described in 1950. In addition to aerobic growth on iron, reduced S and hydrogen, *At. ferrooxidans* is also able to couple the oxidation of sulfur or hydrogen to ferric iron reduction. The higher relative chemical reactivity of iron with oxygen compared to sulfur (Johnson et al. 2012) coupled with a comparatively simple enzymatic pathway results in iron being preferentially oxidized over reduced S. This has an important consequence in biomining operations where polymeric RISCs and elemental sulfur have been suggested to be at least partially responsible for passivation of chalcopyrite mineral surfaces in bioheaps. The accumulation of elemental sulfur from preferential ferrous iron oxidation results in reduced copper dissolution rates and yields.

Iron/sulfur-oxidizing *Acidithiobacillus* spp. have been identified from many biomining and AMD drainage environments, typically when ferrous iron predominates, such as during the early stages of bioheap operations. These were previously often identified in the literature solely as *At. ferrooxidans* but later studies demonstrated several new taxa (Moya-Beltrán et al. 2021). *At. ferrivorans* is an example of an *Acidithiobacillus* species that was previously identified as strains of *At. ferrooxidans*. This acidophile has been tested in the laboratory for low-temperature bioleaching of pyrite, pyrite/arsenopyrite, and chalcopyrite concentrates, identified in the Talvivaara pilot-scale bioheap (Chap. 12) and was identified as the main/sole iron-oxidizer in acidic waters in Antarctica. Several other iron/sulfur-oxidizing *Acidithiobacillus* spp. have been classified including *Acidithiobacillus ferrianus, Acidithiobacillus ferridurans*, and *Acidithiobacillus ferriphilus* (the latter includes psychrotolerant/eurypsychrophilic and mesophilic strains). All of these species catalyze the oxidative dissolution of pyrite and other sulfide minerals. Strains of *At. ferrooxidans, At. ferridurans*, and *At. ferrianus* also use hydrogen as electron donor, which those of *At. ferriphilus* do not, and only one strain of *At. ferrivorans* (Peru 6) has been reported to have this ability. Based upon the growth characteristics of investigated *At. ferriphilus* strains, it has been suggested they may play a role in, for example, biomining operations operated with brackish waters. In addition, *Acidiferrobacter thiooxydans* (previously classified as "*Thiobacillus ferrooxidans*" m-1) is a thermotolerant and moderately osmophilic species often identified in sulfide mineral-impacted environments. Finally, several *Acidihalobacter* spp. have been characterized with *Acidihalobacter prosperus* (originally termed "*Thiobacillus prosperus*"), *Acidihalobacter aeolinanus*, and *Acidihalobacter ferrooxydans* isolated from the Aeolian Islands, Italy along with *Acidihalobacter yilgarnensis* from Western Australia. *Ah. prosperus*, in particular, has been investigated for its ability to catalyze sulfide mineral dissolution in brackish waters and its mechanisms of salt tolerance have been investigated (discussed further in Chap. 14). Recent studies have shown *Ah. yilgarnensis* catalyzes chalcopyrite dissolution in the presence of approximately 0.5 M NaCl, suggesting it may be a valid candidate for biomining using saline waters.

Species of the Firmicutes genus *Sulfobacillus* include *Sulfobacillus thermosulfidooxidans,* which was originally isolated from a copper-zinc-pyrite deposit in Eastern Kazakhstan and subsequently identified in many biomining operations. These include a Chilean operational industrial bioleaching consortium and an Australian spent chalcopyrite/pyrite test bioheap. In contrast to the "strong" iron-oxidizer *L. ferriphilum*, *Sb. thermosulfidooxidans* along with *Sb. acidophilus, Sulfobacillus sibiricus*, and the archaeal strain *Acidiplasma* Fv-Ap have been characterized as "weak" iron-oxidizers and are candidates for efficient chalcopyrite bioleaching. In addition, *Sb. benefaciens* that was originally isolated from a pilot-scale stirred-tank cobaltiferous pyrite concentrate has also been identified from stirred-tank bioreactors treating a chalcopyrite concentrate and soil in the vicinity of coal gangue in China. Finally, *Sulfobacillus thermotolerans* has been identified in industrial biooxidation of refractory gold ores operated at 38–42 °C. A second genus of acidophilic iron/sulfur-oxidizers from the Firmicutes is *Alicyclobacillus*, which includes *Alicyclobacillus ferrooxydans* isolated from solfataric soil, *Alicyclobacillus disulfidooxidans* (originally named *Sulfobacillus disulfidooxidans*) that uses pyrite as its sole energy source, and *Alicyclobacillus tolerans* (reclassified from *Sulfobacillus thermosulfidooxidans* subsp. thermotolerans) and catalyzes pyrite dissolution in the presence of yeast extract. Finally, the moderately thermophilic Actinobacterium, "*Acidithiomicrobium*" strain P2 has the potential to contribute to higher temperature (up to 60 °C) biomining operations due to its abilities to fix carbon dioxide as well as to oxidize both ferrous iron and sulfur, though it appears to have relatively low tolerance for some transition metals, such as copper (Norris et al. 2011).

5.4.3 Elemental Sulfur- and Reduced Inorganic Sulfur Compound-Oxidizers

The ability to oxidize reduced S to sulfate is widely found in acidophilic prokaryotes (Table 5.1; reviewed in Dopson and Johnson 2012) and is of great importance in biomining technologies. The first sulfur-oxidizing acidophile to be described was *(Acidi)thiobacillus thiooxidans* in 1922. Elemental sulfur is thermodynamically stable at acidic pH values under moderate temperatures and pressures and is often subject to microbially accelerated dissimilatory oxidation in natural and man-made environments. Reduced S is more energetically favorable for acidophilic prokaryotes than ferrous iron as an energy source due to: (1) more electrons being available from sulfur oxidation than from iron oxidation and (2) the more negative standard redox potential of sulfate/reduced S than that of ferric/ferrous iron.

The most commonly identified reduced S-oxidizing acidophile found in biomining operations is the thermotolerant bacterium *Acidithiobacillus caldus*. This acidophile has the likely role to remove reduced S during mineral dissolution such that passivating layers do not form and metal dissolution continues. Two examples of *At. caldus* being identified in industrial biomining operations are in a chalcocite heap bioleaching operation in Myanmar and in the Fairview biooxidation tanks for gold recovery, South Africa. The mesophile *At. thiooxidans* has a similar mode of growth as *At. caldus* but has not been so commonly identified in industrial operations. Finally, the reduced S-oxidizing facultative autotroph *Acidiphilium acidophilum* (originally described as *Thiobacillus acidophilum*) has been described with *Acidiphilium* spp. being present in the Rio Tinto (Spain).

5.4.4 Other Species

Other reduced S-oxidizing species include *Aciditerrimonas ferrireducens,* identified in a solfataric field in Japan, and the psychrotolerant/eurypsychrophilic bacterium *Acidobacterium capsulatum* that was isolated from an acidic sphagnum peat bog in West Siberia, Russia.

5.4.5 Iron, Sulfur, and Carbon Metabolism

The most intensely studied acidophile dissimilatory iron oxidation pathway is that in *At. ferrooxidans* while recent genomic and proteomic studies have provided insights into this pathway in other acidophiles (Bonnefoy and Holmes 2012; Ilbert and Bonnefoy 2013). While the redox proteins in most respiratory chains are organized "horizontally" along the cytoplasmic membrane, their topography in Gram-negative iron-oxidizers such as *At. ferrooxidans* is vertical that allows a connection to form between the extracellular matrix and cytoplasm. In the energy providing "downhill" electron-flow pathway, electrons travel from an outer membrane-embedded cytochrome *c*, through a series of electron carriers, to an integral inner membrane cytochrome oxidase that catalyzes oxygen reduction to water. In addition, the electrons can also flow "uphill" in an energy requiring process to reduce NAD^+ via the inner membrane-bound NADH dehydrogenase complex 1. While acidophiles have a "ready-made" transmembrane pH gradient for generating ATP via the inner membrane-bound ATPase, this process would soon result in an acidified cytoplasmic pH without the accompanying export of protons (or import of electrons) during electron transport. These "downhill" and "uphill" electron flows are commonly found in iron-oxidizing chemoautotrophs with an overall conserved organization although the proteins involved can vary. For example, the *rus* operon mediates iron oxidation in, e.g., *Acidithiobacillus* spp., "*Fv. myxofaciens*" and *Ah. prosperus* although an additional *At. ferrivorans* iron oxidation pathway is suggested via the high potential iron–sulfur protein (HiPIP) Iro and possibly an isozyme of rusticyanin, RusB. In addition, the salt-tolerant γ-proteobacterium *Ah. prosperus* redox proteins are similar to those in *At. ferrooxidans* with the exception that the Cyc1 electron carrier was not detected. In contrast, the outer membrane cytochrome *c* Cyc572 is proposed to be the direct oxidant of ferrous iron in *Leptospirillum* spp. Finally, the *Sb. thermosulfidooxidans* genome contains a blue copper protein, sulfocyanin, although further details of the *Sb. thermosulfidooxidans* ferrous oxidation pathway are lacking.

In contrast to iron that has only zero-valent, +2 and +3 oxidation states, sulfur exists in nine different oxidation states, ranging from −2 to +6, which complicates identification of reaction intermediates and component protein molecules involved in its metabolic pathway (Wang et al. 2019). Some reduced S, such as thiosulfate, are also metabolized abiotically in acidic liquors, adding further difficulties to clarifying the mechanism. In addition, electrons deriving from reduced S oxidation enter the electron transport chain at cytochrome *b*. This eliminates the need for the "uphill" electron flow (as is the case for iron oxidation) and thus, electrons are more readily available for the reduction of NAD^+ to NADH. While ferrous iron is the preferred substrate when both iron and reduced S are available, this explains why sulfur-oxidizers are often found in greater numbers than iron-oxidizers in biomining environments (Johnson and Hallberg 2009). Several metabolic pathways for reduced S oxidation are present in acidophiles that have commonalities between the different species and are predicted to involve a number of enzymes, enzyme complexes, and electron carriers located in different cellular compartments (Rohwerder and Sand 2007; Quatrini et al. 2009). However, recent "omics"-based studies have identified variations with multiple genes predicted to encode proteins metabolizing reduced S intermediates. *Acidithiobacillus* spp. including *At. caldus*, *At. thiooxidans*, and *At. ferrooxidans* oxidize a wide range of reduced S, such as sulfide (H_2S), elemental sulfur, thiosulfate, tetrathionate ($S_4O_6^{2-}$), and sulfite (HSO_3^-). Periplasmic sulfide oxidation is mediated by sulfide:quinone oxidoreductase (SQR) to generate elemental sulfur. Only highly reactive thiol-bound sulfane sulfur atoms (R-SS_nH) can be oxidized by sulfur dioxygenase (SDO) and its enzymatic activity has been detected in *At. thiooxidans*, *A. acidophilum*, and *A. cryptum*. Furthermore, extracellular octameric elemental sulfur (S_8) has to pass the outer membrane prior to oxidation by sulfur oxygenase reductase (SOR). In *Acidithiobacillus* and *Acidiphilium* spp., the actual substrate of this enzyme is glutathione persulfide (GSSH) that is formed by glutathione (GSH) and elemental sulfur. In addition, genome analysis of *At. thiooxidans* suggests internally generated sulfur from thiosulfate sulfurtransferase is metabolized by heterodisulfide reductase (Hdr). Thiosulfate can be oxidized by thiosulfate:quinone oxidoreductase (TQR/TQO/DoxD) complex to generate tetrathionate that is metabolized by tetrathionate hydrolase (TetH/TTH) to ultimately form sulfur and thiosulfate. Several *Acidithiobacillus* genomes are predicted to encode both TQR/TQO/DoxD and a truncated Sox complex for thiosulfate oxidation. Sulfite is produced by Hdr and SOR that is predicted to be further oxidized to sulfate by a series of reactions including ATP sulfurylase (SAT). In contrast, little is known about reduced S oxidation in Gram-positive acidophiles although due to the lack of an outer membrane or periplasmic space, reduced S metabolism is thought to be localized in the cytoplasm or possibly the so-called S layer cell wall of *Sulfobacillus* spp. may be involved (Rohwerder and Sand 2007; Johnson and Hallberg 2009).

Based on how they obtain carbon, acidophilic Bacteria can be classified into obligate autotrophs that fix carbon dioxide to assimilate their cellular carbon, obligate heterotrophs that use organic materials as carbon sources, and facultative autotrophs that can do both. In microbial consortia such as found in bioheaps, autotrophic iron- and reduced S-oxidizers function as primary producers to feed organic substrates for heterotrophs. In turn, heterotrophs provide carbon dioxide via oxidation of organic substrates to support the growth of autotrophic acidophiles. The major route for carbon dioxide fixation in acidophilic Bacteria is the Calvin–Benson–Bassham (CBB) cycle. The key enzyme in this cycle, ribulose bisphosphate carboxylase/oxygenase (RuBisCO) has been detected in, for example, *Acidithiobacillus* spp. and "*Ferrovum*" strain JA12 as well as in facultative autotrophic *Sb. thermosulfidooxidans* (Johnson and Hallberg 2009). It was also predicted that *L. ferriphilum* and *L. ferrooxidans* likely fix carbon dioxide via the reductive tricarboxylic acid (TCA) pathway using pyruvate:ferredox oxidoreductase (PFOR). Despite their diversity, there has been relatively little study on the biochemistry of organic carbon metabolism in acidophiles. *At. ferrooxidans* predicted central carbon metabolism involves the pentose phosphate pathway, glycolysis, glycogen metabolism, and an incomplete TCA cycle. In *Acidiphilium* spp., both the pentose phosphate and Entner–Doudoroff pathways are functional but the Embden–Meyerhof–Parnas pathway (glycolysis) is apparently absent. Gram-positive acidophilic heterotrophs include both Firmicutes and Actinobacteria. The first *Sulfobacillus* species to be described, *Sb. thermosulfidooxidans* has an absolute requirement for an inorganic electron donor (e.g., ferrous iron) to grow on organic carbon. *Sb. thermosulfidooxidans* oxidizes sugars via the Embden–Meyerhof–Parnas and/or the Entner–Doudoroff pathway while the TCA cycle and the glyoxylate bypass are inoperative. Finally, although *Am. ferrooxidans* is capable of fixing carbon dioxide it can be more readily cultivated as a heterotroph on yeast extract (Johnson and Hallberg 2009).

5.5 Archaea

The temperature optima for Archaea typically encountered in biomining and sulfide mineral environments range from 35° to ~75 °C for *Ferroplasma acidiphilum* and *Metallosphaera sedula*, respectively (Table 5.2). As for the biomining Bacteria, the majority of species oxidize ferrous iron and/or reduced S although exceptions occur such as the heterotrophic *C. divulgatum* and *Thermoplasma acidophilum*. In addition to iron- and sulfur-oxidizing species, other extremely acidophilic taxa are able to use hydrogen as an electron donor. These include the obligatory hydrogen-oxidizing *Stygiolobus azoricus* along with members of the *Sulfolobus*, *Acidianus*, and *Metallosphaera* genera.

5.5.1 Iron-Oxidizers

The iron-oxidizing Ferroplasmaceae were originally identified from an arsenopyrite/pyrite concentrate pilot biooxidation plant for gold recovery with the type species being named *Fp. acidiphilum*. Since then, this species has been identified from AMD sites and 16S rRNA gene sequences most similar to *Fp. acidiphilum* were identified from a microbial community leaching a low-grade copper ore. *Fp. acidiphilum* has also been tested for treating a copper–zinc concentrate. A second Ferroplasmaceae isolate, "*Ferroplasma acidarmanus*," was identified from AMD where it was demonstrated to be able to survive, but not divide at negative pH values. In addition to its ability to oxidize ferrous iron, the role of *Fp. acidiphilum* in biomining has been suggested to metabolize organic carbon exudates from autotrophic species such as *L. ferriphilum* and *At. caldus* such that they do not reach toxic concentrations. The ability to grow at high metal concentrations and extremely low pH coupled with a heterotrophic lifestyle resulted in *Ferroplasma* isolates constituting $> 99\%$ of the cultivable cells in the last in the series of three pilot bioreactors treating a polymetallic sulfide concentrate at 45 °C. Finally, *Acidiplasma cupricumulans* (reclassified from *Ferroplasma cupricumulans*) was isolated from an industrial chalcocite bioheap and obtains energy from both ferrous oxidation and ferric reduction. *Ap. cupricumulans* has also been identified from laboratory-scale BIOX leaching tanks.

5.5.2 Iron- and Sulfur-Oxidizers

Biomining-related archaeal acidophiles that oxidize ferrous iron as well as reduced S are currently confined to the *Acidianus*, *Metallosphaera*, *Sulfuracidifex*, and *Sulfurococcus* genera from the Sulfobales (Counts et al. 2020). Some species of these genera are also able to mediate dissimilatory hydrogen oxidation coupled to sulfur reduction to sulfide.

Sulfuracidifex metallicus (originally named as *Sulfolobus metallicus*) has been detected in an industrial copper ore test bioheap and chalcopyrite dissolution by an axenic culture of *Sa. metallicus* has been reported. This ability of *Sa. metallicus* to mediate sulfide mineral dissolution at elevated temperatures is of great interest due to increased reaction rates and metal recoveries. However, *Sa. metallicus* is (and most bioleaching Archaea are) reported to be susceptible to the effects of elevated solids loads in biooxidation tanks and the most efficient solid loading is suggested to be a compromise between high pulp densities and rapid growth rates. Nevertheless, *Sa. metallicus* cells have been adapted to pulp densities up to 30% (wt vol^{-1}) with only a 15% reduction in metal dissolution. An additional potential issue with the use of *Sa. metallicus* is the toxicity of chemicals used during ore processing, such as for solvent extraction, if process waters are recycled, as these inhibit ferrous iron oxidation by this archaeon.

Further biomining Archaea are from the *Metallosphaera* genus and both *M. sedula* and *Metallosphaera cuprina* have been demonstrated to solubilize metals from sulfide ores including pyrite and chalcopyrite. The ferrous iron- and reduced S-oxidizing archaeon *Acidianus brierleyi* has been tested for dissolution of chalcopyrite containing concentrates at 70 °C when it was found to constitute up to 98% of the microbial community, with the remainder being *M. sedula*. In addition, *Ad. brierleyi* was more efficient at bioleaching chalcopyrite ores and a concentrate from India than *Sa. metallicus*. An additional candidate species of the *Acidianus* genus, "*Acidianus copahuensis*" has been shown to colonize pyrite and chalcopyrite surfaces and successfully mediated 100% zinc dissolution with the addition of tetrathionate from an ore containing pyrite, sphalerite, pyrrhotite, galena, and chalcopyrite. Another *Acidianus* sp., *Ad. sulfidivorans,* has been utilized to generate scorodite ($FeAsO_4 \cdot 2H_2O$) for arsenic removal from contaminated waters such as AMD. Finally, *Sulfurococcus yellowstonii* has been demonstrated to grow in the presence of sulfide minerals.

5.5.3 Other Species

Tp. acidophilum was originally isolated from a coal refuse pile but members of the Thermoplasmatales are present in AMD biofilms and *Tp. acidophilum* has also been identified from the Rio Tinto and Chilean sulfide mineral mining sites. This species has been described as a "scavenger" as it also likely metabolizes organic carbon released from autotrophic species. Finally, although the mesophilic, heterotrophic, and extremely acidophilic *C. divulgatum* was identified from a streamer type biofilm on copper ore, it has not been shown to have an active role in biomining operations to date.

5.5.4 Iron, Sulfur, and Carbon Metabolisms

In contrast to many iron-oxidizing biomining Bacteria, the acidophilic Archaea have a horizontal electron transfer mechanism from ferrous iron to the terminal oxidase and do not contain rusticyanin identified in many bacterial ferrous iron-oxidizers. Instead, "*Fp. acidarmanus*" is suggested to mediate ferrous iron oxidation via a blue copper-haem sulfocyanin protein that passes the electrons directly to a *cbb*$_3$ terminal oxidase. In addition, a *fox* gene-coded ferrous iron oxidation pathway was first identified in *Sa. metallicus* that has subsequently been identified in all named Sulfolobales genera except the obligatory anaerobic *Stygiolobus* and the heterotrophic *Saccharolobus* (Counts et al. 2020).

Many of the same enzymes that mediate bacterial reduced S oxidation are also present in all or selected species from the Sulfolobales. These include SQR sulfide: quinone oxidoreductase, SOR sulfur oxygenase reductase and Hdr heterodisulfide reductase for sulfur metabolism, TQR/TQO/DoxD thiosulfate:quinone oxidoreductase-mediated thiosulfate oxidation and TetH tetrathionate hydrolase. This reduced S oxidation pathway feeds electrons via the quinone pool to DoxBCE and SoxABCDL terminal oxidases. Of these reduced S metabolizing systems, only Hdr and DoxBCE terminal oxidases are conserved throughout the order. This suggests the archaeal reduced S metabolizing species are similar to their bacterial counterparts in that they have commonalities in their oxidation systems but also variations between different species.

Many acidophilic archaeal species belonging to the Sulfolobales (*Acidianus* spp., *Metallosphaera* spp. *Saccharolobus* spp., *Sulfuracidifex* spp., and *Sulfurococcus* spp.) are facultative autotrophs, while *Sa. metallicus* and *Sg. azoricus* are obligate autotrophs (Table 5.2; Bräsen et al. 2014). On the other hand, biomining-related Archaea belonging to the Thermoplasmatales (*Acidiplasma* spp., *Ferroplasma* spp., *Thermoplasma* spp., and *Thermogymnomonas* sp.) grow either as facultative autotrophs or heterotrophically. The central carbon metabolism in Sulfolobales has been studied intensively using *Saccharolobus solfataricus* (previously *Sulfolobus solfataricus*) as a model organism (Counts et al. 2020). Carbon metabolism pathways reconstructed for Sulfolobales have numerous differences compared to those from Bacteria. In *Sl. solfataricus*, glucose is degraded to pyruvate by the archaeal-type/modified branched Entner–Doudoroff pathway. The presence of the TCA cycle is evident throughout the order, serving to produce biomolecule precursors and as an entry point to the carbon fixation pathway. To thrive in environments with little organic compounds, Sulfolobales are highly dependent upon autotrophy, particularly the 3-hydroxypropionate/4-hydroxybutyrate cycle that is most intensively studied in *M. sedula*. The Thermoplasmatales, *Tp. acidophilum* and *Picrophilus torridus*, likewise metabolize glucose exclusively via the modified Entner–Doudoroff pathway. In contrast to *S. acidocaldarius* that harbors a branched Entner–Doudoroff pathway, the presence of a strictly non-phosphorylative Entner–Doudoroff pathway is suggested in *Tp. acidophilum* and *P. torridus*. Analysis of "*Fp. acidarmanus*" protein levels under heterotrophic versus facultative autotrophic growth conditions identified 15 proteins related to central carbon metabolism including those associated with glycolysis, the pentose phosphate pathway, and the TCA cycle along with a permease potentially involved in organic carbon uptake.

5.6 Biomining of Non-Sulfidic Ores and Wastes

Compared to sulfidic minerals, bioleaching of non-sulfides has received relatively little attention to date. However, metals of economic interest also extensively occur in non-sulfidic ores such as nickel laterites. Among these ores, nickel- and cobalt-enriched limonite deposits should be the most amenable to reductive bioleaching by acidophilic and chemolithotrophic Bacteria. This is because ferric iron minerals are disrupted by reduction to release associated metals. Instead of feeding costly organic substrate to heterotrophic ferric reducers, the use of chemolithotrophic acidophiles, such as *At. ferrooxidans,* has been investigated. These acidophiles couple the oxidation of elemental sulfur to the reduction of ferric iron, thereby accelerating the dissolution of minerals such as goethite (FeOOH) in limonitic deposits, releasing nickel, and other transition metals (cobalt, chromium, and manganese). It has been proposed that the acidophilic microorganisms accelerate the dissolution of goethite nickel laterite by reducing the small amounts of soluble ferric ions produced by acid dissolution of the mineral, thereby causing a shift in the equilibrium between solid phase and soluble ferric ions. More details can be found in Watling (2015) and Chap. 15. In addition, acidophilic microorganisms have been investigated to extract metal values from "urban mine" sources including electronic wastes (e-wastes), spent catalysts or other waste materials. The investigated acidophiles include *Acidithiobacillus* spp. for lithium-ion batteries as well the recovery of rare earth elements from resources such as magnets (described in Chap. 14). Finally, other than the Bacteria and Archaea focused upon in this chapter, filamentous fungal species (e.g., *Aspergillus* and *Penicillium* spp.) have been studied for the production of organic acids, which solubilize metals from non-sulfidic minerals based on the combination of acidolysis and metal–organic acid complexation.

5.7 Future Aspects

High-throughput microbial community profiling based upon markers such as the 16S rRNA gene has shown the presence of a large number of unknown acidophiles that await characterization. These novel taxa have the potential to play important roles in biomining. Particular areas of interest include halotolerant acidophiles that can be used for metal solubilization in brackish water and high temperature, autotrophic iron-oxidizers that have the potential to efficiently leach copper from chalcopyrite in bioheaps. In addition, reconstruction of (near) complete genomes from environmental DNA has resulted in the discovery and description of diverse candidate taxa. Finally, genome sequencing has also enabled the discovery of novel species previously assigned to a single taxon by traditional techniques. An example is the recent analysis of approaching 100 Acidithiobacillia class genomes that resulted in the (re-)classification of 19 lineages at varying taxonomic levels (Moya-Beltrán et al. 2021). These (meta)genomic techniques can be used in the future to generate hypotheses for growth characteristics of biomining Bacteria, such as the ability to grow on a novel substrate that may be exploited in commercial biomining operations.

The identification of complete communities including rare taxa has questioned the concept of a core biomining community and future studies will likely show how each biomining environment has its own acidophile community adapted to the specific conditions. Understanding of the existing acidophile species along with continually described new species will facilitate the "top down" approach to microbial consortia for biomining operations. Finally, understanding the *modus operandi* of biomining prokaryotes will also aid in selecting for communities with desired traits to maintain favorable conditions in bioheaps to increase rates of mineral dissolution. One example is the use of "weak" iron-oxidizers (i.e., those with relatively low affinities for ferrous iron as electron donor) to maintain the ORP in the desired range for optimum chalcopyrite oxidation without the accumulation of passivating layers.

Acknowledgments Mark Dopson acknowledges funding from the Swedish Innovation Agency, Vinnova via the ERA-MIN 2 project Microbial Consortia for enhanced Copper Recovery (MiCCuR). Naoko Okibe acknowledges funding from the Japan Society for the Promotion of Science (JSPS) KAKENHI Grant Number JP20H00647.

References

Bonnefoy V, Holmes DS (2012) Genomic insights into microbial oxidation and iron homeostasis in extremely acidic environments. Environ Microbiol 14:1597–1611. https://doi.org/10.1111/j.1462-2920.2011.02626.x

Bräsen C, Esser D, Rauch B, Siebers B (2014) Carbohydrate metabolism in archaea: current insights into unusual enzymes and pathways and their regulation. Microbiol Mol Biol Rev 78: 89–175. https://doi.org/10.1128/MMBR.00041-13

Counts JA, Willard DJ, Kelly RM (2020) Life in hot acid: a genome-based reassessment of the archaeal order *Sulfolobales*. Environ Microbiol. https://doi.org/10.1111/1462-2920.15189

Dopson M, Holmes DS (2014) Metal resistance in acidophilic microorganisms and its significance for biotechnologies. Appl Microbiol Biotechnol 98:8133–8144. https://doi.org/10.1007/s00253-014-5982-2

Dopson M, Johnson DB (2012) Biodiversity, metabolism and applications of acidophilic sulfur-metabolizing micro-organisms. Environ Microbiol 14:2620–2631. https://doi.org/10.1111/j.1462-2920.2012.02749.x

Dopson M, Ossandon F, Lövgren L, Holmes DS (2014) Metal resistance or tolerance? Acidophiles confront high metal loads via both abiotic and biotic mechanisms. Front Microbiol 5:157. https://doi.org/10.3389/fmicb.2014.00157

Harrison JP, Gheeraert N, Tsigelnitskiy D, Cockell CS (2013) The limits for life under multiple extremes. Trends Microbiol 21:204–212. https://doi.org/10.1016/j.tim.2013.01.006

Ilbert M, Bonnefoy V (2013) Insight into the evolution of the iron oxidation pathways. Biochim Biophys Acta Bioenerg 1827:161–175. https://doi.org/10.1016/j.bbabio.2012.10.001

Johnson DB, Hallberg KB (2009) Carbon, iron and sulfur metabolism in acidophilic micro-organisms. Adv Microb Physiol 54:201–255. https://doi.org/10.1016/S0065-2911(08)00003-9

Johnson DB, Kanao T, Hedrich S (2012) Redox transformations of iron at extremely low pH: fundamental and applied aspects. Front Microbiol 3:96. https://doi.org/10.3389/fmicb.2012.00096

Moya-Beltrán A, Beard S, Rojas-Villalobos C, Issotta F, Gallardo Y, Ulloa R, Giaveno A, Degli Esposti M, Johnson DB, Quatrini R (2021) Genomic evolution of the class *Acidithiobacillia*: deep-branching Proteobacteria living in extreme acidic conditions. ISME J. https://doi.org/10.1038/s41396-021-00995-x

Norris PR, Davis-Belmar CS, Brown CF, Calvo-Bado LA (2011) Autotrophic, sulfur-oxidizing actinobacteria in acidic environments. Extremophiles 15:155–163. https://doi.org/10.1007/s00792-011-0358-3

Parks DH, Chuvochina M, Waite DW, Rinke C, Skarshewski A, Chaumeil P-A, Hugenholtz P (2018) A standardized bacterial taxonomy based on genome phylogeny substantially revises the tree of life. Nat Biotechnol 36:996–1004. https://doi.org/10.1038/nbt.4229

Quatrini R, Appia-Ayme C, Denis Y, Jedlicki E, Holmes DS, Bonnefoy V (2009) Extending the models for iron and sulfur oxidation in the extreme acidophile *Acidithiobacillus ferrooxidans*. BMC Genomics 10:394. https://doi.org/10.1186/1471-2164-10-394

Rawlings DE (2002) Heavy metal mining using microbes. Annu Rev Microbiol 56:65–91. https://doi.org/10.1146/annurev.micro.56.012302.161052

Rawlings DE, Johnson DB (2007) The microbiology of biomining: development and optimization of mineral-oxidizing microbial consortia. Microbiology 153:315–324. https://doi.org/10.1099/mic.0.2006/001206-0

Rohwerder T, Sand W (2007) Oxidation of inorganic sulfur compounds in acidophilic prokaryotes. Eng Life Sci 7:301–309. https://doi.org/10.1002/elsc.200720204

Schippers A, Sand W (1999) Bacterial leaching of metal sulfides proceeds by two indirect mechanisms via thiosulfate or via polysulfides and sulfur. Appl Environ Microbiol 65:319–321. https://doi.org/10.1128/AEM.65.1.319-321.1999

Slonczewski JL, Fujisawa M, Dopson M, Krulwich TA (2009) Cytoplasmic pH measurement and homeostasis in bacteria and archaea. Adv Microb Physiol 55:1–79. https://doi.org/10.1016/S0065-2911(09)05501-5

Wang R, Lin J-Q, Liu X-M, Pang X, Zhang C-J, Yang C-L, Gao X-Y, Lin C-M, Li Y-Q, Li Y, Lin J-Q, Chen L-X (2019) Sulfur oxidation in the acidophilic autotrophic *Acidithiobacillus* spp. Front Microbiol 9:3290. https://doi.org/10.3389/fmicb.2018.03290

Watling HR (2015) Review of biohydrometallurgical metals extraction from polymetallic mineral resources. Fortschr Mineral 5:1–60. https://doi.org/10.3390/min5010001

Chapter 6
Biomolecular and Cultivation Tools

Raquel Quatrini, Elizabeth Lindsay Jane Watkin, and David Barrie Johnson

Abstract Acidophiles are a defined group of extremophilic microorganisms that have distinct physiological features that separate them from the rest of the biosphere. Those that populate biomining operations and environments that are impacted by mining (acid mine drainage waters, etc.) often face additional challenges and stresses, such as elevated concentrations of potentially toxic metals and metalloids. Techniques used to isolate, cultivate, and maintain these (predominantly prokaryotic) microorganisms in laboratories are therefore necessarily different from those used for more "mainstream" life forms. While molecular techniques that are used routinely in biology are also appropriate for studying acidophiles, some protocol modifications are usually required, especially in the sampling and preparation stages, for them to be successfully applied. This chapter describes how both cultivation-based and biomolecular techniques have been developed and applied to study "biomining" microorganisms, and how this has led to major advances in understanding both how they function as pure cultures and in mixed communities.

Keywords Acidophiles · Cultivation · Preservation · Molecular biology · Detection · Identification · Typing

---

R. Quatrini (✉)
Centro Científico y Tecnológico de Excelencia Ciencia & Vida, Santiago, Chile

Facultad de Medicina y Ciencia, Universidad San Sebastián, Providencia, Santiago, Chile

ANID – Millennium Science Initiative Program – Millennium Nucleus in the Biology of the Intestinal Microbiota, Santiago, Chile
e-mail: rquatrini@cienciavida.org

E. L. J. Watkin
Curtin Medical School, Curtin University, Perth, WA, Australia
e-mail: E.Watkin@curtin.edu.au

D. B. Johnson
School of Natural Sciences, Bangor University, Bangor, UK

Faculty of Health and Life Sciences, Coventry University, Coventry, UK

Natural History Museum, London, UK
e-mail: d.b.johnson@bangor.ac.uk

D. B. Johnson et al. (eds.), *Biomining Technologies*,
https://doi.org/10.1007/978-3-031-05382-5_6

6.1 Physiological and Phylogenetic Diversities of Microorganisms Commonly Encountered in Biomining Operations

Biomining operations harness the abilities of microorganisms that are directly or indirectly involved in accelerating the dissolution of (mostly sulfide) minerals. All current full-scale biomining operations operate at low pH, restricting life forms that can thrive in these environments to acidophilic microorganisms. Although temperatures in stirred-tank operations (Chap. 3) are generally maintained at a set value (typically between 40 and 50 °C) those in heaps and dumps cannot be controlled and can vary from near freezing to >80 °C, which has a major determinative impact on the indigenous microflora. Other challenges that acidophiles encounter in biomining environments include elevated concentrations of potentially toxic metals and metalloids, frequently extreme osmotic pressure, and high levels of salt (NaCl) and noxious organic compounds used for extracting target metals from pregnant leach solutions. In addition to prokaryotes that are directly or indirectly involved in mediating sulfide mineral dissolution, other types of organisms are frequently encountered that exploit biomining environments without having any perceptible benefit to the industrial process, especially in relic features (waste rocks, etc.), drainage waters and pit lakes.

A comprehensive account of current knowledge of biodiversity of acidophilic microorganisms is given in Chap. 5. While these are found in all three Domains of the "tree of life" (Bacteria, Archaea, and Eukarya), those with important roles in biomining are exclusively prokaryotic (bacteria and archaea). Eukaryotic fungi encountered in working biomines have only nuisance value, as they can colonise, and frequently clog, irrigation pipes and nozzles, while acidophilic protozoa can also have a detrimental impact by grazing on biomining bacteria (Johnson and Aguilera 2016). However, eukaryotic acidophiles become increasingly rare at elevated temperatures.

As a group, prokaryotic acidophiles exhibit similar broad-range metabolic diversity to the rest of the biosphere, in being able to use solar or chemical (organic and inorganic) energy and being able to obtain the carbon required for their growth from either inorganic or organic sources, and in some cases, both. An important trait that differs between acidophilic prokaryotes and the rest of the biosphere is that chemolithotrophy (the ability, in simplistic terms, to use energy from oxidising inorganic electron donors) is widespread among the former. Both ferrous iron (oxidation state +2) and sulfur (if its oxidation state is $< +6$) are well known in this respect, and a number of species can also utilise hydrogen. Extremely acidophilic nitrifying prokaryotes are currently unknown, and fermentative metabolisms appear to be absent. Molecular oxygen, ferric iron, and (more rarely) manganese (IV) can be used as electron acceptors. Consideration of the metabolic abilities and limitations of species of acidophilic prokaryotes is of central importance when devising strategies and protocols for isolating, cultivating, and maintaining these microorganisms, as set out below. Additional information may be found in Hallberg and Johnson (2007), Johnson and Aguilera (2016), and Johnson and Quatrini (2020).

6.2 Sampling: Sites and Protocols

The type strains of validated species of acidophilic prokaryotes, and sometimes also other strains, can be purchased from national culture collections, such as the *Japanese Collection of Microorganisms* (JCM), the *Deutsche Sammlung von Mikroorganismen und Zellkulturen* (DSMZ), and the *American Type Culture Collection* (ATCC). Many research groups working on biomining projects also maintain, and sometimes allow access to, their own culture collections. In many situations, however, it is both preferable and necessary for researchers to sample from industrial and environmental sites both to recover pure and mixed cultures and to obtain, from molecular analysis, comprehensive datasets of indigenous microbial diversities. Working and abandoned metal mines are appropriate for obtaining metal-tolerant mineral-oxidising prokaryotes, while low-pH geothermal sites can be used to source moderately and extremely thermophilic acidophiles. Low pH, oxygen-depleted environments (e.g., sediments in drainage streams, and anoxic zones in volcanic and pit lakes) are appropriate when facultative and obligate anaerobic acidophiles are sought, and extremely acidic saline sites (which are relatively rare) when targeting halo-acidophiles. Strict adherence to national restrictions and laws and the Nagoya Protocol (https://www.cbd.int/abs/doc/protocol/nagoya-protocol-en.pdf) as well as safety measures, is always essential.

In addition to sampling for microbiological and molecular analysis, it is also important to record GPS locations and the physico-chemical metadata (pH, temperature, redox potentials, dissolved oxygen contents and conductivities of water samples, etc.) on site. Semiquantitative measurements of ferrous iron, other dissolved metals, and solutes (such as sulfate and chloride) can also be done on-site using, for example, Merckoquant test strips. Together these on-site analyses can pinpoint which sites in an area are best suited for a given purpose. Liquid samples used to enumerate and isolate acidophiles should fill sterile containers, and be kept cool (4–10 °C, not frozen); solid samples, such as tailings, should be placed in sealable plastic bags and again kept at low temperature. Water samples also need to be filtered (through sterile 0.2 μm polycarbonate or other suitable membrane filters) on-site, for two distinct purposes: to harvest microbial cells, from which DNA can later be extracted (preferably from membranes frozen on-site in liquid nitrogen), and to obtain sterile samples for quantitative determination of metals and other analytes (e.g., dissolved organic carbon and nitrogen and sulfur species). A small amount of concentrated acid (usually nitric) can be used to help preserve samples for metal analysis. The volumes of water required for the latter are relatively small, but large volumes of water (several litres) are generally needed to obtain sufficient harvested biomass on the membranes themselves. Since cell numbers cannot be assessed accurately on-site (though are often between 10^4 and 10^6 cells mL^{-1}, except in biofilms and microbial streamers), the maximum amount of water that can pass through each membrane, before biomass or particulate materials impede this, should be filtered. Acidity and exceptionally high concentrations of metals, other ions, and precipitates in bioleaching solutions make these samples particularly sensitive to hydrolysis and oxidative degradation once cellular integrity is compromised or cells are fully lysed, and free radical generation due to the presence of iron and copper initiating the Fenton and Haber-Weiss reactions can degrade RNA, DNA, and proteins within hours. Thus, thorough washing steps with deionised water or metal chelator solutions are mandatory to help facilitate the recovery of clean cell pellets once samples are returned to the laboratory.

6.3 Cultivation-Based Approaches

6.3.1 Enumeration, Enrichment, and Cultivation of "Biomining" Microorganisms

6.3.1.1 Microscopy

Microscopes—both light and electron—are as useful and versatile in acidophile microbiology as elsewhere. Phase-contrast microscope observation of cell morphologies (most biomining bacteria are rod shaped while archaea are regular or irregular cocci), motility, and the presence or absence of endospores, can aid in their preliminary identification. Cell numbers in liquid samples can be readily enumerated using gridded counting chambers (e.g., Thoma or Neubauer chambers/slides), though this requires a minimum of about 10^6 cells/mL and is, therefore, more suited to laboratory cultures than environmental samples. This limitation can be resolved by filtering known volumes of samples through polycarbonate membrane filters (0.1–0.2 μm pore size) and fixing and staining trapped cells. Different stains can be used, some, such as DAPI (4′,6-diamidino-2-phenylindole) and SYBR® Green bind preferentially to dsDNA and will label most microorganisms in the sample. Other stains, such as SYTO9 (a membrane permeable dye) and propidium iodide (a membrane impermeable dye) when combined can be used to differentiate living, though not necessarily metabolically active (e.g., viable but non-culturable cells) from non-viable prokaryotes. Microscopic examinations should be accompanied by culturing and molecular techniques in order to achieve identification of acidophiles at various levels (e.g., domain, phylum, or species).

6.3.1.2 Liquid Media

Liquid media are used to enumerate, enrich, cultivate and maintain acidophiles, and formulating appropriate media compositions is critical to the success of these practices. Media compositions vary with the nutritional requirements of the targeted

Table 6.1 Suitable liquid media for enriching acidophilic prokaryotes frequently encountered in biomining operations

Organism	Medium	pH	T (°C)	Notes
L. ferriphilum	10 mM Fe^{2+} + pyrite	1.5	40	*L. ferriphilum* will become increasingly dominant with prolonged incubation
Sulfobacillus spp.	10 mM Fe^{2+} + 0.02% YE	1.8	35–50	Temperature can be adjusted to select for different species
At. caldus	0.5% S^0	2.0	40	pH declines with incubation time; *At. caldus* will thrive at pH ~1
Acidiphilium spp.	0.02% YE (+/− 5 mM glucose)	2.5	30	Very readily enriched for and isolated from samples > pH 2
Acidocella spp.	0.02% YE (+/− 5 mM glucose)	3.5	30	Very readily enriched for and isolated from samples > pH 3
At. ferrivorans	10 mM Fe^{2+} + pyrite	2.0	15	Cold-tolerant species. pH increases during growth
Ferrimicrobium	10 mM Fe^{2+} + 0.02% YE	2.0	30	May also enrich for *Acidiphilium*, etc.
Ferroplasma spp.	10 mM Fe^{2+} + 0.05% YE	1.5	35	Cell morphology is very different from acidophilic bacteria (cell wall-less, round irregular cells)

microorganism(s), as illustrated in Table 6.1. All liquid (and solid) media used for acidophilic prokaryotes need to contain predominantly sulfate salts, rather than chlorides or nitrates which can inhibit growth of acidophiles.

The "9K" medium, so-called because it contains 9 g ferrous iron/L (161 mM) has been used widely to cultivate iron-oxidising acidophiles, and also sulfur-oxidisers, when iron is replaced by elemental sulfur or tetrathionate. This medium, first described in 1958, is flawed in a number of respects: (1) it does not contain sodium or any micronutrients; (2) it contains amounts of nitrogen and phosphorus greatly in excess (>50×) necessary for the biomass it supports; (3) its relatively high pH (~3.5) and phosphate content causes ferrous phosphate to precipitate in newly prepared medium, the turbidity occluding that due to microbial growth. Since most laboratories now use pure-grade (e.g., reverse osmosis) water, the nutrient deficiencies of 9K can be acute, and result in poor or no growth, particularly on successive transfers, as one or more micronutrient present in the initial inoculum/sample are diluted out. Excessive concentrations of phosphate can inhibit the growth of some acidophiles, and both ferrous and ferric iron precipitates are better avoided, as bacteria can attach to these, making harvesting biomass problematic and negatively affecting downstream molecular biology applications. Updated comprehensive formulations of basal salts and trace element mixes that are superior to 9K for cultivating acidophiles have been published (e.g., Ňancucheo et al. 2016).

6.3.1.3 Electron Donors

Ferrous Iron Iron-oxidising acidophiles use ferrous iron and, in many cases, other electron donors. Although ferrous iron is soluble over a wide pH range, it is susceptible to abiotic oxidation above pH 3.0–3.5 (at 30 °C), and is heat-labile, and therefore stock solutions are generally prepared at low pH ($\leq$2), filter-sterilised, and kept cool. Ferrous iron oxidation consumes protons, causing culture pH to increase. Ferric iron is much less soluble than ferrous and hydrolyses at and above pH ~2.3, forming a variety of hydrated, sulfate-containing, amorphous, and crystalline ferric iron minerals, such as schwertmannite ($Fe_8O_8(OH)_6(SO_4)·nH_2O$) and jarosites ($(K,Na,H)Fe_3(OH)_6(SO_4)_2$). This is usually undesirable and can be avoided by lowering the initial amount of ferrous iron in cultures (which reduces biomass yields) or maintaining low pH. The latter is possible in pH-controlled bioreactors but not in shake flask cultures. Buffers are used to counterbalance pH changes in microbial cultures, but many of these (organic and phosphate-based) need to be avoided with acidophiles because of toxicity and/or precipitation problems, whereas the sulfate/bisulfate buffer is highly appropriate. Sulfuric acid has two pK_a values, <0 (H_2SO_4/HSO_4^-) and 1.92 (HSO_4^-/SO_4^{2-}), the latter being close to the pH at which many acidophiles are cultivated. The addition of a sulfate salt to growth media can therefore greatly increase their buffering capacity. For example, adding 50 mM magnesium sulfate to a medium containing 50 mM ferrous sulfate and adjusting to pH 1.5 with sulfuric acid increases its total acidity (proton + bisulfate) from 85 mM to 120 mM, providing sufficient buffering to avoid precipitation of the 50 mM ferric iron produced. Addition of excessive amounts of additional sulfate to culture media should, however, be avoided as this induces osmotic stress and results in prokaryotes diverting energy into producing osmoprotectants (such as trehalose) rather than biomass.

Elemental Sulfur, Sulfur Salts, and Hydrogen Different forms of sulfur can be used to cultivate chemolithotrophic acidophiles, such as *Acidithiobacillus* spp. Elemental sulfur (S^0) is widely used but is insoluble and hydrophobic unless "wetted" by microbial surfactants. Thiosulfate ($S_2O_3^{2-}$) is a soluble sulfur salt used routinely to cultivate neutrophiles but is not stable in acidic liquors and is oxidised by ferric iron. In contrast, tetrathionate ($S_4O_6^{2-}$) is soluble and stable at low pH, though it is toxic to many acidophiles at ~5 mM. Oxidation of S^0 and sulfur oxyanions generates acidity, and culture pH can fall to the point where microbial growth is inhibited or ceases. Again, the sulfate/bisulfate buffer can be used to alleviate this problem in shake flask cultures ($SO_4^{2-} + H^+ \rightarrow HSO_4^-$).

Hydrogen is used by a wide range of species and strains of chemolithotrophic acidophiles, though pragmatic problems relating to mass transfer of the gas into aqueous phases and working with a highly explosive gas require different cultivation protocols. One major advantage of using hydrogen as an electron donor is that its oxidation, coupled to oxygen, is a pH-neutral reaction, and cell densities of >10^9/mL can be obtained in shake flask cultures.

Organic Electron Donors A variety of organic electron donors can be used to cultivate obligate and facultative heterotrophic acidophiles, though these are mostly noncomplex compounds such as low molecular weight alcohols (e.g., glycerol) and monomeric and dimeric sugars. Polymeric and aromatic compounds tend not to be utilised, though there are exceptions (e.g., *Acidocella aromatica*, which can degrade aromatic compounds such as phenol). Mono- and dicarboxylic acids are better avoided as they can be highly toxic to acidophiles even in very low concentrations, but tricarboxylic citric acid can be used by some *Acidiphilium* and other species. Yeast extract is often included in growth media to enhance microbial growth by providing growth factors, though some bacteria, such as *Acidiphilium cryptum*, can grow well on single organic compounds.

6.3.1.4 Anaerobic and Microaerobic Cultivation of Acidophiles

While laboratory cultures of acidophilic microorganisms tend to be grown routinely under aerobic conditions, many can also grow in the absence of oxygen or under oxygen-limiting conditions, using alternative electron acceptors. Both anaerobic and microaerobic atmospheres can be generated using atmosphere-generating systems available from a range of suppliers. The electron acceptor most commonly used by acidophiles, apart from oxygen, is ferric iron (more species of acidophiles can reduce than can oxidise iron). It can be coupled to the oxidation of sulfur, hydrogen, or organic compounds. Again, pH constraints need to be borne in mind: the reduction of soluble ferric iron is an acid-generating reaction, while the reductive dissolution of ferric iron minerals generates alkalinity. Acidophilic prokaryotes that can grow via the dissimilatory reduction of sulfate and/or sulfur are mostly archaea and are less well-known than iron reducers. The dissimilatory reduction of sulfate at low pH is strongly acid-consuming at low pH, and the hydrogen sulfide generated can be used to precipitate, and thereby recover, metals present in process liquors and drainage streams.

6.3.1.5 Solid Media

While solid media are used routinely to cultivate neutrophilic microorganisms, early attempts to use them to grow chemolithotrophs either failed all together or were non-reproducible. This situation was remedied by the development of an "overlay technique", where a double layer gel is prepared, the bottom layer of which is inoculated with an active culture of an acidophilic heterotroph that is able to detoxify the medium during plate incubation, by removing organic materials (chiefly pyruvic acid and galactose) released by acid hydrolysis of the gelling agent. For most applications, this is agarose or agar (which need to be sterilised separately from acidic components of solid media) though for extreme thermo-acidophiles Gelrite™ is more appropriate as this can form thermo-stable gels. A strain of *A. cryptum* (SJH) has been used widely in various overlay plate formulations as this is a highly

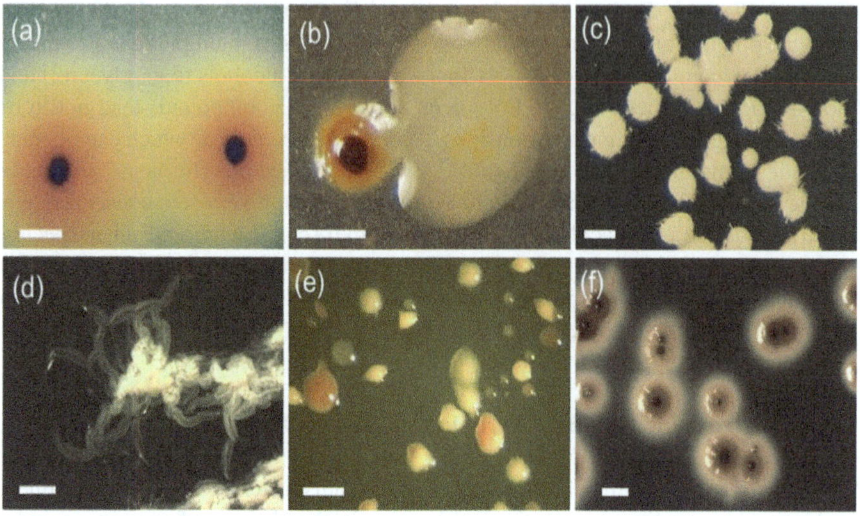

Fig. 6.1 Colonies of acidophilic bacteria grown on overlay plates: (**a**) *Leptospirillum ferriphilum*^{T} (on ferrous iron), (**b**) *Acidithiobacillus ferrivorans* strain Peru 6 (on ferrous iron/tetrathionate), (**c**) *Sulfobacillus* sp. (on tetrathionate), (**d**) *Acidithiobacillus ferrooxidans*^{T} (on hydrogen), (**e**) mixed population of *Acidiphilium* and *Acidocella* spp. (on yeast extract), and (**f**) *Desulfosporosinus acididurans*^{T} (on glycerol/yeast extract; anaerobic)

versatile heterotrophic acidophile that can grow as low as pH 0.8, tolerate >500 mM sodium chloride, and at elevated concentrations of many transition metals. Preadaptation and cultivation of *A. cryptum* in a medium with similar chemical composition to that of the solid media allows it to be used in overlay plates that can have a similar generic makeup (e.g., containing ferrous iron and/or tetrathionate, and elevated salt concentrations for halophilic acidophiles) to liquid media. However, since *A. cryptum* SJH metabolises most organic compounds that might be used to isolate and cultivate other heterotrophic acidophiles, a different species needs to be used in overlay plates. *Acidocella aromatica* has the unusual characteristic of being able to degrade many aliphatic acids, including pyruvic, but does not grow on yeast extract or, with the exception of fructose, monosaccharides, or glycerol. Overlay media incorporating *Ac. aromatica* have been used to isolate and cultivate a wide range of heterotrophic acidophiles, including sulfate-reducers (Ňancucheo et al. 2016). Colonies of acidophilic microorganisms grown on different overlay medium formulations are shown in Fig. 6.1.

6.3.2 Microbial Activity Measurements

6.3.2.1 Specific Rates of Oxidation and Reduction of Electron Donors and Acceptors

Measurement of specific rates of oxidation and reduction of iron and sulfur are useful for comparing pure and mixed cultures and understanding how these are influenced by growth histories, in laboratory and environmental samples (Johnson et al. 2012). Cultures grown under defined conditions are harvested and resuspended in a small volume of basal salts. Biomass (e.g., as protein) in the suspension is measured and aliquots are used to record rates of iron or sulfur oxido-reduction at defined pH and temperature over a period (typically 20 min to 2 h) which needs to be much less than the doubling time of the culture in question, to prevent any significant increase in biomass. In the case of iron oxido-reduction, measurements of ferrous iron concentrations can be used (e.g., using the Ferrozine colorimetric assay) while for sulfur measuring changes in sulfate concentrations (e.g., using ion chromatography) are appropriate. Results are expressed as μg iron/sulfur oxidised/reduced $minute^{-1}$ mg $protein^{-1}$. These data are not only useful to compare species and strains but can also be used in the design of bioreactors, for example, to calculate the magnitude of a microbial population required to oxidise (and precipitate) iron in mine drainage waters.

6.3.2.2 Microcalorimetry

The microbially catalyzed oxidation of sulfide minerals such as pyrite is exothermic and this can be used to assess the activity of chemolithotrophic acidophiles using a microcalorimeter capable of recording very small (microwatt) changes in thermal energy (e.g., Hedrich et al. 2016). A representative sample (ca. 10 g) is placed in an oxygen-containing vessel, which is sealed and placed in the microcalorimeter. After equilibration, the heat output over a 2- to 4-h period is recorded. Knowledge of which mineral is mainly being oxidised allows its rate of oxidation to be calculated, by reference to published datasets.

6.3.2.3 ATP Measurements

Generation of ATP is an ongoing process in metabolically active biomass; levels decline and eventually become zero as cells become moribund and die. Measurements of ATP can therefore be used to determine the "health" of a microbial population, which may or may not correlate with cell densities. Measurements are facilitated by being carried out using commercially available kits and relatively inexpensive equipment (a luminometer). Results as "relative light units" since the assay involves ATP fuelling emission of photons are obtained within minutes of sampling. The technique is sensitive (able to detect 10^3 metabolically active cells mL^{-1}) and can readily be used on-site by operators at mine sites. However, ATP measurements do not give information on the identity of organisms present in a sample and, being an enzyme-based technique, are prone to inhibition by acidity and dissolved metals that characterise mineral leachates. A protocol developed by Okibe and Johnson (2011) eliminated these problems and was used successfully to measure ATP in both laboratory cultures of acidophiles and leachates from biomining operations.

6.3.3 *Maintenance and Preservation of Biomining Microorganisms*

Numbers of viable cells of some acidophiles decline rapidly once their substrate is depleted. For example, numbers of viable *Leptospirillum* spp. can fall from ~10^7 to ~10^1 mL^{-1} within several days of ferrous iron in the culture media being depleted. While regular subculturing can be used to maintain cultures of acidophiles, this can be highly demanding of time when multiple cultures are involved, and so short- to long-term methods of maintaining acidophiles in an active state are useful. For iron- and sulfur-oxidising species, viable cultures can be maintained by first growing cells either in ferrous iron or sulfur liquid medium, adding sterile coarse grain pyrite or elemental sulfur, further incubation of up to 1 week, and then storage at 4 °C, pyrite and sulfur acting as a slow-release substrate for the acidophiles during this time. This strategy can be effective for up to at least 2 years for strains of *Acidithiobacillus* spp. and *L. ferrooxidans*; strains of *L. ferriphilum* tend to lose viability when held at low temperatures and are better stored (for up to 6 months) in pyrite-containing medium at ca. 20 °C.

For long-term preservation, most acidophiles can be frozen and stored at −70 °C using a suitable cryoprotectant. Glycerol (used routinely for neutrophiles) is effective for heterotrophic acidophiles, while 7% (v/v) dimethylsulfoxide (DMSO) is more universally successful for all acidophiles. The initial pH of the suspensions used (which need to contain ~10^9 cells mL^{-1}, as many inevitably die during freezing and thawing) should not be too great as acid concentrations increase during the freezing process, which needs to be rapid. Freezing cells onto sterile glass beads allows stored cultures to be accessed several times. Problems with attempts to store *Leptospirillum* spp. at −70 °C, means that they are often maintained and released as active liquid culture in national collections, such as the DSMZ.

6.4 Biomolecular Techniques

Biomolecular techniques are cost-effective and fast alternatives, or complementary approaches, to traditional culture-dependent methods. As in many other areas of application a variety of molecular tools have been useful to detect, identify, type, enumerate, and trace acidophiles in different industrial environments and along bioleaching progress (Fig. 6.2). Most frequently, the tools used for these purposes target DNA, with fewer applications directed to RNA or proteins, even though the latter are far more abundant in biomass than DNA. This is due to RNA and proteins varying with the general treatment conditions and the specific sample collection, handling, and storage conditions, while DNA-based procedures are more robust, stable, and resistant to most of these changes. Cell lysis is frequently challenging when working with acidophiles, since their membranes have been selected to endure not only extreme acidity, but often also extreme temperatures and oxidant

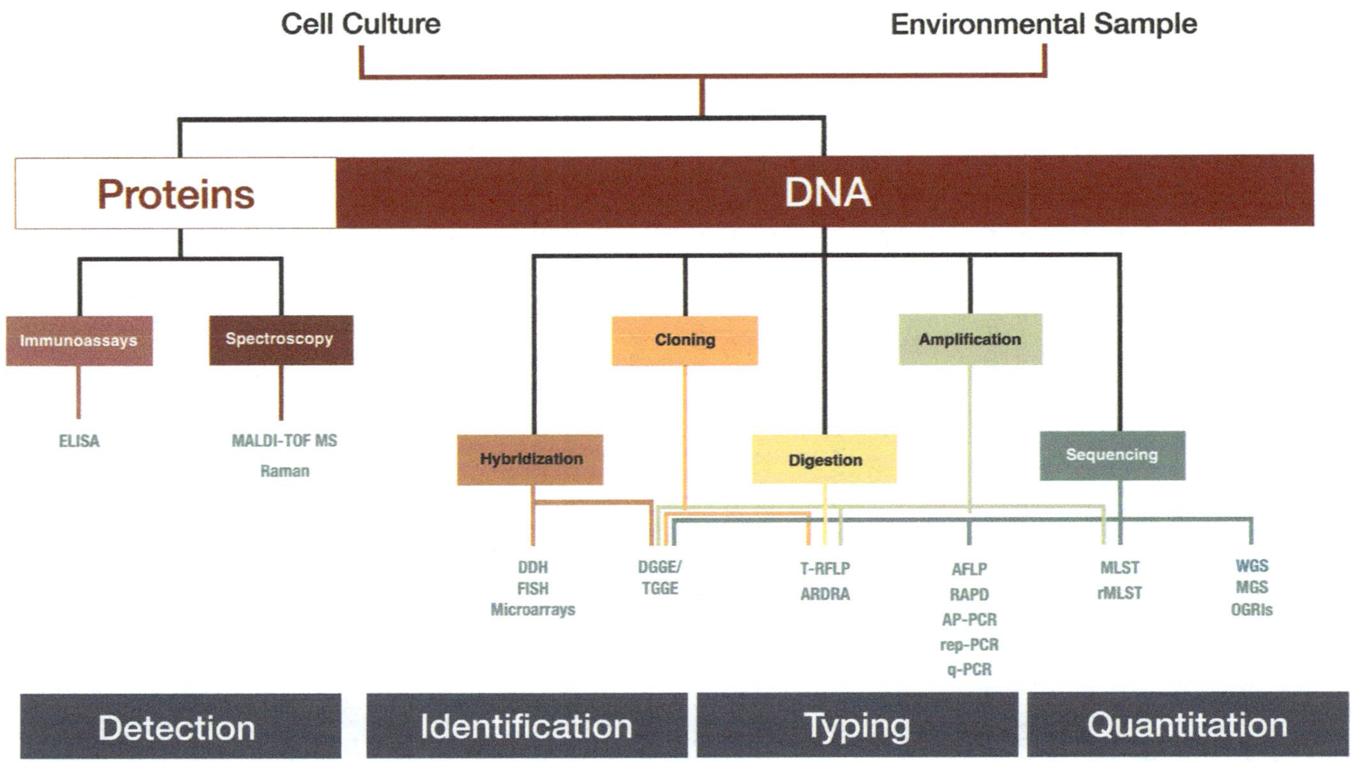

Fig. 6.2 Molecular typing techniques used in the study of bioleaching acidophiles. The technologies are displayed in colored boxes. The techniques are shown in green. Technique acronyms are explained in Table 6.2 and/or in the text

conditions, and their sensitivity to lysis is quite variable. For example, DNA can be extracted with ease from Gram-negative acidophiles such as *Acidithiobacillus* spp. but extraction is far more difficult with *Sulfobacillus* spp. and other Gram-positive acidophiles. Commercial kits are therefore not always well suited to extract nucleic acids from pure cultures of acidophiles and environmental samples, and it is often necessary to use a combination of physical (freeze/heat pulses), mechanical (sonication, bead-beating, cryo-milling), and/or chemical (detergents, enzymes) treatments depending on the target biomolecule and the integrity of the sample required. In the sections below, we cover some of the most common molecular techniques (categorised into different principal categories, according to the technology they rely on to resolve or detect inherent differences between target molecular markers) used for bacterial typing in general, and applied to the detection, identification, and study of the different acidophiles in industrial systems in particular.

6.4.1 Molecular Markers Used in Molecular Detection, Identification, and Typing

Molecular markers are DNA sequences that can be used to identify a particular genotype; they may be taxonomic or functional in nature. Regardless of its nature, in order to be a reliable indicator, a molecular marker must meet a number of criteria. Namely, it must be present and functionally constant in all organisms of interest (ubiquitous), it must display an appropriate level of sequence conservation and, finally, be detectable. Not only are molecular markers a useful means by which to identify species, but they can also be utilised to determine the similarity between microorganisms, to reveal the structure of microbial communities, and to detect the presence of microorganisms in the environment.

The most widely used molecular marker for the determination of phylogenetic relationships as well as typing is the 16S ribosomal RNA gene, which is highly conserved and occurs in all bacteria and archaea. It is composed of both conserved and variable regions and as such provides the ideal vehicle for comparison between isolates. In addition, there are many 16S rRNA gene sequences in public databases facilitating comparison with known species of bacteria and archaea. Its resolution, however, is insufficient to differentiate reliably beyond species level. Additionally, as microorganisms can contain different numbers of copies of the gene, there is limited benefit in the use of this gene region for purposes of quantification. Due to these limitations, several other markers such as functional (e.g., *amoA*, *nifH*, and *dsrAB*) or conserved single-copy housekeeping genes (*recA*, *atpD*, *rpoB*) have been proposed; a number of these are used conjunctly in Multi Locus Sequence Analysis (MLSA). By relying on five to seven informative loci, MLSA provides greater power of resolution than the use of a single gene for the identification, typing, and determination of relatedness of isolates. This strategy has been used to determine relationships between closely related iron- and sulfur-oxidising microorganisms

(Nuñez et al. 2014). Bacterial housekeeping genes (HKGs) and ribosomal protein loci are frequently utilised in this technique as they meet the above-stated criteria of suitable molecular markers (Khaleque et al. 2020).

6.4.2 *Fingerprinting Techniques*

Several experimental procedures, collectively known as molecular fingerprinting techniques, have been devised to help identify similarities and dissimilarities between microbial strains, populations, or communities, based on the unique and differential characteristics of a given molecule (e.g., repeat pattern at the whole-genome level) or molecular marker (e.g., DNA restriction profile of the 16S rRNA gene). Several of these techniques have been successfully applied to discriminate closely related acidophiles, to assess the genetic variability of industrial strains from different sources, or to trace changes in the cultured and uncultured diversity profiles of given communities during the progress of bioleaching or upon changes in operational procedures. Depending on the particular focus, and the required level of resolution, different types of fingerprinting techniques can be applied (Table 6.2). To map variations in DNA, resulting from changes in the targeted DNA molecule size (due to insertions, deletions, and rearrangements) or sequence (due to nucleotide differences or variations in the repeat patterns), either total genomic DNA or PCR-amplified genes or DNA regions can be utilised. The alluded variations are then exposed by different profiling techniques based on the use of low-specificity primers (PCR-based fingerprinting), restriction enzymes (Restriction-based fingerprinting), or denaturant agents or conditions (Conformational polymorphism fingerprinting), and resolved using gel or capillary-based electrophoresis techniques. Regardless of which technique is used, the resulting banding patterns are generally unique and serve as a "fingerprint" for strain identification, or population and community comparative analyses. In PCR-based fingerprinting applications, arbitrary, low-selectivity, or repeat-specific, short oligonucleotide primers (8–12 bases in length) are used to initiate amplification at multiple target sites in the genome and produce distinct genomic fingerprints. Low annealing temperatures (35–45 °C) are used to permit promiscuous pairing of the primers and to allow single mismatches.

For their ease of use and low cost, several variant methods have been successfully applied in biomining-related studies, including Randomly Amplified Polymorphic DNA analysis (RAPD) using multiple random decamers, Arbitrary Primed-PCR (AP-PCR) using single primers, Enterobacterial Repetitive Intergenic Consensus Sequence (ERIC-PCR), BOX-PCR, and Repetitive Extragenic Palindromic–PCR (REP-PCR) using repeat directed single primers. However, due to the promiscuous nature of these PCR-based techniques, in which banding patterns will vary according to reaction settings and conditions on a user-to-user basis, reproducibility of the assays is frequently an issue and their value is limited to intra-experiment comparisons. When samples are processed in parallel under standardised conditions reliable, informative and discriminant fingerprints can be obtained. In turn,

Table 6.2 Molecular typing techniques and markers used in the study of bioleaching acidophiles and their communities

	Technique	Marker	Discriminatory power[a]	Application		Advantages	Disadvantages
				Pure cultures	Communities		
Restriction-based fingerprinting	AFLP	gDNA, rDNA	High—Strain level	X		High reproducibility	Complex, difficult interpretation
	RFLP	gDNA (HKG)	Mid—Species level	X		Rapid, low cost, high reproducibility	Laborious set-up
	T-RFLP	rDNA (HKG)	Mid—Species level	X		Rapid, high sensitivity and reproducibility, HT	PCR-derived biases
	ARDRA	rDNA	Mid—Species level	X		Rapid, low cost, high reproducibility	Limited resolution
Anonymous fingerprinting	RAPD	gDNA (random targets)	High—Strain level	X		Rapid, low cost, no previous sequence knowledge required, HT	Limited reproducibility, not universal
	AP-PCR	gDNA (random targets)	High—Strain level	X			
	ERIC-PCR	gDNA (ERIC-type repeats)	High—Strain level	X			
	BOX-PCR	gDNA (BOX-type repeats)	High—Strain level	X			
	REP-PCR	gDNA (Palindromic repeats)	High—Strain level	X			
Conformational polymorphism fingerprinting	SSCP	rDNA, ssDNA	Mid—Species level	X	X	Rapid, semi-quantitative	Complex, low reproducibility, laborious phylogenetic identification
	DGGE	rDNA, dsDNA	Mid—Species level	X	X	Rapid, semi-quantitative	
	TGGE	rDNA, dsDNA	Mid—Species level	X	X		

Hybridisation dependent	DDH	gDNA	Mid—Species level	X		Universal standard. Genome level comparison	Complex implementation. No central database. High experimental error
	FISH	rRNA (HKG)	Mid—Species level	X	X	Phylogenetic identification, semi-quantitative	Dependent on probe sequences, unable to identify unknown species
	Microarray	rRNA (multiple genes)	Mid—Species level	X	X	Phylogenetic identification, semi-quantitative, rapid	Cross hybridisation, PCR-derived biases, low-abundance species are difficult to detect
Real-time PCR dependent	qPCR	rDNA, gDNA, rRNA, mRNA	High—Strain level	X	X	Quantitative, high sensitivity and reproducibility, rapid	Genetic information required for probe design, unable to identify unknown species, cost
Sequence dependent	Sequencing	rDNA	Mid—Species level	X	X	Phylogenetic identification, quantitative, rapid. Discovery of unknown bacteria	Cost, HGT that may distort relationships
	MLSA	gDNA (HKG)	High—Strain level	X	X	High reproducibility	Genetic information required for selection of informative genes, cost, laborious
	NGS	rDNA, gDNA, mgDNA	Mid to high—Species to strain level	X	X	High sensitivity, rapid	Cost, laborious data management

Table adapted from Nuñez et al. (2016)

[a]Levels of discriminatory power, based on the tools' capacities to discriminate between genera (low), species (mid), and strains (high)

Abbreviations: AFLP: amplified fragment length polymorphism; RFLP: restriction fragment length polymorphism; T-RFLP: terminal restriction fragment length polymorphism; ARDRA: amplification of ribosomal DNA restriction analysis; RAPD: randomly amplified polymorphic DNA analysis; AP-PCR: arbitrary primed-PCR; ERIC-PCR: enterobacterial repetitive intergenic consensus sequence primed-PCR; BOX-PCR: BOX elements primed-PCR; REP-PCR: repetitive extragenic palindromic primed-PCR; SSCP: single-strand conformation polymorphism; DGGE: denaturing gradient gel electrophoresis; TGGE: temperature gradient gel electrophoresis; DDH: DNA–DNA hybridisation; FISH: fluorescent in situ hybridization; q-PCR: quantitative PCR: MLSA: multilocus sequence analysis; NGS: next generation sequencing analysis; HKG: housekeeping genes; HT: high-throughput; HGT: horizontal gene transfer; gDNA: genomic DNA; rDNA: ribosomal RNA genes; rRNA: ribosomal RNA; mRNA: messenger RNA; mgDNA: metagenomic DNA restriction-based methods which produce molecular fingerprints by enzymatic digestion of reference and test DNA samples (generally combining a frequent and an average cutter restriction endonuclease) are much more stable and reproducible. A number of variant techniques, namely Amplification of Ribosomal DNA Restriction Analysis (ARDRA), Amplified Fragment Length Polymorphism (AFLP), and Terminal-Restriction Fragment Length Polymorphism (T-RFLP), have been used in the analysis of bioleaching acidophiles and their communities, providing fair estimates of the overall genetic similarity of diverse types of samples and treatments under evaluation. Denaturing Gradient Gel Electrophoresis (DGGE) is by far the most widely used application in conformational polymorphism fingerprinting techniques. Due to its greater resolution (DNA fragments of similar length but different sequences can be resolved on a gel, as band patterns of varying complexity), this technique is very useful for analyzing and comparing whole microbial communities and their spatial and/or temporal variations. Detection and identification of species of interest can also be achieved, either by hybridisation of specific probes or by sequencing of excised gel bands (e.g., Nuñez et al. 2016).

6.4.3 Hybridisation-Dependent Approaches

A number of approaches that rely on the hybridisation of nucleic acids are used routinely to detect, identify, trace, and quantify microbial species in samples of diverse complexity. These methods involve using one or multiple probes to hybridise with the test sample and a label, usually fluorescent, to allow visualisation of the hybridised products, using microscopy, flow cytometry, or other specialised scanning devices.

One of the first such methods used in microbiology was DNA–DNA hybridisation (DDH), which measures the sequence relatedness between a given pair of genomes using the DNA of one organism as the labelled probe, and the other as the unlabelled target. The denatured DNAs are mixed and incubated to allow the DNA strands to anneal, forming a hybrid double-stranded DNA. The degree of hybridisation is quantified (e.g., using the increment of melting temperature) to obtain a measure of relatedness. Early molecular studies used DDH to group bioleaching acidophiles in genospecies, some of which have survived the passage of time while others have been reassigned to novel species using more sensitive molecular and genomic techniques. The DDH technique was long considered as the gold standard in the delineation of species (where a 70% DDH value corresponded to the 97% 16S rRNA gene sequence identity), but its many experimental pitfalls have caused it to be replaced by sequencing based approaches (Sect. 6.4.5).

Other hybridisation-dependent approaches are better suited for tracing and quantitating microorganisms in their ecological contexts. Fluorescence In Situ Hybridization (FISH), and its technically improved assay variants like Catalysed Reporter Deposition-FISH (CARD-FISH), can be applied to detect and quantify microorganisms of interest in microbial communities of considerable complexity. Genus- and species-specific probes, targeting 16S rRNA genes of different bioleaching acidophiles have been used to define the community structures of mine-impacted environments (Hallberg and Johnson 2007), including the underground AMD system of the Richmond mine in California and Mynydd Parys in north Wales, and the Rio Tinto in Spain. Applications in the biomining industry have also been reported, proving the usefulness of CARD-FISH to visualise metabolically active cells within microbial communities attached to mineral surfaces and to follow changes in the abundance of particular acidophiles in bioreactors or in mine tailing dumps.

DNA microarrays can also serve these purposes, though they provide only relative quantitation measures. Depending on the type of microarray (phylochips, functional gene arrays, and whole-genome coverage arrays) a few taxonomically (e.g., 16S rRNA gene) or metabolically informative marker genes (e.g., *nifH*) from different taxa, or alternatively the whole gene complement of a single or group of microorganisms, can be utilised for parallel detection and identification of diverse microorganisms within a community (hybridising DNA) or their activities (hybridising RNA). A number of dedicated microarrays were designed in the mid-2000s for screening acidophiles in biomining environments, but this approach has been mostly abandoned in favour of metagenomic sequencing applications (reviewed in Nuñez et al. 2016).

6.4.4 End-Point and Real-Time PCR Approaches

The PCR-based fingerprinting techniques described above are dependent on end-point PCR products and as such are only able to deduce the relative abundance of detected species. The use of quantitative real-time PCR (qPCR), however, enables quantification of species in microbial systems. qPCR detects and measures amplified DNA products during amplification resulting in increased sensitivity and reliability, and quantification is achieved through the comparison with a known standard. The 16S rRNA gene is most often targeted for this, but since many prokaryotes contain multiple copies of this gene this can lead to the potential for either over or underestimation of the true abundance of target species, and other markers that occur in a single copy, such as *gyr*B and *rpo*B may be more suitable for enumeration procedures (e.g., Zammit et al. 2008). qPCR has been used across a range of bioleaching environments including stirred tanks (e.g., Wang et al. 2014) and bioheaps (reviewed in Nuñez et al. 2016) to evaluate the presence and to assess the roles of different acidophilic prokaryotes.

While qPCR is a powerful tool in the analysis of microbial community dynamics, a number of caveats must be considered for its use in biomining systems. The reliance on the design and use of specific primers for target organisms may result in overlooking significant or novel players in the microbial system. Additionally, the extraction of DNA from bioleaching systems often results in the co-extraction of PCR inhibitors that are found in mineral- and metal-rich samples. Therefore, care must be taken in order to minimise the downstream effects of these inhibitors on

PCR amplification and quantification in these systems. Additionally, DNA extraction protocols do not discriminate between active and inactive cells and therefore no information about the activity status of the microorganisms can be determined. This can in part be addressed by targeting RNA rather than DNA.

6.4.5 Sequence-Dependent Approaches

Nucleic acid sequencing is probably the most widely used molecular genotyping technique in modern days. Traditional Sanger-based sequencing has been used for decades in the field of biomining to assess the presence and the degree of relatedness between microorganisms, based upon the level of identity of one or a few marker genes. More recently shotgun sequencing-based approaches have been applied to obtain the genomic blueprints of well-known bioleaching microorganisms and also native acidophiles recovered in culture from very diverse environments. Advances in genomic sequencing have provided relevant insights into the metabolic potential of these microbes and the ecosystem-level interactions they establish between them that are relevant to mineral processing (summarised in Cárdenas et al. 2016). With recent advances in high-throughput sequencing techniques, the ability to assess microbial communities at a taxonomic and functional level has expanded significantly and, as such, the understanding of how these ecosystems operate. Metagenomics can be applied to diverse natural habitats resulting in the identification of new genes and gene products from uncultured microbes, assembly of whole genomes, as well as comparisons of community gene content from microbial assemblages of different origins.

The most frequently utilised application of high-throughput sequencing techniques is that of targeted metagenomics, where PCR-based amplification of a target gene (most often 16S rRNA gene) is combined with next generation sequencing, hence known also as amplicon sequencing. Detailed community profiles can be generated after extensive bioinformatic analysis of the generated data. This approach has been applied to the study of microbial ecology within acidophilic communities such as bioleaching heaps, bioreactors, and acid mine drainage waters (reviewed in Zhou et al. 2018). Following the seminal work by Banfield and colleagues (Baker and Banfield 2016), an increasing number of whole metagenomic analyses from AMD, biomining-related, and natural acidic environments have been performed, enabling the establishment of relevant links between the occurrence of certain microbes, their activities and the geochemistry of cognate sites (reviewed in Nuñez et al. 2016; Quatrini and Johnson 2018). In addition, high-throughput sequencing technologies targeting environmental RNA samples have been used to characterise the functional response of microbial communities to changing conditions and treatments (e.g., different energy sources, operational temperatures, and bioleaching technologies), leading to a greater understanding of biological processes in bioleaching microorganisms than can be determined by metagenomics alone (e.g., Christel et al. 2018). While it is possible for researchers to study natural microbial communities' structure and gene expression profiles through analysis of nucleic acids directly extracted from the environment, there are limitations on the interpretation of sequence-based/generated data due to potential biases introduced during the extraction process of nucleic acids from environments as well as PCR-based biases. Nevertheless, this powerful tool is used widely to gain an understanding of the ecology and community dynamics within acidophilic communities.

6.4.6 Protein-Dependent Approaches

Several methods for microbial classification and identification that are based on either the detection of proteins or peptides have been described. Protein-based techniques have the advantage of yield and stability over nucleic acids-based methods, and under particular setups they can achieve similar levels of resolution.

Immuno assays, such as the enzyme-linked immunosorbent assay (ELISA), targeting proteins (and other macromolecules) of the external surface of either bacteria or archaea have been used to detect and enumerate specific acidophiles immobilised on membrane filters, solid supports, or mineral particles. Even if accurate and easy to use, immunological identification systems are expensive, require availability of pure cultures for antibody development, entail cumbersome preparative and standardisation procedures, and are difficult to use when numbers of target microorganisms are small, as is the case in many bioleaching samples. Furthermore, comprehensive libraries of ELISA antibodies required to detect a wide range of different acidophilic species are unavailable.

During the past 20 years or so, numerous studies have shown the applicability of mass spectrometry (MS) for microbial identification, taxonomy and strain typing. In particular, matrix-assisted laser desorption/ionisation time-of-flight mass spectrometry (MALDI-TOF MS) has become established as a widely used technology in clinical and food microbiology based on its speed, simplicity, and high resolution (up to strain-level variations). Given its high-throughput capacity, MALDI-TOF MS is currently replacing other existing biochemical and molecular identification methods for large-scale "culturomic" and microbiome analysis. The method detects the mass-to-charge ratio (m/z) of the bioanalytes (proteins/peptides) and uses the spectral information derived from most abundant extracellular or intracellular proteins (primarily, highly abundant ribosomal proteins) in a sample as a fingerprint for the identification of particular microorganisms. Despite its numerous advantages, applicability of this approach is possible only when there is spectral information available for reference microorganisms. Such information is currently limited for many environmental and industrially relevant microbial groups. While some data has been published and some applications reported (e.g., Kantor et al. 2017), the lack of sufficient reference spectra restricts the application of MALDI-TOF MS in the biomining industry.

Other vibrational spectroscopy technologies, such as Raman or Fourier transform infrared (FT-IR), hold the promise of achieving comprehensive and cost-effective "whole-organism" fingerprints (spectra) that enable the identification and classification of relevant bioleaching acidophiles and improve understanding of the intrinsic biochemical composition and variations of intact microbial cells (phenotype) as recovered from specific industrial bioleaching environments.

6.5 Closing Remarks

This chapter has described the various techniques and approaches that have been developed over many years to study and understand the microbiology of biomining environments. While often used in isolation, cultivation-based and molecular methodologies are, however, better used in tandem, and can provide the impetus to further develop both areas, such as the discovery, using a biomolecular approach, of a new genus or species of acidophile that cultivation techniques had previously overlooked, as was the case with *Ferrovum myxofaciens* (Johnson and Aguilera 2016).

Acknowledgements This work was supported in part by ANID (under Grants FONDECYT 1221035 y Centro Ciencia & Vida, FB210008, Financiamiento Basal para Centros Científicos y Tecnológicos de Excelencia) and by ANID—Millennium Science Initiative Programme—NCN17_093, granted by the Ministry of Economy, Development and Tourism from Chile.

References

Baker B, Banfield JF (2016) Metagenomics of acid mine drainage at Iron Mountain California, expanding our view from individual genes and cultures to entire communities. In: Quatrini R, Johnson DB (eds) Acidophiles: life in extremely acidic environments. Caister Academic Press, London, pp 221–232. https://doi.org/10.21775/9781910190333.13

Cárdenas JP, Quatrini R, Holmes DS (2016) Genomic and metagenomic challenges and opportunities for bioleaching: a mini-review. Res Microbiol 167:529–538. https://doi.org/10.1016/j.resmic.2016.06.007

Christel S, Herold M, Bellenberg S, El Hajjami M, Buetti-Dinh A, Pivkin IV, Sand W, Wilmes P, Poetsch A, Dopson M (2018) Multi-omics reveals the lifestyle of the acidophilic, mineral-oxidizing model species *Leptospirillum ferriphilum*^{T}. Appl Environ Microbiol 84:e02091–e02017. https://doi.org/10.1128/AEM.02091-17

Hallberg KB, Johnson DB (2007) Isolation, enumeration, growth and preservation of acidophilic prokaryotes. In: Hurst CJ, Crawford RL, Garland JL, Lipson DA, Mills AL, Stetzenbach LD (eds) Manual of environmental microbiology, 3rd edn. ASM Press, Washington, DC, pp 1155–1165. https://doi.org/10.1128/9781555815882.ch95

Hedrich S, Guézennec AG, Charron M, Schippers A, Joulian C (2016) Quantitative monitoring of microbial species during bioleaching of a copper concentrate. Front Microbiol 7:2044. https://doi.org/10.3389/fmicb.2016.02044

Johnson DB, Aguilera A (2016) The microbiology of extremely acidic environments. In: Yates MV, Nakatsu CH, Miller RV, Pillai SD (eds) Manual of environmental microbiology, 4th edn. ASM Press, New York. https://doi.org/10.1128/9781555818821.ch4.3.1

Johnson DB, Quatrini R (2020) Acidophile microbiology in space and time. Curr Issues Mol Biol 39:63–76. https://doi.org/10.21775/cimb.039.063

Johnson DB, Kanao T, Hedrich S (2012) Redox transformations of iron at extremely low pH: fundamental and applied aspects. Front Microbiol 3:96. https://doi.org/10.3389/fmicb.2012.00096

Kantor RS, Huddy RJ, Iyer R, Thomas BC, Brown CT, Anantharaman K, Tringe S, Hettich RL, Harrison ST, Banfield JF (2017) Genome-resolved meta-omics ties microbial dynamics to process performance in biotechnology for thiocyanate degradation. Environ Sci Technol 51: 2944–2953. https://doi.org/10.1021/acs.est.6b04477

Khaleque HN, González C, Johnson DB, Kaksonen AH, Holmes DS, Watkin EL (2020) Genome-based classification of *Acidihalobacter prosperus* F5 (= DSM 105917= JCM 32255) as *Acidihalobacter yilgarnensis* sp. nov. Int J Syst Evol Microbiol 70:6226–6234. https://doi.org/10.1099/ijsem.0.004519

Ñancucheo I, Rowe OF, Hedrich S, Johnson DB (2016) Solid and liquid media for isolating and cultivating acidophilic and acid-tolerant sulfate-reducing bacteria. FEMS Microbiol Lett 363: fnw083. https://doi.org/10.1093/femsle/fnw083

Nuñez H, Loyola D, Cárdenas JP, Holmes DS, Johnson DB, Quatrini R (2014) Multi locus sequence typing scheme for *Acidithiobacillus caldus* strain evaluation and differentiation. Res Microbiol 165:735–742. https://doi.org/10.1016/j.resmic.2014.07.014

Nuñez H, Covarrubias PC, Moya-Beltrán A, Issotta F, Atavales J, Acuña LG, Johnson DB, Quatrini R (2016) Detection, identification and typing of *Acidithiobacillus* species and strains: a review. Res Microbiol 167:555–567. https://doi.org/10.1016/j.resmic.2016.05.006

Okibe N, Johnson DB (2011) A rapid ATP-based method for determining active microbial populations in mineral leach liquors. Hydrometallurgy 108:195–198. https://doi.org/10.1016/j.hydromet.2011.04.008

Quatrini R, Johnson DB (2018) Microbiomes in extremely acidic environments: functionalities and interactions that allow survival and growth of prokaryotes at low pH. Curr Opin Microbiol 43: 139–147. https://doi.org/10.1016/j.mib.2018.01.011

Wang Y, Zeng W, Qiu G, Chen X, Zhou H (2014) A moderately thermophilic mixed microbial culture for bioleaching of chalcopyrite concentrate at high pulp density. Appl Environ Microbiol 80:741–750. https://doi.org/10.1128/AEM.02907-13

Zammit CM, Mutch LA, Watling HR, Watkin ELJ (2008) Evaluation of quantitative real-time polymerase chain reaction for enumeration of biomining microorganisms in culture. Hydrometallurgy 94:185–189. https://doi.org/10.1016/j.hydromet.2008.05.034

Zhou S, Gan M, Zhu J, Liu X, Qiu G (2018) Assessment of bioleaching microbial community structure and function based on next-generation sequencing technologies. Fortschr Mineral 8: 596. https://doi.org/10.3390/min8120596

Chapter 7
Microbial Ecology of Bioheaps, Stirred Tanks, and Mine Wastes

Christopher George Bryan and Susan Therese Largier Harrison

Abstract Bioleaching systems, from unmanaged acid rock drainage-generating mine wastes to highly controlled tank leaching, involve complex interplay between physico-chemical conditions and microbial communities. Environmental conditions influence microbial community composition but can themselves be the result of microbial activity. Advances in molecular and classical microbiological methods and the application of ecological tools have greatly improved our understanding of the microbial ecology of mineral bioleaching and biooxidation processes. The early discoveries that microbial processes contributed to dissolution of metal sulfides at low pH in the early 1950s were followed by the isolation and characterisation of relatively few "biomining bacteria" for the next 20 or so years, though this situation has changed radically since the 1980s. The subsequent evolution of industrial operations and advanced microbiological methods has begun to reveal the full extent of microbial diversity and function in these systems. This chapter discusses the microbial ecology of mine wastes, heap, dump, and tank bioleaching, general approaches for their study as well as current questions and future directions.

Keywords Bioheaps · Tank leaching · Mine wastes · Microbial ecology · Sampling techniques

C. G. Bryan (✉)
BRGM, Orléans, France
e-mail: c.bryan@brgm.fr

S. T. L. Harrison
Department of Chemical Engineering & DVC: Research and Internationalisation, Centre for Bioprocess Engineering Research (CeBER), University of Cape Town, Cape Town, South Africa
e-mail: Sue.Harrison@uct.ac.za

D. B. Johnson et al. (eds.), *Biomining Technologies*,
https://doi.org/10.1007/978-3-031-05382-5_7

7.1 Microbiology and Biomining

Prior to the 1930s, it was assumed sulfide mineral oxidation (and by extension dump and in place leaching) was an entirely chemical process. Research into the phenomenon of acid rock drainage (ARD) from mine wastes led to the identification of the biological role in mineral oxidation. The first acidophilic bacterium, *Acidithiobacillus thiooxidans,* had been described around 1920 and Colmer and Hinkle (1947) confirmed the presence of this sulfur-oxidiser in mine water, as well as an iron-oxidising bacterium, later named as *Thiobacillus* (now *Acidithiobacillus*) *ferrooxidans*. An increasing number of studies during this period demonstrated the abundance of such bacteria in mine waters, yet while their dependence on sulfur and/or iron oxidation was proven, it was not until the 1950s that the extent and scale of their role in pyrite oxidation and ARD formation began to be understood.

Consequently, the earliest commercial biomining systems were not designed to specifically promote microbial activity. An early study of dump leaching (Bhappu et al. 1969) showed that dumps were not conducive to microbial activity. The authors used most probable number (MPN) analysis to show that mesophiles were only active at the dump surface and theorised that the heaps were oxygen-depleted lower down. Studies such as this led to heaps being engineered to better promote microbial activity. However, there is still a tendency to "black-box" the microbiological processes that are fundamental to successful biohydrometallurgical applications. Without understanding these highly complex interactions, it is difficult to identify and target key operational parameters needed to maximise the potential of biohydrometallurgy. One of the pioneers of this field, Giovanni Rossi, was among the first to stress the intricate and important interrelationships between multiple scientific disciplines involved in biohydrometallurgy, and the need for scientists with expertise in these areas to work together (Rossi 1990).

7.2 Microbial Ecology of Biological Mineral-Oxidising Systems

Mineral-oxidising microorganisms may be found, and are active, in a wide variety of places (Chap. 1). The consequences of their actions are either desirable—and to be encouraged and optimised—or undesirable. The principal role of the microbial component of biohydrometallurgical processes is simple: the modification of the reduction–oxidation (redox) potential to manipulate the oxidation state of sulfur and some of the metals (chiefly iron) in the mineral, thereby influencing their solubility. In oxidative leaching, this is mediated through the (re)generation of lixiviant solutions containing ferric iron and sulfuric acid.

Bioleaching systems may be laboratory scale, industrial installations, or natural or anthropogenic environments. They may be saturated, such as acid rock/acid mine drainage (ARD/AMD) and stirred-tank reactors or unsaturated such as waste rock dumps and tailings ponds, heap- and dump leaching operations, or small-scale simulation columns. The microbiology of these systems is very different, and each presents different levels of complexity in terms of sampling and analysis, and how to extrapolate the data.

Bioleaching environments are generally considered to be extreme, with elevated concentrations of dissolved metals and acidic pH (typically <2, but more variable in heaps and dumps). Such conditions present challenges to both classical and molecular microbiology methods (Chap. 6). Advances in both cultivation techniques and DNA (as well as RNA and protein) extraction and analysis have led to step-changes in the understanding of the major microorganisms involved in bioleaching. To fully understand the role of the bioleaching microbiota, it is important to know not only who is present, but where they are and the roles that they play—both in the leaching of the metals and in interacting with other members of the microbial consortium. To have confidence in the data produced from these often heterogeneous and dynamic systems, it is critically important to have a robust and reliable sampling and sample processing strategy. Further, the relevance of standardised metabolic tests needs to be tested and validated under the extreme environments of biomining.

The challenges of sampling, sample processing, test work, and data analysis vary, depending on the system under study. Microbial communities are in a constant state of flux, with the growth cycles of different species overlaid on one another. Therefore, "snapshots" of microbial community composition need to be interpreted with this in mind. Overarching this, is the community response to environmental conditions. For example, if a sample spends several days or longer in transit prior to DNA extraction and analysis, there is the possibility that the environmental signature (of the heap or reactor) has been lost or modified, and therefore proper stabilisation of the sample during transportation and storage is critical. The geographical and relative isolation of the system is important. For example, commercial operations are often in remote locations and may have limited sample processing and microbiological facilities on site. As such, the sample transit time and conditions may need to be taken into consideration when interpreting the relevance of results.

Microorganisms involved in current biomining operations are predominantly acidophilic. Furthermore, they can be classified as those which can oxidise iron and/or sulfur, and whether they are capable of autotrophic or heterotrophic growth (obligate or facultative; Chap. 5). Further to this, their likelihood of dominating or thriving in a particular leaching environment is driven by their propensities for iron oxidation, e.g., their affinities for ferrous iron (or ferric iron in reductive bioprocessing) and sensitivities to ferric iron. Similarly, their ability to scavenge inorganic carbon, nitrogen, and sulfur or to grow rapidly when each is plentiful impacts the community, as does tolerance to potentially inhibiting solutes. Further to this, the location of the microorganisms in the reaction system and their planktonic or sessile nature, the mixing patterns and contacting in the system and the degree to which the system is open to flow together with the time period required for renewal of the liquid volume within it (retention time) impact culture dynamics. The dynamic conditions within closed or slowly renewed systems (such as heaps and dumps) lead to changing culture dominance in response to ability to thrive in the changing environment. Conversely, while conditions may be more stable in flow-through systems, small perturbations may lead to rapid changes in the microbial ecology due to wash out from the system of species and strains that thrive less well following perturbations.

Despite the sometimes, though misplaced, reported difficulties in cultivating acidophiles, biohydrometallurgy is somewhat unique in that, as far as is known, the majority of the microorganisms which regularly constitute greater than 95% of the total diversity in active bioleaching systems (i.e., heaps and tanks) have been cultivated in pure culture. Therefore, it is possible to work with them and determine important physiological features such as growth rates, substrate affinities, and inhibition constants. Further, their preferred carbon and energy sources and the metabolic by-products from each system can be determined. The interaction of these biokinetic traits and the interactive roles in the metabolic cycles are critical in determining the microbial community dynamics. Indeed, in an in silico age of bioinformatics, molecular biology, and high throughput sequencing, it is important not to overlook the essential in vitro work needed to determine the effects of environmental factors on microbial activity and limits.

Previously, measuring rates and amounts of, for example, ferrous iron oxidised or CO_2 fixed, have been proposed as proxies for biomass production. However, it has been shown that there is no cardinal relationship between these measurements and biomass, but rather that operating conditions affect the relationships between biomass production and substrate utilisation. Indirect methods such as microcalorimetry and respirometry may be better measures of microbial activity and biomass production though they are generally carried out using small samples which may not be representative of the entire heap or waste material.

Several studies have suggested that some microorganisms have a greater propensity to attach to mineral and other surfaces than others, while elsewhere the microbial ecology of continuously operated commercial biooxidation tanks has found similar microbial communities in the planktonic (aqueous) phase and those attached to surfaces. These data suggest that the tracking of microbial ecology in the tank biooxidation system is sufficiently, although not absolutely, represented by the study of planktonic cells, thus simplifying routine monitoring approaches to provide lead (aspirational) and lag (current) indicators of the "health" of microbial communities in stirred tanks.

In biological mineral-oxidising systems, there are three primary microbial functions: the oxidation of iron, the oxidation of elemental sulfur, and sulfur oxy-anions, and the metabolism of organic carbon. In some cases, all three roles can be fulfilled by a single organism (e.g., *Sulfobacillus* spp.) but mostly several species work in concert and it is generally accepted that mixtures of organisms (consortia) perform optimally. While in biomining processes, nutrients (nitrogen, phosphorus, and potassium) are usually provided, in mine waste environments this is not the case, and there is even greater importance of other metabolic functions such as nitrogen cycling. In any case, the required ecological functions must be provided either by indigenous microorganisms or through inoculation. "Top-down" and "bottom-up"

Table 7.1 Major described bioleaching microorganisms frequently detected in mine wastes, heap/dump bioleaching, and commercial/industrial pilot bioreactor systems (modified from Chap. 5)

Genus and species	Wastes	Heaps/ Dumps	Reactors
Numbers of unique taxa typically detected using high throughput sequencing methods (per site)	1000–10,000	~100	1–10
Acidimicrobium ferrooxidans	✓	✓	
Ferrimicrobium acidiphilum	✓	✓	
Leptospirillum ferriphilum	✓	✓	✓
Leptospirillum ferrooxidans	✓	✓	
Acidiplasma cupricumulans		✓	✓
'Ferroplasma acidarmanus'			✓
Ferroplasma acidiphilum	✓	✓	✓
Acidiferrobacter thiooxydans	✓		✓
Iron-oxidising *Acidithiobacillus* spp.	✓	✓	
"*Acidithiomicrobium* P2"	✓		
Alicyclobacillus spp.	✓	✓	
Sulfobacillus acidophilus	✓		✓
Sulfobacillus benefaciens			✓
Sulfobacillus thermosulfidooxidans	✓		✓
Sulfobacillus thermotolerans		✓	
"*Sulfolobus*-like" strain ICHT3[a]			✓
Acidianus brierleyi			✓
"*Acidianus*-like" strain ICHT4[a]			✓
Metallosphaera sedula		✓	✓
"*Metallosphaera*-like" strain ICHT2[a]			✓
Sulfuracidifex metallicus		✓	✓
Acidiphilium acidophilum	✓		
Acidithiobacillus caldus			✓
Acidithiobacillus thiooxidans	✓	✓	
Acidiphilium cryptum	✓	✓	
"*Stygiolobus*-like" strain ICHT1[a]			✓
Thermoplasma acidophilum	✓		

[a]Not isolated in pure culture and undescribed. Included due to their industrial relevance (BioCOP)

approaches have been proposed to determine the most effective bioleaching consortia for different applications (Rawlings and Johnson 2007) (Table 7.1).

7.2.1 Heaps and Dumps

Mineral heaps, from engineered bioleaching heaps, to dumps and waste heaps of historical mine waste are much more difficult to study than well-mixed tank processes. In the laboratory and at pilot level, they are mimicked by small and large-scale studies, across a wide range of scales from a few grams of low-grade ore to tons of solid material. These are all typically unsaturated systems, with microorganisms found both free-swimming in flowing and interstitial liquid phases and attached (either electrostatically or within a biofilm) to the mineral surface. Brierley (2001) noted that both chemical and physical conditions changed radically within mineral heaps from their construction to the point at which they were decommissioned. Understanding the changes in chemical and biological parameters that occur provides immense scope for process optimisation.

Due to their economic importance, there are generally more studies covering the microbiology of heap and dump leaching operations than mine wastes and there are several comprehensive reviews available in the literature (e.g., Watling 2006). Increased process control (in heap leaching), and the usually greater concentrations of reactive minerals means that heap and dump leaching operations generate more extreme conditions. Traditionally, effluents draining heaps and other unsaturated ore beds have been analysed for both their chemistry and microbiology, and these data used to infer microbial ecology, activity, and physico-chemical conditions within the heap. However, increasingly, this is recognised to be an indirect approach which may identify trends, but not actual conditions, in the heap, and it may not be a valid approach when used as a sole index of microbial activity.

Sampling of large-scale heaps, dumps, and tailings dams is typically carried out by taking core samples, and data have been published of cores taken from active heaps and recently spent heap residues. However, there are disadvantages in adapting this as a routine approach. Coring is expensive and often not practical for production heaps, let alone mine wastes. Furthermore, coring disturbs the bed structure and may lead to channelling and non-ideal fluid flow. The size and location of samples must also be considered to address how representative they are. This leaves many questions on how best to analyse microbiology within heaps, including whether it is possible to interpret data from effluents draining heaps and dumps to infer what is happening within them.

The relative activities (for example, iron oxidation, sulfur oxidation, and gene expression) of the key microorganisms implicated in bioleaching in the different phases (planktonic or attached) is difficult to evaluate. Early work demonstrated that microbial attachment is not essential for bioleaching, since ferric iron produced by planktonic cells can oxidize sulfide minerals. While it might seem logical that attached cells might play a more important role in mineral dissolution owing to their spatial proximity, reduced diffusion barriers, and the ability to concentrate leach agents into the biofilm phase, this is not easily proven.

Heap and dump leaching systems are essentially unmixed batch processes (Chap. 2) and subject to temporal and spatial variation in conditions and microbial biodiversity as well as inherent variability in mineralogy. Heaps may be inoculated during agglomeration or through irrigation, either just once or continually, though indigenous microorganisms will still be present and active. Comprehensive studies using high throughput sequencing (HTS) methods may reveal the presence of hundreds of taxa, but typically at least 90% of the abundance (in terms of biomass)

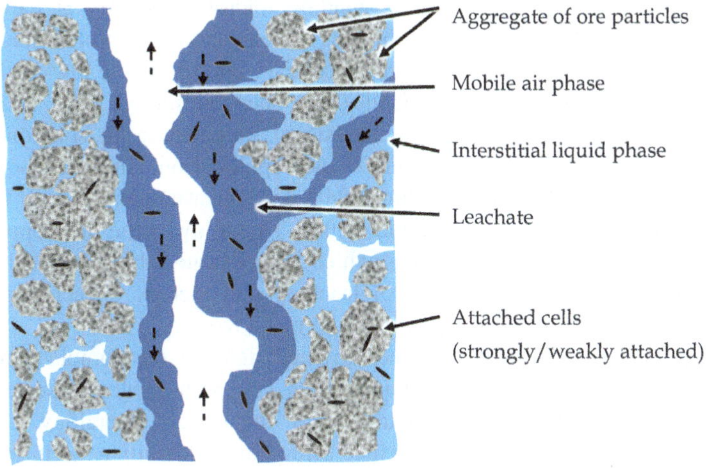

Fig. 7.1 Schematic representation of the four phases within the ore bed (modified after Govender et al. 2013)

is made up of less than ten individual taxa at most, the majority of which are at least partially described. The potential roles of the remaining rare taxa are unclear. They may be remnants of the indigenous community that are just managing to survive as the conditions are not so extreme as to eliminate them, without necessarily contributing to ecological function in any meaningful way.

At the same time, heap and dumps contain many different microenvironments, within which these minor taxa may dominate, but which bulk sampling procedures homogenise. Furthermore, it is important not to conflate abundance with activity. For example, studies have shown that while *Leptospirillum* spp. may not numerically dominate a heap leaching system, they can account for the majority of the heap metatranscriptome, i.e., they are the most active component of the ecosystem in terms of gene expression.

Unlike bioreactors, in bioheaps there appears to be a significant difference between the microbial communities associated with the mineral (attached to the surface, or free-swimming in the interstitial liquid within and between agglomerates) and the flowing solution (leachate liquors; Fig. 7.1). Laboratory-based studies have indicated that the concentrations of cells and metals in the interstitial fluid may be orders of magnitude greater than in the flowing leachate, and that the microbial community structure is significantly different as well with acidithiobacilli more abundant in the interstitial phase and species such as *Leptospirillum* spp. and *Acidiferrobacter* spp. reported to be more abundant in the percolating liquors. Most of the biomass (over 90%) in a heap is thought to be associated with the ore (either attached or in the stagnant interstitial fluid) and that this biomass experiences solution physico-chemical conditions (pH, redox potentials, concentrations of soluble metals, etc.) that are very different to those inferred from the leachate chemistry.

This may have implications in understanding microbial activity at the mineral surface and in developing ecological models of heap and dump bioleaching.

Bioleaching of full-scale run-of-mine secondary copper sulfide low-grade heaps operated at the Escondida Mine (Chile) has been monitored since 2006 (e.g., Demergasso et al. 2018). Monthly analysis of pregnant leach solutions (PLS) draining the heap using MPN counts of acidophilic iron- and sulfur-oxidisers, Q-PCR and activity tests were combined with daily measurement of PLS temperature, pH, redox potentials, and soluble metals. This allowed the development and implementation of data analysis tools which demonstrated that PLS analysis can indicate the stage in the life cycle of leaching and thereby expected dominant microbial community as well as prediction of metal release.

Such predictive studies do not seek to shed light on the complete microbial consortium or on its optimisation but provide key performance indicators. This approach has since been extended to include the initial mineralogy, expanded daily data collection on the PLS to include acidity and alkalinity as well as concentrations of soluble copper and iron, among others. The data analysis algorithms have been expanded to allow empirical pattern recognition to predict recoveries and also to determine rules for stacking heaps in order to maximise their performance.

7.2.1.1 Microbial Succession and Thermal Gradients

Initially heaps and dumps will be at ambient temperature. In general terms, studies tend to suggest that during these early stages sulfur-oxidising organisms such as *Acidithiobacillus* spp. are the most abundant. This may be due to the initial abundance of elemental sulfur and sulfur oxy-anions resulting from chemical dissolution of sulfide minerals and the relatively low redox potentials. Biological ferrous iron oxidation causes redox potentials to become more positive and concentration of total soluble iron increases, favouring iron-oxidisers such as *Leptospirillum* spp. and *Ferroplasma* spp. For example, Wakeman et al. (2008) monitored microbial community succession in leachates from laboratory-scale columns. The community was dominated initially by *At. ferrooxidans*, transitioning to *At. ferrivorans* (referred to as "*Acidithiobacillus* sp. NO37" at the time) and then *L. ferriphilum*, all of which were included in the inoculum. Indigenous taxa including *Alicyclobacillus* spp. were significant members of the leachate communities initially but their numbers declined as bioleaching progressed.

The oxidation of reduced forms of sulfur is exothermic and, depending on heap design and operation as well as mineralogy, the interior temperature typically increases, sometimes to above 60 °C. These changes in temperature cause temporal and spatial changes in microbial community structure. Over time, temperature in most parts of the heap will increase, with consequent shifts toward thermo-tolerant and moderately thermophilic microorganisms such as *Sulfobacillus* spp. (and some other iron-oxidising Gram-positive bacteria) and *Ferroplasma* spp. and eventually to more extremely thermophilic archaea such as *Metallosphaera* spp., *Sulfolobus* spp., and *Thermoplasma* spp.

While mesophilic bioleaching organisms are often indigenous to mined ore, the same is not considered true for thermophiles or, if they are present, they tend to be far fewer in number and therefore, it is desirable to add them to the heap in order to exploit elevated temperatures. This may be done during heap construction but would then need to remain viable and not washed out during temperature ramping, or they may be introduced in irrigating solutions.

Depending on the design and operating parameters, heaps will have spatial variation in temperature (Chaps. 2 and 10), leading also to variations in microbial communities. The implications of these heterogeneous communities on heap performance are important. In very general terms, iron- and sulfur-oxidising mesophiles have efficient CO_2 fixation and assimilation pathways and are the primary producers in biomining systems. The acidithiobacilli are particularly good at scavenging CO_2 at very low dissolved concentrations. As temperature increases above 50 °C, communities become dominated by organisms such as *Sulfobacillus* spp. and acidophilic archaea which are less efficient at fixing CO_2. Some are considered to be obligate heterotrophs and most display better growth in laboratory cultures that are supplemented with organic carbon. Therefore, optimising microbial activity in these higher temperature phases requires a different approach to that at mesophilic temperatures.

7.2.2 Mine Wastes

The microbiology of acid mine/rock drainage waters are more amenable to study than mineral-rich bioleaching ecosystems, and has been the focus of a number of reviews (e.g., Nordstrom et al. 2015). These communities are often relatively simple and structurally different to those found within the waste dumps that generate the ARD. In contrast, solid mine wastes represent the most heterogeneous mineral leaching systems, with widely varying ages, mineralogical composition, sulfide mineral content, pore/interstitial water pH, and other geochemical parameters such as bioavailable and total metal concentrations. There have been few systematic studies of mine waste microbial ecology and it is difficult to form generalised conclusions from published data.

The most comprehensive studies of mine wastes have combined molecular biology and cultivation-based approaches to elucidate not only the detectable biodiversity but also its potential for mineral oxidation and thus metal dissolution. The biomolecular and cultivation methodologies that have been used successfully to elucidate the microbiology biomining and mine-impacted environments are described in Chap. 6. Among examples of the importance of this approach for elucidating the microbial biodiversity of biological mineral-oxidising systems was the pioneering work done at the former Richmond mine at Iron Mountain, California (e.g., Edwards et al. 2000). Such work demonstrated how combined molecular and classical microbiology can be used to study the links between geochemical and environmental factors and microbial ecology in the ARD and the subsurface, ultimately improving understanding of key factors influencing microbial community structure and function.

In the early days of molecular microbiology, the methods were less able to detect the full breadth of microbial diversity. Consequently, it was assumed that the majority of taxa in mine wastes and ARD could be readily cultivated. However, advances in high throughput sequencing methods have radically altered this perception. In most cases, and in contrast to active bioleaching systems, the commonly isolated bioleaching microorganisms from weathered mine wastes make up a relatively small fraction of the total number of taxa (often less than 5%). Despite this, enrichments from such wastes can usually oxidise pyrite, though some enrichment cultures are better than others. As soon as wastes are put into leaching conditions (e.g., shake flasks for enrichment cultures) diversity decreases rapidly (within a single subculture) and significantly, from thousands of taxa to less than ten (Sbaffi et al. 2017). Such cultures are usually dominated by well-documented bioleaching organisms such as *Leptospirillum* spp., iron-oxidising acidithiobacilli, *Sulfobacillus* spp. (and other iron- and/or sulfur-oxidising Firmicutes), with varying abundances of other mineral-oxidising organisms such as *Acidiferrobacter*, *Thermoplasma* spp., and *Ferroplasma* spp. and non-mineral-oxidising heterotrophs such as *Acidiphilium* spp. Enrichment cultures are completely unlike their source waste in terms of community composition and structure, more so where the waste is particularly aged. Therefore, cultivation-based methods can be misleading in terms of both the microbial abundance and diversity revealed. Nevertheless, they are essential in determining the potential for such communities to catalyse mineral oxidation, given the right conditions, and in the goal of bioprospecting for novel strains and consortia with potentially exploitable traits.

Work done by Axel Schippers and co-workers using combinations of classical (e.g., MPN) and molecular microbiology methods began to elucidate some general trends in the microbial succession in mine wastes (e.g., Korehi et al. 2014; Schippers et al. 2010). When wastes are first deposited, they have neutral pH or even slightly alkaline (especially flotation tailings). They are initially colonised by a wide variety of microorganisms, including pioneer sulfur-oxidising Proteobacteria such as *Thiobacillus* spp. Due to the activities of these prokaryotes, the environment becomes more acidic. Closest to active sulfide oxidation zones (oxidation fronts or areas rich in reactive sulfides), biodiversity is more limited and communities tend to be dominated by mineral-oxidising taxa such as *Acidithiobacillus* spp., *Leptospirillum* spp., *Sulfobacillus* spp., and iron-oxidising archaea. However, in deeper parts of the waste, particularly saturated zones, these microorganisms may be present but in low numbers, and mineral oxidation is limited by oxygen supply.

Historic, unmanaged wastes often can still produce ARD, even after hundreds of years. These wastes typically have greater microbial diversity than more recent waste dumps, with often thousands of taxa routinely detectable by HTS. Communities can vary from mostly bacterial to predominantly archaeal. The most common bacterial phyla reported in weathered wastes include Proteobacteria, Acidobacteria, Chloroflexi, Planctomycetes, and candidate division AD3, with the Euryarchaeota and Crenarchaeota the most frequent archaeal phyla (e.g., Sbaffi et al. 2017). While most of these organisms represent poorly characterised, deeply branching phylogenetic groups, there is a general notion that they represent systems in transition from purely lithotrophic to more "heterotrophically-inclined" microbial communities. While many of these taxa represent novel lineages, similar organisms can be found in aging volcanic deposits or environments impacted by mine wastes and ARD. Currently, it is difficult to speculate as to their role in these systems.

It is clear that microbial diversity increases with mine waste age and that older waste deposits tend to be dominated by microorganisms that have yet to be isolated and characterised. However, it is not age, but pH that is by far the most significant factor that controls microbial diversity in mine wastes. Diversity decreases with decreasing pH and neither mineralogy nor chemistry have an effect to the same extent. While there does seem to be a possible link with the total metal(loid) content this is not entirely independent of pH itself.

Sulfidic ARD-generating mine wastes may be blended with amendments and planted with grasses and other plants during site rehabilitation or phytostabilisation. Interestingly, this does not necessarily lead to the displacement of the lithotrophic, mineral-oxidising community (e.g., Hottenstein et al. 2019). Even where a system is no longer net acid-generating these organisms are still found, juxtaposed with heterotrophic (non-mineral-oxidising) communities. This suggests that litho-autotrophic niches may still exist even in apparently stabilised waste dumps, and that these may be reactivated rapidly should conditions change, for example, through re-exposure of sulfidic material to both oxygen and water.

7.2.3 Bioreactors

Stirred-tank reactors (STRs; Chap. 3) are well-mixed systems and in theory, allow representative samples to be taken at any point within the reactor or at the outlet of a continuous STR. Typically, direct counting (using phase contrast microscopy) of planktonic cells in liquid phase is possible in order to estimate biomass and to follow growth rates, although this can be complicated due to the presence of fine particles. However, it is much more difficult to quantify microorganisms attached to the mineral surface and to determine what proportion of the total biomass is either attached or planktonic. Early work in laboratory-scale tanks suggested that planktonic cells accounted for about 66% of the microbial population present in the tank.

Attachment studies performed on fine-grained concentrates, using both flow-through systems in which cells are contacted with fine-grained mineral on coated beads and also submerged slurry systems, demonstrated rapid attachment of *L. ferriphilum*, *At. ferrooxidans*, and *Metallosphaera hakonensis*. *L. ferriphilum* and *At. ferrooxidans* previously adapted to a copper sulfide concentrate showed rapid attachment levels (75–79%) in a flow-through system, with *M. hakonensis* showing 30–45% attachment to mineral concentrate at 65 °C. Conversely, attachment to low-grade ores and quartzite gangue material was much lower (~25%) as was contacting at temperatures not optimal for metabolic growth. Data from growing and metabolically active, well-agitated slurry reactor systems at scales of 1 litre to 500 m^3 and at solids concentrations of 10–25% have suggested that enumerating planktonic cells could be a reasonable proxy for the bioreactor as a whole. Shear stress, and slurry densities of fine particles of 10–15% are also important factors in controlling microbial attachment.

Bioreactor systems, whether laboratory-, pilot-, or commercial-scale, are the most homogeneous bioleaching environments spatially, and, at least in the case of continuous stirred-tank reactors (CSTR) at steady state, temporally. The operating conditions are tightly controlled, designed to maximise mineral dissolution and often push the microbial consortia to their limits. As a result, the microbial ecology is relatively noncomplex and usually comprise between two and four dominant taxa. Where other taxa have been identified through cultivation or HTS approaches, they are far more limited in diversity and abundance than in heaps and dumps.

At the time of early bioreactor development for bioleaching in the latter part of the twentieth century, the acidithiobacilli were the most widely studied bioleaching organisms and were assumed to be the most important in stirred tanks. Moreover, bioleaching cultures were routinely enriched from ARD and mine wastes, where these organisms often dominated. As a result, early reactors were designed to run at mesophilic temperatures of around 35 °C. An example of this is the development of the cobaltiferous pyrite bioleaching operation in Kasese, Uganda, which began during the early 1990s (Morin and d'Hugues 2007) where the original inoculum used in laboratory studies, derived from a gold-bearing arsenopyrite bioleaching CSTR, was dominated by *Acidithiobacillus*-like bacteria. This was in line with the early BIOX development work (Chap. 4), with early reactors thought to comprise similar microbial consortia. Such results are incongruent with subsequent work which showed that even at such temperatures *Leptospirillum* spp. would be expected to be the dominant iron-oxidisers due to their ability to sustain higher rates of iron oxidation in high redox (<840 mV) environments (Rawlings et al. 1999). While it is theoretically possible that *Leptospirillum* spp. were absent from the inocula as well as the (unsterilised) concentrates, it is generally accepted that the inferred microbial community composition of these early gold biooxidation systems were erroneous observations due to limitations of the techniques available at the time, and the dominant acidophiles in the full-scale commercial tanks at the Kasese plant were found to be *L. ferriphilum, At. caldus, Sb. benefaciens*, and a *Ferroplasma*-like archaeon. However, the stirred tanks used in the BIONORD process developed by Polyus JSC (Chap. 11) which are typically maintained between 35 °C and 40 °C are dominated by *Ferroplasma acidiphilum, Acidiferrobacter thiooxidans, At. thiooxidans*, and *Acidiphilium multivorum* while *L. ferriphilum* accounts for less than 7% of the total biomass.

The oxidation of sulfide minerals such as pyrite and arsenopyrite is exothermic, and much of the capital and operating expenditure of tank bioleaching is associated with cooling. Moreover, increasing temperature increases reaction rate, and during the optimisation of operating parameters it is desirable to push the operating temperature to the highest possible before system instability. As a result, both the BIOX and Kasese plants were ultimately run above 40 °C, and most commercial biooxidation systems operate between 40 and 45 °C. It has been frequently reported that such bioreactor communities comprise *At. caldus* and *L. ferriphilum* (usually as the dominant prokaryotes) with *Sulfobacillus* spp. (frequently *Sb. thermosulfidooxidans*) present in smaller numbers. Acidophilic archaea (mostly *Ferroplasma* spp.) have also been detected, and their numbers tend to increase in downstream tanks as organic carbon concentrations increase.

Although it is not proven why *Sb. thermosulfidooxidans* is the dominant *Sulfobacillus* spp. in certain situations and *Sb. benefaciens* in others, it appears to be linked with CO_2 availability and concentrations of organic carbon, with *Sb. benefaciens* better able to scavenge CO_2 than *Sb. thermosulfidooxidans* (Johnson et al. 2008), and also their different degrees of tolerance to transition metals (e.g., *Sb. thermosulfidooxidans* is more resistant to copper). It is also interesting to note in this context that the Mintek bioleaching culture, used in the Mondo bioleaching process (Chap. 12) operates at 45 °C and is dominated by *L. ferriphilum* and *At. caldus* despite 45 °C being outside the accepted temperature range of many (though not all) strains of *L. ferriphilum*.

Whether *L. ferriphilum* is the principal iron-oxidising organism in all tank bioleaching systems that operate at 40–45 °C has been questioned. Analysis of a long-term continuous bioleaching system processing a nickel–copper concentrate at 40 °C indicated that it was dominated by *Sb. thermosulfidooxidans* and another *Sulfobacillus* sp. with very low levels of *L. ferriphilum*, despite having been inoculated with the BRGM-Kasese culture (Bryan, unpublished data). Moreover, analysis of the microbial communities of several commercial BIOX® plants in different countries showed archaeal dominance by organisms such as the iron-oxidiser *Ferroplasma acidiphilum* and a *Thermoplasma* spp. (Smart et al. 2017).

For a long time, ~45 °C has been considered the upper limit for tank bioleaching before stepping up to 60–80 °C for archaeal processes (MesoTHERM®, BioCOP, etc.). However, the IBCCO (Iranian Babak Copper Company) commercial bioreactors in Iran processing copper concentrate operate between 45 and 60 °C and are dominated by *Sulfobacillus* spp. suggesting that they can fulfil the iron oxidation demand in this system (Manafi et al. 2021). One of the challenges associated with *Sulfobacillus*-dominated bioleaching systems is that they are relatively poor at assimilating CO_2. At lower temperatures, primary producers such as *Leptospirillum* spp. and *Acidithiobacillus* spp. efficiently fix and assimilate CO_2, providing sulfobacilli with organic carbon in the form of cellular debris and exudates. A novel facultatively autotrophic, acidophilic actinobacterium ("*Acidithiomicrobium* P2") has shown some promise in co-culture with *Sb. thermosulfidooxidans* and *At. caldus* in the leaching of nickel concentrate in continuous laboratory-scale culture at 49 °C (Norris 2017). However, its sensitivity to copper may limit its role in bioleaching operations.

Currently there are no commercial high-temperature biomining operations, although the development of the MesoTherm® process and the pilot-scale BioCOP process are discussed in more detail in Chaps. 3 and 4, respectively. Generally, commercial interest in thermophile bioleaching has centred around three culture types: *Sulfuracidifex* (previously *Sulfolobus*) *metallicus* at 64–68 °C, *Acidianus brierleyi* at ~70 °C (used in pilot work by Mintek for example), and the ICHT culture at ~78 °C (used in the BioCOP and BRGM HIOX pilot plant; Norris et al. 2013). The ICHT culture comprises 4–5 novel *Sulfolobus*-, *Stygiolobus*-, *Acidianus*-, and *Metallosphaera*-like archaeal species (Table 7.2).

7.3 Challenges and Future Directions

Generally, in unsaturated systems, pH seems to be the factor that correlates most closely with greater microbial diversity. As pH increases, so does microbial diversity. However, it is not clear whether this is a direct or indirect effect, as pH will cause other parameters to change also, such as the solubility of transition metals, CO_2 solubility, and organic acid protonation, which complicates the interpretation of data. In saturated systems where pH and temperature are constant, biodiversity is much more restricted. These systems are much more sensitive to changes in feed materials and resulting fluctuations in soluble metal(loid) concentrations and redox potentials. It is possible that as these systems are operating at their biological limits, the lack of biodiversity affects their resilience; they lack a biological buffer which may allow the system to adapt to changes. The links between environmental parameters like pH, system biodiversity, function, and resilience require further study.

The monitoring of heap and dump leaching operations remains a challenge. On the one hand, it is probable that the majority of heap biomass is found in interstitial (stagnant) fluid, or attached to the mineral surface, and that the microbial community of the leachate is probably not representative of the microbial community as a whole. On the other hand, microbiological and chemical analyses of heap leachates can be used to indicate heap performance. Nevertheless, better understanding of the microbial community of a heap and the conditions that it operates in (pH, solute concentrations, redox, etc.) will improve the ability to model heap performance and understand operational limits.

Understanding differences between microbial performance and adaptation is crucial to successful biomining operations. However, the best-adapted culture is not necessarily the best performing. Investigations have been carried out on top-down, designed cultures for bioleaching (particularly tank leaching), but organisms that have the most desirable traits may not be able to colonise a system where an indigenous community is present as these organisms may be better adapted (even if they are not the most efficient from an operational point of view). In other words, during either a bottom-up or top-down approach, it is important allow the more effective microbial consortia (in terms of accelerating mineral oxidation and solubilising metals) sufficient opportunity to adapt to the system (geochemistry, mineralogy, operating conditions, etc.). On the other hand, it has been suggested that, if during primary succession most environmental niches are highly colonised, it could be difficult for subsequent organisms to displace these, even if they are better adapted. Other approaches used to attempt to effect ecological control include

Table 7.2 Summary of typical dominant microbial ecology of biological mineral-oxidising systems

Environment	Typical microbial ecology
Mine wastes and tailings	***Freshly deposited/unoxidised (pH ~ 3–8):*** Typically, higher diversity with higher pH. Relative absence of common mineral-oxidising taxa such as *Acidithiobacillus* spp. and *Leptospirillum* spp. Initial colonisation by pioneer organisms, often dominated by neutrophilic Proteobacteria including *Thiobacillus* spp. leading to acidification and decrease in pH. Increasing occurrence of major mineral-oxidising taxa as pH decreases.
	Actively leaching wastes/oxidation front (pH < 3): Generally restricted diversity. Common mineral-oxidising taxa dominate, typically iron-oxidising *Acidithiobacillus* spp., *Leptospirillum* spp., and *Sulfobacillus* spp.
	Weathered/well-oxidised wastes (pH 2–6): High biodiversity, generally more heterotrophically inclined. Common mineral-oxidising taxa present, but not dominant. Proteobacteria, Acidobacteria, Chloroflexi, Planctomycetes, candidate division AD3, Euryarchaeota, and Crenarchaeota most commonly detected phyla. Diversity controlled by pH and, to a lesser extent, overall metal(loid) load.
Commercial heap/dump leaching	***Initial colonisation phase:*** Primarily determined by composition of inoculum applied during construction/irrigation. *Acidithiobacillus* spp. tend to dominate.
	Active oxidation phase: Transition to increased dominance of iron-oxidising taxa such as *Acidithiobacillus* spp. and *Leptospirillum* spp.
	During thermal ramping: Heaps may be inoculated during construction and/or during irrigation with a specific (moderately) thermophilic culture as temperature ramps up. Transition to *Sulfobacillus* spp. and *Ferroplasma* spp. through to *Sulfolobus* spp. and *Thermoplasma* spp.
Commercial and industrial pilot tank leaching	***35–40 °C:*** Typically dominated by *Ferroplasma acidiphilum, Acidiferrobacter thiooxydans, At. thiooxidans, Acidiphilium multivorum*, and (to a lesser degree) *L. ferriphilum* (BIONORD).
	40–45 °C: Simple, stable communities almost entirely dominated by *At. caldus*, *L. ferriphilum* and *Sb. thermosulfidooxidans*/*Sb. benefaciens* (KCC, Mondo); **or** *Ferroplasma acidiphilum* and a *Thermoplasma* spp. (e.g., BIOX Fairview and Suzdal). The relative dominance of each species varies depending on operating conditions and tank position (primary, secondary, etc.).
	45–60 °C: Single commercial plant operating at this temperature range, dominated by two *Sulfobacillus* spp. (IBCCO). Laboratory-scale pilot data suggest *At. caldus* may be codominant with *Sulfobacillus* spp. at temperatures up to ~50 °C.

(continued)

Table 7.2 (continued)

Environment	Typical microbial ecology
	60–70 °C: *Acidiplasma cupricumulans*; increasing dominance of *Metallosphaera* spp. with increasing temperature (MesoTHERM).
	70–80 °C: Single *Acidianus* sp. at 70 °C (MINTEK). Consortium of novel *Sulfolobus*-, *Stygiolobus*-, *Acidianus*- (to a lesser extent), and *Metallosphaera*-like species (BioCOP).

continuous inoculation with appropriate microorganisms (e.g., thermophiles for self-heating bioheaps).

Ecological engineering of mine wastes is desirable for many reasons. From an environmental perspective, colonising mine waste with microorganisms that do not accelerate the breakdown of minerals, leading to acid generation and solubilisation of metals, could provide a useful method to limit ARD formation. How to achieve such control is less evident. Sterilising, or otherwise inhibiting the indigenous (or best-adapted) community is likely to be prohibitively expensive for most operations. Alternatively, continuous application of readily accessible organic matter has been shown to sustain the dominance of heterotrophic populations that not only do not oxidise residual sulfide minerals but can also immobilise soluble metals and generate alkalinity (Johnson 2014).

Bioleaching systems are complex, in terms of bio- and geochemical processes and interactions, sampling, analysis, and interpretation. Understanding microbial diversity, its function, and response to environmental conditions are key in improving process design, and system modelling and prediction. Improvements in microbiological tools such as DNA (and RNA) extraction, sequencing and ecological analytical approaches are now being applied more routinely to these systems. It is essential to properly consider sampling and experimental design with microbiology in mind. Important to this is not just knowing which microorganisms are present in a system, but their locations, activities and responses to environmental conditions.

References

Bhappu RB, Johnson P, Brierley J, Reynolds D (1969) Theoretical and practical studies on dump leaching. AIME Trans 244:307–320

Brierley CL (2001) Bacterial succession in bioheap leaching. Hydrometall 59(2):249–255. https://doi.org/10.1016/S0304-386X(00)00171-7

Colmer AR, Hinkle ME (1947) The role of microorganisms in acid mine drainage: a preliminary report. Science 106:253–256. https://doi.org/10.1126/science.106.2751.253

Demergasso C, Véliz R, Galleguillos P, Marín S, Acosta M, Zepeda V, Zeballos J, Henríquez F, Pizarro R, Bekios-Calfa J (2018) Decision support system for bioleaching processes. Hydrometall 181:113–122. https://doi.org/10.1016/j.hydromet.2018.08.009

Edwards KJ, Bond PL, Druschel GK, McGuire MM, Hamers RJ, Banfield JF (2000) Geochemical and biological aspects of sulfide mineral dissolution: lessons from Iron Mountain, California. Chem Geol 169(3):383–397. https://doi.org/10.1016/S0009-2541(00)00216-3

Govender E, Bryan CG, Harrison STL (2013) Quantification of growth and colonisation of low grade sulphidic ores by acidophilic chemoautotrophs using a novel experimental system. Miner Eng 48:108–115. https://doi.org/10.1016/j.mineng.2012.09.010

Hottenstein JD, Neilson JW, Gil-Loaiza J, Root RA, White SA, Chorover J, Maier RM (2019) Soil microbiome dynamics during pyritic mine tailing phytostabilization: understanding microbial bioindicators of soil acidification. Front Microbiol 10:1211. https://doi.org/10.3389/fmicb.2019.01211

Johnson DB (2014) Recent developments in microbiological approaches for securing mine wastes and for recovering metals from mine waters. Minerals 4:279–292. https://doi.org/10.3390/min4020279

Johnson DB, Joulian C, d'Hugues P, Hallberg KB (2008) *Sulfobacillus benefaciens* sp. nov., an acidophilic facultative anaerobic Firmicute isolated from mineral bioleaching operations. Extremophiles 12:789–798. https://doi.org/10.1007/s00792-008-0184-4

Korehi H, Blöthe M, Schippers A (2014) Microbial diversity at the moderate acidic stage in three different sulfidic mine tailings dumps generating acid mine drainage. Res Microbiol 165:713–718. https://doi.org/10.1016/j.resmic.2014.08.007

Manafi Z, Kargar M, Kafilzadeh F (2021) Tank bioleaching of a copper concentrate using the moderately thermophilic microorganisms *Sulfobacillus thermosulfidoxidans* KMM3 and *Sulfobacillus acidophilus* KMM26. Rev Metal 57:e207. https://doi.org/10.3989/revmetalm.207

Morin DHR, d'Hugues P (2007) Bioleaching of a cobalt-containing pyrite in stirred reactors: a case study from laboratory scale to industrial application. In: Rawlings DE, Johnson DB (eds) Biomining. Springer, Heidelberg, Germany, pp 35–55

Nordstrom DK, Blowes DW, Ptacek CJ (2015) Hydrogeochemistry and microbiology of mine drainage: an update. Appl Geochem 57:3–16. https://doi.org/10.1016/j.apgeochem.2015.02.008

Norris PR (2017) Selection of thermophiles for base metal sulfide concentrate leaching, Part II: nickel-copper and nickel concentrates. Miner Eng 106:13–17. https://doi.org/10.1016/j.mineng.2016.12.006

Norris PR, Burton NP, Clark DA (2013) Mineral sulfide concentrate leaching in high temperature bioreactors. Miner Eng 48:10–19. https://doi.org/10.1016/j.mineng.2013.01.001

Rawlings DE, Johnson DB (2007) The microbiology of biomining: development and optimization of mineral-oxidizing microbial consortia. Microbiology 153:315–324. https://doi.org/10.1099/mic.0.2006/001206-0

Rawlings DE, Tributsch H, Hansford GS (1999) Reasons why '*Leptospirillum*'-like species rather than *Thiobacillus ferrooxidans* are the dominant iron-oxidizing bacteria in many commercial processes for the biooxidation of pyrite and related ores. Microbiology (UK) 145:5–13. https://doi.org/10.1099/13500872-145-1-5

Rossi G (1990) Biohydrometallurgy. McGraw-Hill

Sbaffi T, Buckling A, Bryan CG (2017) Microbial Community Composition of Mine Wastes in Cornwall and West Devon (UK), vol 262. Solid State Phenomena, Trans Tech Publications, Switzerland, pp 290–293. https://doi.org/10.4028/www.scientific.net/SSP.262.290

Schippers A, Breuker A, Blazejak A, Bosecker K, Kock D, Wright TL (2010) The biogeochemistry and microbiology of sulfidic mine waste and bioleaching dumps and heaps, and novel Fe(II)-oxidizing bacteria. Hydrometall 104:342–350. https://doi.org/10.1016/j.hydromet.2010.01.012

Smart M, Huddy RJ, Edward CJ, Fourie C, Shumba T, Iron J, Harrison STL (2017) Linking microbial community dynamics in BIOX® leaching tanks to process conditions: integrating laboratory and commercial experience. Solid State Phenomena, Trans Tech Publications 262:38–42. https://doi.org/10.4028/www.scientific.net/SSP.262.38

Wakeman K, Auvinen H, Johnson DB (2008) Microbiological and geochemical dynamics in simulated-heap leaching of a polymetallic sulfide ore. Biotechnol Bioeng 101:739–750. https://doi.org/10.1002/bit.21951

Watling HR (2006) The bioleaching of sulphide minerals with emphasis on copper sulphides–a review. Hydrometall 84:81–108. https://doi.org/10.1016/j.hydromet.2006.05.001

Chapter 8
Biomining in China: History and Current Status

Guanzhou Qiu, Xueduan Liu, and Ruiyong Zhang

Abstract While the earliest biomining activities in China were documented in 6th ~ 7th century BC fundamental research and biomining applications started relatively late in this country. Rapid development, from phenotypic to genotypic characterisation of biomining microorganisms, as well as from theoretical to practical applications, has been made in China since the 1950s. The central government has attached great importance to biohydrometallurgy and is supporting integrated applications of bioleaching technology in copper, gold, and uranium extraction to ensure China's economic reserves of strategic mineral resources. Examples include a bioleaching plant at the Zijinshan copper mine with an annual processing capacity of 60 million tons of copper ore, the first demonstration project of biooxidation of refractory gold ore developed by the Changchun Gold Research Institute (CCGRI), and the leaching of a sulfide-entrained uranium ore where yields have been increased to 96%. This chapter describes the history of biomining in China, and introduces the development and application of biomining technologies (e.g., recovery of copper, zinc, gold, and uranium). The main challenges and future directions are also discussed.

Keywords Biomining · Acidophiles · Heap leaching · Tank leaching · Acid rock drainage (ARD) · China

G. Qiu · X. Liu
Central South University, Changsha, People's Republic of China
e-mail: qgz@csu.edu.cn

R. Zhang (✉)
Key Laboratory of Marine Environmental Corrosion and Bio-fouling, Institute of Oceanology, Chinese Academy of Sciences, Qingdao, People's Republic of China

Open Studio for Marine Corrosion and Protection, Pilot National Laboratory for Marine Science and Technology (Qingdao), Qingdao, People's Republic of China
e-mail: ruiyong.zhang@qdio.ac.cn

D. B. Johnson et al. (eds.), *Biomining Technologies*,
https://doi.org/10.1007/978-3-031-05382-5_8

8.1 History of Biomining in China

The earliest record of biomining activities in China can be traced back to the 6th ~ 7th century BC in the book titled *Shan-hai Ching*, where it was written that "the LuoShui River flows out and into WeiShui River in SongGuo mountain, which contains significant amounts of copper". Later on, Liu An, the Huainan King, wrote a book titled *Huai Nan Wan Bi Shu* in the 2nd century BC and recorded copper extraction from acid mine drainage (AMD): "copper was obtained when iron was put into Baiqing solution". During the Tang and Song Dynasty (600–960 AD), factories producing copper using hydrometallurgy were established, and annual copper production reached more than 1000 t (Qiu and Liu 2019). During the Song Dynasty (960–1279 AD) the pioneering hydrometallurgist Qian Zhang wrote a book *Synopsis of Copper Leaching* to introduce the copper extraction by hydrometallurgy (Golas 1995).

The first laboratory focusing on biohydrometallurgy in China was built by Professor Fuxu He in Central South Institute of Mining and Metallurgy (now Central South University, CSU, Changsha, China) in 1958. Characterisation of bioleaching microorganisms and bioleaching of metal ores were started in China at that time. In 1960, industrial applications were conducted at the Tongguanshan Copper Mine (Tongling, Anhui, China) by the Institute of Microbiology of Chinese Academy of Sciences (CAS). Bioleaching technology was subsequently applied in a demonstration plant to extract uranium from a low-grade uranium ore heap comprising 700 t of ore. In 1995, CSU exploited and tested copper extraction from low-grade copper ore wastes by biomining at the Dexing Copper Mine, Jiangxi Province. Two years later, a bioleaching plant with an annual production of 2000 t of cathode copper was successfully established at the mine (Fig. 8.1).

In 1999, the group headed by Prof. Guanzhou Qiu (CSU) started to cooperate with the Oak Ridge National Laboratory (USA) to carry out ecological and genomic research of bioleaching microorganisms, initiating the era of genomic studies in the field of biomining in China. In 2000, the first pilot-scale plant for biooxidative pre-treatment of refractory gold ore with annual treatment of 50 t gold concentrate was officially launched. In 2005, a biomining plant with a capacity of 30,000 t of cathode copper was built in Zijin Mining Company, and the cathode copper purity reached LME grade A.

Currently, the main research activities on biomining in China are performed in CSU, GRINM GROUP (formerly known as General Research Institute for Nonferrous Metals), Institute of Process Engineering of CAS, Institute of Microbiology of CAS, Institute of Oceanology of CAS, Shandong University, Changchun Gold Research Institute, Beijing Research Institute of Chemical Engineering and Metallurgy of China, Northeastern University, University of Science and Technology Beijing, Kunming University of Science and Technology, East China University of Technology, University of South China, and Jiangnan University, etc. The research carried out on bioleaching includes mainly three aspects: (i) microbiology of bioleaching (Guo et al. 2013; Wang et al. 2016; Zhang et al. 2010); (ii) microbial–mineral interactions (Li et al. 2020; Xia et al. 2020; Yu et al. 2018; Yin et al. 2020);

Fig. 8.1 The biomining plant set in 1997 in Dexing Copper Mine, Jiangxi Province, China

(iii) multiple factors that strongly influence bioleaching efficiency (Feng et al. 2021; Gan et al. 2019; Hao et al. 2021; Huang et al. 2022; Liu et al. 2016; Qiu et al. 2011; Ruan et al. 2006; Shang et al. 2021; Wang et al. 2018). The Chinese government has given major financial support for fundamental research and application of biohydrometallurgy, and has established a number of national science and technology plans, including "National Basic Research Program of China (973 Program)", "National High Technology Research and Development Program (863 Program)", and "National High Technology Industrialisation Demonstration Project." The National Natural Science Foundation of China (NSFC) has also provided funding (approximately 1 billion Yuan; approx. $155 M USD). For instance, the number of approved projects in its Engineering and Materials Department increased from 10 in 2000 to 50 in 2010, and the total budget increased tenfold.

8.2 Biomining Development in China

8.2.1 Macroscopic to Microscopic Views of Biohydrometallurgy

In China, the early research in biohydrometallurgy focused mainly on the macro level. Metallurgists often used acid mine drainage (AMD) for improving the bioleaching efficiency of metal ores, without knowing how the bioleaching process worked. Since the discovery of bacteria associated with AMD in 1947 by Colmer et al. (Colmer et al. 1950; Temple and Colmer 1951), microbiologists started to search for, isolate and select strains of microorganisms that were more effective in biomining applications. In 2004, CSU participated in whole genome sequencing of the type strain of *Acidithiobacillus ferrooxidans*, which was the first sequencing of a biomining microorganism. Based on the data, a national standard (GB/T 20929—2007) entitled "Methods for the detection of *At. ferrooxidans* and its oxidation activity by microarray" was established by CSU. The establishment of a national standard enabled rapid and accurate screening of bioleaching microorganisms with high iron- and/or sulfur-oxidising abilities. The full map and annotation of the genome of *At. ferrooxidans* laid the foundation for studying bioleaching mechanisms at the molecular level and realising the orientation of microbial leaching research from phenotypic to genotypic level. In Shandong University, extensive activities on genetic modification of biomining microorganisms for understanding iron and sulfur metabolisms in acidophiles, and improvement of biomining efficacy have been ongoing (Li et al. 2010; Hao et al. 2012; Gao et al. 2020).

8.2.2 From Qualitative to Quantitative Analysis

Biomining microorganisms usually comprise acidophiles which perform iron and/or sulfur oxidation activities (Johnson 2014). It is essential to find a way to quantitatively analyse microbial composition and function for the clarification of multifactors influencing bioleaching efficiency. With the rapid development of molecular methods, genetic, genomic, and metagenomic technologies are increasingly being applied in biomining applications. In particular, the application of genomics has led to significant progress in quantitative analysis of bioleaching systems, such as community structure and function. The development of microbial functional gene array and community genomic array technologies has led the research level from single function of a single population to whole functions of a single population and whole functions of a microbial community. Based on these technologies, the dynamics of microbial community structures and leaching functions can be detected quantitatively and may be used to analyse the effect of leaching parameters on microbial growth and iron-/sulfur-oxidation ability. The established microbial function gene array developed in CSU was used to study the microbial structure and function, allowing the simultaneous detection of the microbial community structure and function in bioleaching system. Based on results from studying the succession mechanism of bioleaching microbial community structure and function, the microbial consortium was optimised through combining different species and strains of microorganisms. The new optimised consortium was successfully applied for the bioleaching of low-grade copper sulfide at the YuShui Copper Mine, Guangdong Province, leading to enhanced levels of copper extraction and recovery.

8.2.3 *From Theory to Practice*

8.2.3.1 Biomining of Copper Ores

China is currently the top global consumer of copper, but its present domestic production accounts for only about 20% of its copper demand. Biomining technologies facilitate the exploitation of low-grade copper resources in China as elsewhere and can be therefore enhance global copper production. Below, several biomining industry applications for copper production are listed and described.

(1) Dexing Copper Mine

The heap bioleaching plant in Dexing Copper Mine is an example of a successful application of biomining technology. More than 3.5 Gt of waste ores have been produced during the life of the mine. These contain 0.05%–0.25% (by wt.) copper, so there are approximately 600,000 t of residual copper metal in this material. Since the waste ores are mainly composed of primary copper sulfides such as chalcopyrite, it has proved difficult to obtain high leaching rate using conventional biomining. In order to improve bioprocessing of the Dexing copper waste ores, two research projects "Studies on bioleaching of low-grade sulfide ore with selected bacterial consortium" and "Studies on the catalytic mechanism and strengthening bioleaching strains isolated from Dexing copper mine, and their industrial application" were carried out. Using the quantitative analysis technology developed by CSU, strains of *Acidithiobacillus* spp. and *Leptospirillum* spp. and other biomining prokaryotes, with high growth rate, high oxidation ability, and high resistance to metal ions were obtained by microarray screening and used to improve copper extraction. Copper recovery was further enhanced by upgrading the SX-EW plant at the mine (Fig. 8.1).

(2) Zijinshan copper mine

The Zijinshan copper mine was the first example of a successful industrial application of biomining in China. This mine is located in Shanghang County (a subtropical region) in Fujian province, and the copper sulfide deposit contains 240 Mt. of ore averaging 0.063% copper. Chalcocite and covellite are the copper minerals comprising the secondary sulfides. Since the copper grade is very low and the deposit contains significant amounts of arsenic, the traditional flotation and smelting process cannot be applied to extract copper economically and effectively, and in 1998 the mine operators began extracting copper using heap bioleaching (Ruan et al. 2006). Due to the relatively warm climate (average atmospheric temperature at the mine is 16–20 °C), heap bioleaching is favourable. Several steps, from shake flask tests to column tests and pilot tests combined with solvent extraction-electrowinning (SX-EW), were initially trialled to improve microbial efficacy and copper recovery. Not surprisingly, *Acidithiobacillus* spp. and *Leptospirillum* spp. appeared to be the dominant leaching organisms in the bioleaching process (Yin et al. 2018; Chen et al. 2020). From these early

Fig. 8.2 Heap bioleaching industrial application in Zijinshan Copper mine, Fujian Province, China

experimental studies (Fig. 8.2), the Zijinshan Copper Mine established a heap bioleaching factory with an ore processing rate of 60 Mt. y^{-1} and an annual cathode copper production of 10 Mt. The exploitable copper reserves have increased from 2.7 to 3.1 Mt. by using biomining technology. The copper recovery for heap bioleaching reaches 80% in a leaching period of approximately 200 days.

(3) Chambishi Copper Mine (Zambia)

Biomining technology developed in China was used to extract and recover copper in Zambia. In 2010, a strategic framework agreement was signed between the Zambian Ministry of Mines and Minerals Development and CSU. Based on the agreement, the "China Nonferrous Metal Mining (Group) Co., LTD—CSU—Zambia biohydrometallurgy technology industrialisation demonstration base" was established. Chambishi Copper Mine contains ~7 million tons of copper, with the chief copper minerals in the ore being bornite and chalcocite, with minor amounts of chalcopyrite. In March 2011, the Zambia Chambishi Copper Company cooperated with CSU to exploit the low-grade copper ore in Chambishi by heap bioleaching. Firstly, the indigenous microorganisms were screened, enriched, and adapted to the heap environment. Then, the adapted microorganisms were sub-cultured in 10 L, 70 L, 2 m^3, 28 m^3, and 150 m^3 stirred tank reactors, successively (Fig. 8.3). The microbial consortium was inoculated into ore heap by irrigation and spraying. The cell numbers were maintained at approx. 10^8 cells/mL in the leachate liquors. In the heap bioleaching of 600,000 t of low-grade copper ore, copper extraction reached up to 50% in 2 months. Bioleaching solution was processed using SX-EW, producing cathode copper at a rate greater than 10,000 t y^{-1}. Biomining technology was estimated to increase copper recovery by 20%, and to reduce acid consumption by at least 35%, compared to the acid leaching process using sulfuric acid. This was a clear demonstration of how low-grade copper resources can be exploited by using biomining technology.

Fig. 8.3 Scale up of the adaption process of bioleaching microorganisms from shake flasks to 150 m^3 stirred reactors

8.2.3.2 Biomining of Uranium Ores

In order to keep pace with the increasing demand of uranium for nuclear power generation, China's uranium production has been oriented to the exploitation of low-grade or refractory uranium ore, and other mineral resources associated with the processing of uranium. During (indirect) bioleaching of uranium ore, U(IV) is oxidised to U(VI) by ferric iron which is regenerated by iron-oxidising acidophiles, thereby maintaining the leaching reaction. In terms of uranium resources, biohydrometallurgy can enable an efficient use of a large number of idle or abandoned uranium sulfide resources in China and is expected to become increasingly important with projected decline in the grade of uranium resources from 0.1% to 0.03%. Bioprocessing has the potential to significantly lower the cut-off grade of uranium ores and thereby increase the economic mining exploitation of low-grade uranium deposits.

The Institute of Microbiology of CAS started biomining technology for uranium recovery, in the 1970s, with a pilot-scale study of heap leaching was conducted in Uranium Mine 711, (Hunan province). A total of 2 t concentrated uranium was enriched from the surface ore containing 0.02% ~ 0.03% of uranium by biomining for 8 years in the Bofang Copper Mine, Hengyang, Hunan province. In the 1980s, heap biomining of uranium gained rapid development and was applied at the Chaotaobei Uranium Mine (Ganzhou, Jiangxi province), a uranium mine in Xinjiang province and Xiangshan Uranium Mine (Jiangxi province). Using the optimised mixed culture in heap leaching at Fuzhou 721 mine in Jiangxi province, up to 96.8% extraction of uranium was achieved in 97 days.

Most of the examples cited are pilot-scale tests. The promotion and application of biohydrometallurgy could make a large number of idle or abandoned uranium sulfide resources available in China. It is anticipated that biomining can improve the exploitable uranium grade from the current limit of 0.1% to 0.03%.

8.2.3.3 Biomining for the Pre-treatment of Gold Ores

Refractory gold deposits are considered to account for about two-thirds of known global gold reserves, but these are not readily processed using conventional technologies. Biooxidation, however, can be used as a pre-treatment of refractory gold ores, allowing the previous metal to be accessed and solubilised by lixiviants such as cyanide (Chap. 4). China has become the world's largest producer of gold (Fig. 8.4) and incorporating biooxidation technology for refractory deposits will help to keep China at the forefront of global gold production.

The Shaanxi Provincial Authority of Land and Mines conducted a pilot-scale study on bioleaching pre-treatment of 2000 t of pyritic gold ore (containing 0.54 g Au t^{-1}) in 1994. Following biooxidation, gold recovery reached 58%. Direct extraction of gold by cyanidation of an arsenic-containing concentrate achieved only 35% gold recovery, whereas this was increased to 93% after 5-day pre-biooxidation. In Xinjiang Province, the Baogutu gold mine used biooxidation pre-treatment and gold leaching reached 92% ~ 97%.

The first biooxidation plant for processing gold ore in China was built at the Zhenyuan Gold Mine, Yunnan province. In 1998, Shaanxi Zhongkuang Technology Co. Ltd. established a biooxidation plant for pre-treatment of gold ore concentrates with a processing capacity of 10 t d^{-1}. In 2001, Tiancheng Gold Co., Ltd. imported

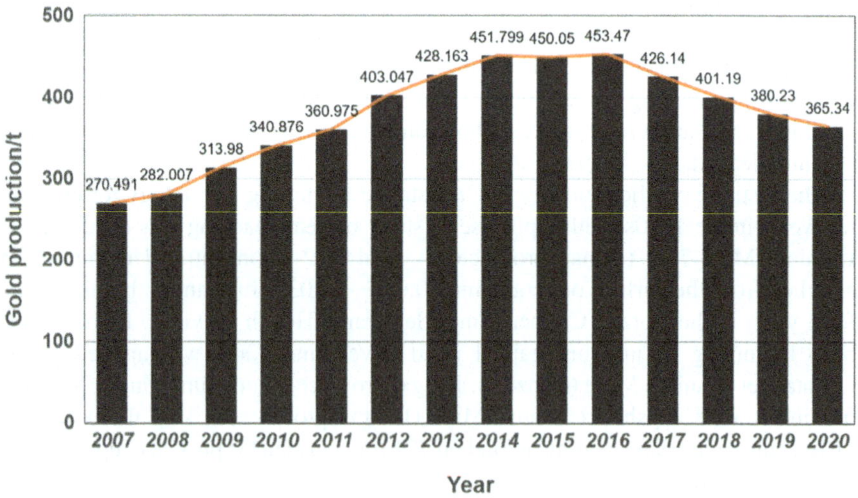

Fig. 8.4 Gold production in China from 2007 to 2020

the BacTech technology from Australia and built a biooxidation plant with a concentrate processing capacity of 100 t d^{-1}, which is no longer operating. In July, 2000, the construction of the first commercial biooxidation plant started at the Yantai Gold Smelter. The plant went into operation at the end of 2000 (Yang et al. 2002). The China National Gold Group Corporation works on the exploitation of arsenic-refractory gold concentrates at the Tianli Gold Company, Liaoning province, China. They have been extensively working on biooxidation/cyanidation technology and developed the "CCGRI" biotechnology in 2005. The processing capacity reached 150 t d^{-1}. The microorganisms (the "HY series" bacteria) employed in this technology are active at 35–52 °C and tolerate up to 22 g As L^{-1}. The HY bacteria were shown to grow in the presence of gold concentrate with a pulp density of 25–27% and 13–15% arsenic content.

Recently, biooxidation of refractory gold ores has developed rapidly and has reached an internationally advanced level. China has built more than 10 biooxidation-cyanide gold plants and currently has the largest number of biooxidation gold plants worldwide. This biotechnology is estimated to contribute ~8% of gold production in China in the near future.

8.3 Future Perspectives

Biomining technologies have undergone significant development in China, and the application of biomining for metal recovery has been industrialised for copper, uranium, gold, and nickel. However, biomining still needs to overcome some detractions in order to facilitate further expansion. For instance, copper production by biomining accounts for less than 8% of the total copper produced in China. This number is still lower than the estimated global level of 10–20% (Chap. 1). Microbial strains that are more effective in industrial applications (e.g., tolerance to high ore pulp density and/or toxic metals) could lead to improvements, and these may to be enriched and selected from both natural and anthropogenic environments. For instance, heap biomining applications in northern China could benefit from using consortia that contain psychrophilic/psychrotolerant strains that remain active at low temperatures. Fresh water is scarce in many places in China (as in many parts of Chile and Australia) and bioleaching using saline and/or brackish waters would help the expansion of biomining, though this requires salt-tolerant, mineral-degrading acidophiles (Chap. 13).

The vast amount of marine mineral resources, such as polymetallic nodules, marine manganese crusts, and massive sulfide deposits on the seafloor, could help to meet the expanding demand for metals in China and other countries, as could the recovery of base and precious metals from e-wastes (Chap. 14). Biomining technologies can be adapted to process these materials.

Last but not least, the contribution of biomining autotrophs which fix CO_2 to the carbon sequestration in a global scale has to be discussed and considered and estimated and this may give additional advantages for the expanding of this green technology in China and the rest of the world, though this carbon sequestration is transient and not likely to be a useful means of carbon sequestration like managed forestry (Chap. 1).

References

Chen J, Zhong S, Tang D, Kuang C (2020) Practical experience in large-scale development of Zijinshan low-grade gold-copper mine. Mining, Metallurgy & Exploration 37(4):1339–1347. https://doi.org/10.1007/s42461-020-00237-2

Colmer AR, Temple KL, Hinkle ME (1950) An iron-oxidizing bacterium from the acid drainage of some bituminous coal mines. J Bacteriol 59(3):317–328. https://doi.org/10.1128/JB.59.3.317-328.1950

Feng S, Yin Y, Yin Z, Zhang H, Zhu D, Tong Y, Yang H (2021) Simultaneously enhance iron/sulfur metabolism in column bioleaching of chalcocite by pyrite and sulfur oxidizers based on joint utilization of waste resource. Environ Res 194:110702. https://doi.org/10.1016/j.envres.2020.110702

Gan M, He P, Gu C, Zheng Z, Zhu J, Zhou S, Liu X, Qiu G (2019) Graphene and visible light enhance pyrite-based Cr(VI) reduction in the presence of *Acidithiobacillus ferrooxidans*. Int Biodeterior Biodegrad 137:78–87. https://doi.org/10.1016/j.ibiod.2018.11.010

Gao XY, Liu XJ, Fu CA, Gu XF, Lin JQ, Liu XM, Pang X, Lin JQ, Chen LX (2020) Novel strategy for improvement of the bioleaching efficiency of *Acidithiobacillus ferrooxidans* based on the AfeI/R quorum sensing system. Minerals 10(3):16. https://doi.org/10.3390/min10030222

Golas PJ (1995) A copper production breakthrough in the song: the copper precipitation process. J Song-Yuan Stud 25:153–168

Guo X, Yin H, Cong J, Dai Z, Liang Y, Liu X (2013) RubisCO gene clusters found in a metagenome microarray from acid mine drainage. Appl Environ Microbiol 79(6):2019–2026. https://doi.org/10.1128/AEM.03400-12

Hao L, Liu X, Wang H, Lin J, Pang X, Lin J (2012) Detection and validation of a small broad-host-range plasmid pBBR1MCS-2 for use in genetic manipulation of the extremely acidophilic *Acidithiobacillus* sp. J Microbiol Methods 90(3):309–314. https://doi.org/10.1016/j.mimet.2012.06.003

Hao X-D, Liu X-D, Zhu P, Liang Y-L, Qiu G-Z, Ma H-Q, Liu Y, Liu Q-J, Ren L-Y, Sarkodies EK, Liu H-W (2021) Responses of bacterial community to potential heap construction methods of fine-grained copper tailings in column bioleaching system. Trans Nonferr Metal Soc China 31(6):1796–1805. https://doi.org/10.1016/S1003-6326(21)65617-0

Huang MQ, Zhang M, Zhan SL, Chen L, Xue ZL (2022) Saturated dissolved oxygen concentration in in situ fragmentation bioleaching of copper sulfide ores. Front Microbiol 13:821635. https://doi.org/10.3389/fmicb.2022.821635

Johnson DB (2014) Biomining—biotechnologies for extracting and recovering metals from ores and waste materials. Curr Opin Biotechnol 30:24–31. https://doi.org/10.1016/j.copbio.2014.04.008

Li B, Lin J, Mi S, Lin J (2010) Arsenic resistance operon structure in *Leptospirillum ferriphilum* and proteomic response to arsenic stress. Bioresour Technol 101(24):9811–9814. https://doi.org/10.1016/j.biortech.2010.07.043

Li Q, Zhu J, Li S, Zhang R, Xiao T, Sand W (2020) Interactions between cells of *Sulfobacillus thermosulfidooxidans* and *Leptospirillum ferriphilum* during pyrite bioleaching. Front Microbiol 11:44. https://doi.org/10.3389/fmicb.2020.00044

Liu HC, Xia JL, Nie ZY, Ma CY, Zheng L, Hong CH, Zhao YD, Wen W (2016) Bioleaching of chalcopyrite by *Acidianus manzaensis* under different constant pH. Miner Eng 98:80–89. https://doi.org/10.1016/j.mineng.2016.07.019

Qiu GZ, Liu XD (2019) Biotech key to unlock mineral resources value. Chin J Nonferrous Met 29(9):1848–1858. (in Chinese)

Qiu G, Li Q, Yu R, Sun Z, Liu Y, Chen M, Yin H, Zhang Y, Liang Y, Xu L, Sun L, Liu X (2011) Column bioleaching of uranium embedded in granite porphyry by a mesophilic acidophilic consortium. Bioresour Technol 102(7):4697–4702. https://doi.org/10.1016/j.biortech.2011.01.038

Ruan RM, Wen JK, Chen JH (2006) Bacterial heap-leaching: practice in Zijinshan copper mine. Hydrometallurgy 83(1–4):77–82. https://doi.org/10.1016/j.hydromet.2006.03.048

Shang H, Gao W-C, Wu B, Wen J-K (2021) Bioleaching and dissolution kinetics of pyrite, chalcocite and covellite. J Cent South Univ 28(7):2037–2051. https://doi.org/10.1007/s11771-021-4751-5

Temple KL, Colmer AR (1951) The autotrophic oxidation of iron by a new bacterium, *Thiobacillus ferrooxidans*. J Bacteriol 62(5):605–611. https://doi.org/10.1128/JB.62.5.605-611.1951

Wang Y, Chen X, Zhou H (2018) Relationships between galvanic interaction, copper extraction and community dynamics during bioleaching of chalcopyrite by a moderately thermophilic culture. Bioresour Technol 265:581–585. https://doi.org/10.1016/j.biortech.2018.07.017

Wang ZB, Li YQ, Lin JQ, Pang X, Liu XM, Liu BQ, Wang R, Zhang CJ, Wu Y, Lin JQ, Chen LX (2016) The two-component system RsrS-RsrR regulates the tetrathionate intermediate pathway for thiosulfate oxidation in *Acidithiobacillus caldus*. Front Microbiol 7:1755. https://doi.org/10.3389/fmicb.2016.01755

Wang X, Sun Z, Liu Y, Min X, Guo Y, Li P, Zheng Z (2019) Effect of particle size on uranium bioleaching in column reactors from a low-grade uranium ore. Bioresour Technol 281:66–71. https://doi.org/10.1016/j.biortech.2019.02.065

Xia JL, Liu HC, Nie ZY, Fan XL, Zhang DR, Zheng XF, Liu LZ, Pan X, Zhou YH (2020) Taking insights into phenomics of microbe-mineral interaction in bioleaching and acid mine drainage: concepts and methodology. Sci Total Environ 729:139005. https://doi.org/10.1016/j.scitotenv.2020.139005

Yang SR, Xie JY, Qiu GZ, Hu YH (2002) Research and application of bioleaching and biooxidation technologies in China. Miner Eng 15(5):361–363. https://doi.org/10.1016/S0892-6875(02)00019-5

Yin S, Wang L, Kabwe E, Chen X, Yan R, An K, Zhang L, Wu A (2018) Copper bioleaching in China: review and prospect. Minerals 8(2):32. https://doi.org/10.3390/min8020032

Yin L, Yang HY, Tong LL, Ma PC, Zhang Q, Zhao MM (2020) Arsenopyrite bio-oxidization behavior in bioleaching process: evidence from laser microscopy, SEM-EDS, and XPS. Front Microbiol 11:1773. https://doi.org/10.3389/fmicb.2020.01773

Yu RL, Hou CW, Liu AJ, Peng TJ, Xia MC, Wu XL, Shen L, Liu YD, Li JK, Yang F, Qiu GZ, Chen M, Zeng WM (2018) Extracellular DNA enhances the adsorption of *Sulfobacillus thermosulfidooxidans* strain ST on chalcopyrite surface. Hydrometallurgy 176:97–103. https://doi.org/10.1016/j.hydromet.2018.01.018

Zhang RY, Xia JL, Peng JH, Zhang Q, Zhang CG, Nie ZY, Qiu GZ (2010) A new strain *Leptospirillum ferriphilum* YTW315 for bioleaching of metal sulfides ores. Trans Nonferrous Met Soc China 20(1):135–141. https://doi.org/10.1016/s1003-6326(09)60110-2

Chapter 9
Copper Bioleaching Operations in Chile: Towards New Challenges and Developments

Mario Vera Véliz, Alvaro Videla Leiva, and Patricio Martínez Bellange

Abstract The Chilean mining industry, after consolidating copper production through bioleaching in copper ore heaps, is now facing major challenges and changes in the exploitation of its copper reserves. The depletion of copper oxide ores, as well as a decrease in copper grade, have generated a significant amount of idle capacity in solvent extraction and electrowinning plants. The percentage of the total copper mined in Chile and produced as a concentrate is projected to increase to ~90% by 2027, with a very small increment in the total amount of copper per year. These changes, plus issues of efficiency and complexity, have contributed to a reduction in application of bioleaching in the Chilean mining industry. Several mining companies have transformed their bioleaching operations to chloride leaching, in search of more efficient technologies for leaching primary copper sulfides such as chalcopyrite, but with future environmental consequences that are still uncertain. These changes have impacted the R&D sector, where personnel have had to reorient their efforts to develop sustainable technologies in line with scarcity of water, complex mineralogy, and recent strict regulations. New products of bioleaching technology will include the revalorisation of tailings, technologies to stabilise dumps, as well as improvements in bioleaching of run-of-mine (ROM) ores and compatibility with new leaching technologies.

Keywords Bioleaching · Microorganisms · Chile · Mining · Codelco · BioSigma · Thin layer leaching · Chloride

---

M. Vera Véliz
Instituto de Ingeniería Biológica y Médica, Escuelas de Ingeniería, Medicina y Ciencias Biológicas, y Departamento de Ingeniería Hidráulica y Ambiental. Escuela de Ingeniería. Pontificia Universidad Católica de Chile, Santiago, Chile
e-mail: mariovera@ing.puc.cl

A. Videla Leiva
Departamento de Ingeniería de Minería, Pontificia Universidad Católica de Chile, Santiago, Chile
e-mail: avidela@ing.puc.cl

P. Martínez Bellange (✉)
Independent Biomining Consultant, Santiago, Chile

D. B. Johnson et al. (eds.), *Biomining Technologies*,
https://doi.org/10.1007/978-3-031-05382-5_9

9.1 Development and Current Status of Copper Bioleaching in Chile

Although Chile has been considered a mineral-rich territory since colonial times, its copper mining industry only started to grow at the beginning of the twentieth century, transforming the country's economy. Since these early years, several new technologies such as from the early process of vat percolation leaching, electrowinning (EW), and copper cementation to, in the 1970s, solvent extraction (SX), "thin layer" (TL) bioleaching processes (1980s) which use mostly mineral layers of 0.5–1 m (greater in some cases), and chloride copper leaching (2000s) have been adapted concomitantly, in parallel with several changes in the copper industry (Domic 2007, Gentina and Acevedo 2016).

The combination of TL-SX-EW technology in Chile started by the technological developments achieved at Lo Aguirre mine, Minera Pudahuel, which was located close to Santiago. In 1975, the company decided to apply thin layer heap bioleaching, previously patented by Holmes and Narver (H & N) in California (US patent number 4017309A). Improvements of the TL process, led to its application in 1980, concomitantly with a second patent for the process in 1981. In addition to its applicability for copper oxides, the TL process allowed secondary copper sulfides, such as chalcocite, covellite, and bornite to be processed. Microbial oxidative action on ores increased both the concentration of ferric iron (the main metal sulfide-oxidising agent), and the oxidation of sulfur and reduced inorganic sulfur compounds (RISCs), ensuring the acidic pH required by mineral-oxidising acidophilic prokaryotes. The ore initially contained mainly copper oxides, but over the years, a growing proportion of the feed consisted of chalcocite (Cu_2S). A nominal 15,000 t y^{-1} of copper cathodes was produced from the time of start-up until the site was shut down in 2001 due to the total depletion of the deposit (Domic 2001).

A very positive scenario for the expansion of biohydrometallurgical projects occurred during the 1990s, fuelled by a combination of several factors such as rising copper prices, depletion of higher-grade ores, increment of costs, and environmental concerns. These, together with an increase of foreign investment in Chile, led to significant changes in the implementation of new projects for copper recovery. Several projects based on the more environmentally friendly TL-SX-EW processes started, leading to production of high-quality copper cathodes at competitive costs. The development of TL-SX-EW copper cathode production in Chile since 1990 is shown in Fig. 9.1. Rapid expansion occurred during 1995–2010, and production of 2088 kt of SX-EW copper cathodes was obtained in 2010. During that period, bioleaching also attracted interest and reached a significant level of maturity. By 2010, out of the 2088 kt of total TL-SX-EW copper production, the Chilean copper bioleaching production contribution was 556 kt per year, led by the Minera Escondida, Minera Spence, Chuquicamata, and Quebrada Blanca operations.

By 2012, bioleaching became the standard method for low-grade copper sulfide processing, allowing processing of copper feed grades as low as 0.4%. Table 9.1 shows a benchmark of the main operations by 2012 (Enrique Roman, personal

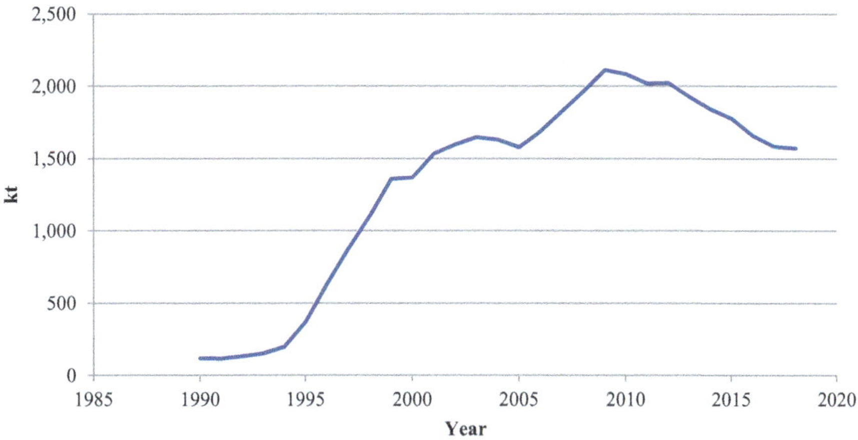

Fig. 9.1 Total production of copper cathodes in kt per year, by TL-SW-EW in Chile between 1990 and 2020. Adapted from Cochilco Annual Statistics Report, 2020 (https://www.cochilco.cl/Paginas/Estadisticas/Publicaciones/Anuario.aspx)

communication). This table includes different types of processes, among these Lo Aguirre, Carmen Andacollo, and Spence used permanent heaps, while Quebrada Blanca, Cerro Colorado, and Zaldívar used "on-off" heap leaching (intermittent irrigation), the latter with run-of-mine (ROM) material. Other operations also including ROM material were Escondida, Los Bronces, and Chuquicamata. It is interesting to note that the latter was the first investment in hydrometallurgy operations made by the Chilean state-owned company, Codelco, which applied dump leaching to ROM ore with low recovery (<25%) of copper. The SX–EW units began operations in mid-1994, with a design production level of 12,500 t y^{-1} of copper cathodes (Domic 2007). During this time, the feasibility of applying bioleaching at a new mine (Ministro Hales), with ores rich in arsenic sulfides, was studied by a joint venture between Codelco and BHP Billiton, using the high temperature BioCop™ sulfide concentrate leaching technology. The process employed thermophilic microorganisms in agitation tanks, operating at temperatures between 65 and 80 °C. A benefit of the copper dissolution was the possibility of precipitating and removing soluble arsenic as inert scorodite ($FeAsO_4$), which is highly stable in aerobic environments. At the time, the operational costs were much higher than those associated with conventional concentrate smelting and refining, and the joint venture ended. BHP sold its part of the pilot plant facility to Codelco, who transformed the industrial facility into Ecometales (www.ecometales.cl), a subsidiary dedicated to the treatment of solution impurities, particularly arsenic by chemical oxidation followed by generation of scorodite.

The Quebrada Blanca operation is situated at an altitude of 4000 m above sea level. It was the first TL process application on a chalcocite ore body, containing about 1% total copper. The plant started in 1994 with a designed capacity of 75,000 t y^{-1} of copper cathode, and by 2012 it had reached 80,000 t y^{-1}, exceeding the

Table 9.1 Benchmarking between bioleaching operations, 2012 (Roman E, personal communication)

Mine	Lo Aguirre	Carmen Andacollo	Spence	Quebrada Blanca	Cerro Colorado	Zaldívar	LGS Chuquicamata	SBL Escondida	Los Bronces
Major copper minerals	Chalcocite Bornite	Chalcocite	Chalcocite Covellite Chalcopyrite	Chalcocite Covellite	Chalcocite Covellite	Chalcocite Covellite	Chalcopyrite	Chalcocite Covellite	Chalcocite Covellite Chalcopyrite
Copper grade (%)	1–1.4	0.77	2	0.73	0.83	0.67	0.35	0.3–0.7	0.45
Copper production capacity (kt y^{-1})	14	21	70	80	125–135	150	22.5	180	45
Heap type	Permanent Heap	Permanent Heap	Dynamic Heap	"On-off" ROM	"On-off" ROM	"On-off" ROM	Dump ROM	ROM	Dump ROM
Height (m) Lift number	4 4	8 10	10	10	6–8	6.5	35	7 18	40 10
Particle size, P80 (mm)	6	19	12	305	152	508	203–381	203–381	203–381
Leaching cycle (d)	1095	280	650	450	490	300	548	316	770
Copper recovery (%)	75	80	70	80	80	80	20	38	35

designed production. Since the oxides and secondary sulfides were almost completely mined out, the annual production dropped to 21,100 t y^{-1} in 2019. Quebrada Blanca is currently switching to production of copper concentrates, with a projected 28-year life-of-mine expansion, by the installation of a 140,000 t day^{-1} concentrator mill plant, to produce 316,000 t y^{-1} of copper.

The Zaldívar mine is located in the Antofagasta region of Chile and operations there started in mid-1995, reaching a maximum production of 150,000 t y^{-1} of copper cathode in 2002, which fell to 116,000 t y^{-1} by 2019. By early 2021, the Zaldívar mine operation had switched to chloride leaching to reduce cycle times and improve copper recovery.

The Escondida operation belongs to an international consortium composed of BHP Billiton (57.5%), Rio Tinto (30%), Mitsubishi (10%) and International Finance Corporation (2.5%). In addition to the production of copper concentrates for direct export, Escondida also produces copper cathodes via TL–SX–EW operation. Copper sulfides are mainly composed of low-grade chalcocite and covellite (Table 9.1). Due to the increment in primary sulfide proportion, the process has recently switched to chloride leaching.

Spence is operated exclusively by BHP, and has produced copper cathodes by microbial leaching. The deposit contains reserves of copper oxides (1.14% Cu, mainly as atacamite; $Cu_2Cl(OH)_3$) and copper sulfide ores (1.12% Cu, primarily supergene chalcocite and some minor amounts of covellite; CuS). These are mined and processed separately, as atacamite generates solutions that are rich in chloride, and oxides leach about twice as fast as the sulfides (Domic 2007). The sulfide ore fraction contains chalcocite, covellite, and chalcopyrite.

Within the last 10 years, Chile has been facing a serious change in the exploitation of its copper ore reserves. The production of copper by TL-SX-EW has decreased, due to the depletion of copper oxides and secondary sulfide ores, as well as a decline in their copper grades, leading to a significant amount of idle SX-EW capacity (Lagos et al. 2018). Despite previously being widely accepted, the long operational cycles and low recoveries associated with bioleaching of ROM have become of increasing concern. In addition, some companies have switched their operations to leaching with high chloride-containing solutions. Projections indicate that the amount of copper production by heap leaching will decrease from a current (2020/2021) estimate of ~30% to approximately 10% by 2027. Concomitant with this decline in heap (bio)leaching, production of copper concentrates will increase to around 90%, with a very small increase in the total amount of copper produced per year (Cochilco 2017).

A comparison of the state of the most prominent bioleaching operations in 2009 and 2019 is shown in Table 9.2. Current main operations using biohydrometallurgy include Cerro Colorado, Quebrada Blanca, Escondida, Zaldívar, and Spence. The Cerro Colorado operation is located in the Antofagasta region, and commenced in 1993, with a starting production capacity of 45 kt y^{-1} of copper cathodes. By 2013, Cerro Colorado reached a total annual copper production of 130 kt y^{-1}. Since then, the annual production has declined until reaching current levels of 70 kt y^{-1}. The environmental permits are expected to expire in 2023, and therefore a down-scaling

Table 9.2 Major bioleaching operations in Chile: production comparison between 2009 and 2019

Bioleaching operations	2009 TL-SX-EW Production ($t\ y^{-1}$)	2019 TL-SX-EW Production ($t\ y^{-1}$)	Company	Copper grade (%)	Situation
Quebrada Blanca	82,000	21,100	Teck	0.3	1994–Present
Chuquicamata	85,000	0	Codelco	0.3	Closed
Carmen Andacollo	22,500	4700	Teck	0.58	1996–Present
Dos Amigos	10,000	0	Cemin	2.5	Closed
Los Bronces	46,400	0	Anglo-American	0.45	Closed
Zaldívar[a]	137,000	58,100	Antofagasta PLC/Barrick gold	0.67	Switched to chloride leaching
Escondida[a]	182,000	250,200	BHP	0.3–0.7	Switched to chloride leaching
Spence[a] Cerro Colorado[a]	128,000 93,700	193,000 71,700	BHP BHP	1.12 1.2–0.6	Switched to chloride leaching Switched to chloride leaching
Total net bio-copper production	649,600	25,800			
% Total Chilean production	10.4	0.4%			

[a]Operations switched to chloride leaching have not been considered for the net amount of bio-copper produced

process has begun. In 2016, under the control of BHP, the operation switched to a full chloride leaching process, to improve secondary sulfide ore leaching kinetics and copper recovery.

Codelco, at the Radomiro Tomic division, has been applying the BioSigma process (see Sect. 9.2.1), for almost a decade. The first industrial test was performed between 2012 and 2014, by using two pilot heaps of 25.000 t of ore each. Biomass was produced in an external bioreactor, which was periodically removed and directly inoculated into the leaching solution and the heap. Usually, the copper ore grade is less than 0.4% and more than 70% consisted of primary copper sulfides, mainly bornite and chalcopyrite. The testing facility has been used to design the ROM bioleaching process at Radomiro Tomic operation, which will allow production of 10,000 $t\ y^{-1}$ of copper, with an estimated recovery of 25% Cu.

Another project currently applying bioleaching is the Sociedad Punta del Cobre (Pucobre), located in the Atacama region. This mine started operations in 1989, with a copper mill concentrator and since 1993 has operated a bioleaching process.

Sulfide ore is mined from the 3.5 Mt. y^{-1} Punta del Cobre underground mine and transported 5 km for its treatment at San Jose plant. The plant is located 18.5 km from Copiapó. Today, the plant can produce over 110,000 t y^{-1} of copper concentrates, grading ~29% copper. In recent years, bioleaching operations have been focused in copper concentrates, reaching 75 t y^{-1} of cathodes production using bioleaching technology (Pucobre team, personal communication).

The most prominent events that have occurred in Chile in relation to the implementation of the TL-SX-EW, applied to copper, are listed in Table 9.3.

9.2 Research, Development, and Biomining Applications in Chile: Industrial Cases

Over the past 30–35 years, there has been a major contribution made by universities in Chile in devising and communicating the results of research projects with potential application to mineral bioprocessing. Some of these projects developed in partnership with large and medium-sized mining companies have generated some interesting results, though these have often faced problems with follow through. The continuous structural changes that the mining industry in Chile faces have had major impact on the application of new technologies, as their continuity is affected by management decisions primarily responding to short-term economics. This has resulted in a shortening of time for long-term development of technologies, with consequent impairment to show their ultimate potential. Nevertheless, at least two companies have significantly contributed to advances in the development and application of bioleaching in Chile, as described below.

9.2.1 BioSigma-CodelcoTech

In 2002, the BioSigma company was launched as a joint venture between the mining companies Codelco (Chile) and JX Nippon Mining and Metals (Japan). The main aim of BioSigma was to commercialise comprehensive biotechnological solutions for sustainable mining, with a commitment to open innovation in Chile and worldwide. At the beginning of 2017, BioSigma was merged with other technological subsidiaries from Codelco, under the name of CodelcoTech, a subsidiary finally closed in 2020 due to corporate decisions. At its peak (2012) BioSigma employed about 100 personnel that had a multidisciplinary profile with its highly qualified technical and professional staff qualified in different specialties. During its 15 years of operation, BioSigma achieved important scientific and technological advances, such as isolating different microorganisms and selecting microbial consortia with improved action and differentiated activities such as oxidation of iron and/or elemental sulfur and inorganic sulfur compounds. This included microbial consortia

Table 9.3 Prominent events in the implementation of the TL-SX-EW technology in Chile (updated from Domic 2007)

1969–1970: First SX-EW pilot trials in Chile, at Chuquicamata, Antofagasta region, testing of vat leach solutions, from the Exótica (currently Mina Sur) mine
1980: Minera Pudahuel Lo Aguirre mine startup, Región Metropolitana; first commercial application of the TL leaching system and of solvent extraction-electrowinning in Chile. Bioleaching pilot plant studies were successful, and it was decided to implement a full-scale operation as soon as the presence of sulfide ore in the feed justified it.
1981: Minera Pudahuel granted a 15-year patent for the TL leaching process (in essence: agglomeration and acid curing, followed by non-flooded heap leaching, with capacities for heap leaching of copper oxide and copper secondary sulfide ores).
1984: First commercial plant using SX-EW, with diluted leach solutions; in situ bioleaching at the El Teniente block-caving mine (solutions contained 1–1.5 g Cu L^{-1}).
1985: Commercial TL bioleaching commenced at Pudahuel's Lo Aguirre facility, with a second solvent extraction circuit entirely engineered in Chile. This was the first controlled fully engineered bioleaching process for copper sulfide ores commercially implemented in Chile.
1987: First commercial plant using SX-EW, using high tenor solutions from the leach of "ripios" (old vat tailings) retreatment plus current vat-leach of the Mina Sur oxide ores at Chuquicamata (leach solutions contained 10–12 g Cu L^{-1}).
1991: First leaching–SX-EW plant using Pudahuel's TL leaching technology and license using seawater for leaching started at Lince (currently Michilla Mine, Antofagasta region).
1993–1994: Two SX-EW plants using Minera Pudahuel microbial TL technology and license, leaching exclusively copper sulfide ore feed, operated at the Cerro Colorado and Quebrada Blanca mines at elevations of 3300 and 4200 m above sea level, respectively.
1994: First flotation of concentrates and ammonia chemical leach–SX-EW, at Coloso plant, La Escondida, Antofagasta region, following the principles of the arbiter process; the plant was closed in 1998 due to failures in the construction materials that negatively influenced overall economics.
2004: First thermophilic microbial leaching of mixed copper sulfide/arsenide flotation concentrates, at semi-commercial scale in a prototype plant for 20,000 ton y^{-1} of copper cathodes; using the BHP Billiton BioCop technology in a joint venture with Codelco, at Alliance copper, for treating the high arsenic copper concentrates from Mansa Mine (currently Ministro Hales), near Chuquicamata, Antofagasta region.
2005: BioSigma launches the SBP Process (a bioleaching process, based on consortia of native microorganisms inoculated into heaps to recover copper from secondary and primary copper sulfides).
2006: BHP Escondida starts a 180,000-t y^{-1} bioleaching operation to process oxides and secondary copper sulfides.
2007: BHP Spence starts a 200,000-t y^{-1} bioleaching operation to process oxides and secondary copper sulfides.
2007: First copper cathode obtained using BioSigma's bioleaching technology at Codelco's Andina division (at 3000 m elevation) in Valparaíso region.
2010: Construction and commissioning of an industrial biomass plant, for bioleaching microorganism production and heap inoculation in Codelco's Radomiro Tomic division, Calama, Antofagasta region.
2012–2014: Industrial testing of crushed ore and ROM material in heaps using BioSigma inoculation technologies at the Radomiro Tomic Division of Codelco.

with greater resistance to cations and anions typically present in elevated concentrations in mining solutions, with a concomitant improved action on chalcopyrite ores. Additionally, from sequencing and annotation of the genomes of isolated microorganisms and the use of high-performance technologies such as transcriptomics, proteomics, and metabolomics, it was possible to establish databases on the genes, proteins and metabolites that participate in the process of bioleaching of copper sulfide mineralogical species, with the corresponding development of biomarkers, the understanding of the importance of certain nutrients and the negative effect produced by certain toxic compounds inherent to the operation process (Bobadilla-Fazzini et al. 2014, 2017; Martínez et al. 2013, 2015). All of this allowed a comprehensive analysis of bioleaching processes, and to develop and patent different analysis and prediction tools, such as phenomenological models of bioleaching heap operation and bioreactor performance, as well as the design of efficient biomass production systems, with the correct microbial activity necessary for the process (US patent number US7837760B2). The latter was reflected in the construction of a pilot biomass plant at the Radomiro Tomic division. This plant was planned to reach a potential operation for inoculating 7.2 Mt. of mineral y^{-1}. It is also worth mentioning the importance of the development of bio-characterisation technologies for the in situ monitoring of the bioleaching process. For this, several technologies were adapted over time, from Denaturing Gradient Gel Electrophoresis (DGGE), through PCR, qPCR, DNA microarrays, up to the use of cutting-edge technologies, such as massively parallel sequencing (NGS) and Q-TOF mass spectrometry. All of this fundamental science development, involving different universities, technology centres and companies, generated relevant information that allowed progress for scaling-up bioleaching of primary copper sulfides in heaps. In addition, the foundations for an improved biomining process, specifically to achieve significant copper recoveries from low-grade sulfides, were established. In 2005, results of laboratory studies initiated validation tests in pilot-scale bioleaching operations using 2500 ton heaps of sulfide minerals located in the Chuquicamata division, Codelco Norte. In 2007, along with continuing the pilot works to validate its technology at the Codelco Norte Division, BioSigma began the industrial prototype validation in 50,000 t heaps containing low-grade primary sulfide mineral at the Andina division (Region de Valparaíso, Los Andes). In this process, the first copper cathodes were obtained with BioSigma's technology (Codelco 2007). Continuing with the scaling and validation on an industrial scale of the BioSigma bioleaching process under actual operating conditions, an industrial-scale plant to produce bioleaching microorganisms for heap inoculation was constructed, and began operation towards the end of 2010, in the Radomiro Tomic division (Codelco 2009). With this capability, an industrial test of crushed ore and ROM material in heaps was carried out between 2012 and 2014 with the application of BioSigma inoculation technology, using consortia of acidophilic microorganisms, and compared with conventional acid treatment leaching technologies. The results were encouraging, with copper recoveries between 30% and 50% greater than competing technologies, and achieved within in a much shorter period of time. With these positive results, Codelco, at Radomiro Tomic division decided to integrate BioSigma bioleaching technology into its production plans during the second half of 2014. This first project was designed to treat 3.6 Mt. of low-grade sulfur ROM ore. The planned amount was increased to 5 Mt. of low-grade mineral for the year 2015 (Codelco 2014). Additionally, a patent related to a solid-phase inoculation technology, "BioSigma Bioleaching Seeds" (BBS; Patent number US10131961B2), that enhanced the BioSigma liquid inoculation technology by improving homogeneous inoculation of the ore, greater microbial resistance to potential toxins and elevated solute potentials, and improved cell viability during manipulation, transport, and storage of the microorganisms, was developed. The BBS technology was combined with an integrated biofilter or "BioSigma Bioleaching Filter" (BBF; INAPI applicant number 201903901), which allowed the use of direct high toxic process solutions (raffinate) with a low fresh water consumption. This conferred several advantages to the inoculation process, such as providing protection of microorganisms against high ionic loads, reduction of biomass volume, possibility for long-term storage, safe transport as well as ensuring a more homogeneous inoculation process (Martínez and Parada 2013).

At the start of 2017, the fusion of three Codelco technological subsidiaries, including BioSigma, resulted in the formation of CodelcoTech. Its main aim was to continue with development of technologies promoted by the former subsidiaries, integrating them under the wing of the current Codelco's corporate roadmap. The remaining BioSigma personnel, now under CodelcoTech, continued to work on technologies (referred to as "2.0"), by facing the challenge of promoting already validated technologies but adapting them to current scenarios. These included problems such as water scarcity, new environmental regulations, and changes in mining operations. Major efforts were focused on technologies for increasing the efficiency of bioleaching through conjugated and sequential processes with oxidants, such as chloride leaching, and the development of mobile modular reactors with higher performance and lower investment costs, among others. Additionally, new lines of development for bioleaching and biotechnology were opened. Among these, the feasibility of exploiting secondary resources such as tailings represents a potentially enormously valuable resource, considering the huge amount of tailings deposits in Chile (over 750 tailing dumps, Sernageomin, https://www.sernageomin.cl/datos-publicos-deposito-de-relaves). In this context, the extraction capacity through bioleaching of critical elements for international industry, such as rare earth elements (REE), nickel, cobalt, among others has gained increasing attention. In addition, there are challenges in physically and chemically stabilising these environmental mining liabilities, bearing in mind their potential use in other industrial applications such as construction. Bioleaching applications have opened an opportunity to use a sustainable circular economy model that can help develop a new way of mining. The CodelcoTech team directed a national project that tackled all these issues with interesting results for the recovery of valuable elements, such as cobalt, using bioleaching technologies, between 2017 and 2020. In parallel, thanks to collaboration with universities and technology centres, other mining issues related with treatment of sulfate-laden waters and acid mine drainage were addressed, through biotreatment in reactors and directed microbiological control (Schwarz et al. 2020; Suárez et al. 2020).

In 2020, the decision was made to close CodelcoTech, with the consequent freezing of all technological developments initiated by BioSigma and continued by the CodelcoTech team. Even so, at the Radomiro Tomic division, the biomass plant referred to above continues to operate, being integral to the optimisation stage to feed mineral to be treated by bioleaching.

The BioSigma-CodelcoTech case in Chile was associated with development and use of bioleaching of low-grade primary sulfides, mainly in large-scale mining (copper production of between 100,000–500,000 t y^{-1}), as was the case of Codelco. It is also interesting to mention that there are other private efforts, where advances have been made in biotechnological developments aimed at medium-sized mining (copper production of less than 50,000 t y^{-1}). In these cases, this type of disruptive technologies may have a greater opportunity of being integrated into the process and developed over time.

9.2.2 *Pucobre: LIAP*

Another interesting case is LIAP, a translated acronym for "Applied Research Laboratory", a company born as a spin-off of the mining company Pucobre, a medium-sized mining company located in the Atacama region, which has been in operation for more than 30 years. LIAP was created in 2011 under the challenge of using bioleaching technology to extract copper at the Biocobre cathode plant, near the city of Copiapó. This challenge generated the development of a laboratory with capabilities to develop technology. It has also allowed tackling some other challenges such the treatment of complex concentrates, a topic that is currently of cross-cutting interest in Chilean mining industry. Many mining sites contain arsenic minerals, and copper concentrates containing arsenic are penalised in their commercialisation in the international market, and cannot be processed pyrometallurgically, due to current regulations. These developments have allowed the company to develop patented technologies, such as the Biocobre technology (US patent 10036081B2), that allows bioleaching of concentrates and tailings material agglomerated with plastic, in a circular economy model that includes waste from different industries. Currently, LIAP-Pucobre is completing a bioleaching pilot program with this technology in confined heaps with 300 t of complex concentrates with high arsenic content. Recoveries of over 90% of copper and other metals, such as gold and silver, have been reported within in a period of operation of 200 days (LIAP team, personal communication).

9.3 Conclusions and Future of Bioleaching in Chile

Chile, as the largest global copper producing-country, faces the need to advance in the generation of more sustainable and environmentally friendly mining methods, to contribute to the construction of a new economy that takes into account the challenges faced by humanity. In this regard, the international community is demanding the use of latest-generation tools and new ways of mining. While Chile's potential in mining production, particularly with copper, has remained relatively intact over time, conditions enabling the development of the mining industry are changing at both national and global scales.

Copper extraction is becoming every year more difficult and energy demanding. The decline in the quality of the geological resources, specifically the decrease in copper grades of current ore deposits, has created a scenario under which mining companies must make great efforts to maintain their current production levels (Lagos et al. 2018). The factors influencing the downturn in productivity include, among others, long haulage distances due to the deepening of ore reservoirs requiring increased movement of ore material. The deterioration of ore quality requires also more complex processing schemes, due to the greater hardness of the rock and the presence of more complex mineralogy and complex contaminants. In addition, the depletion of copper oxides has generated surplus capacity at SX-EW plants, which leads to an opportunity for the processing of complex and low-grade sulfide minerals through hydrometallurgy to fill this capacity, including other metals as cobalt. All these factors are stimulating the need for a strong development of innovative and sustainable technologies to ensure copper production compatible with environmental regulations.

The exponential advances experienced in biotechnology within the last 40 years have captured the attention of the mining industry, in order to explore how to solve some of the above-mentioned challenges. Chile, as a mining country, has been generating knowledge in the last 35 years applied to the development of bioleaching. Although some of the private research and development initiatives have not continued due to multiple factors, the knowledge and progress achieved so far could form the basis for new discoveries and developments to be applied in the near future at new Chilean mining projects. Future mining will target the processing of secondary resources, such as tailings and gravel as well as ROM, and it is increasingly positioning bioleaching as a natural solution perfectly compatible with social demands of sustainability.

One of the keys for the application of innovative technologies that have less impact on global ecosystems, such as bioleaching and other applications in mining associated with applied biotechnology, has been R&D performed mainly in technology centres, universities and mining companies. The latter have played an important role in Chile, although applied research projects have not been without problems, especially with regard to their long-term continuity. As has been observed during the last two decades, these initiatives have been affected by how metal mining is carried out in Chile. Clearly there is a need to promote future changes in project development and planning, so that not only production variables but also sustainability variables are taken into consideration when devising future R&D projects in biomining. Although taking into account the setbacks described, bioleaching shows several advantages compared to other technologies that guarantee its application in some specific areas, and a significant growth could be achieved if R&D is scheduled to fill the current gaps in years to come. For instance, it is well-accepted that bioleaching has a lower cost over alternatives in the treatment of very low-grade sulfidic ores, especially ROM. This economical advantage makes feasible the economical extraction of marginal copper grade ore and therefore increases the mining reserves. Therefore its application with this kind of material is a natural niche to continue its development. Another application of bioleaching may focus on the re-processing of fresh or abandoned tailings, as it allows valorisation of current environmental liabilities. This approach links with circular economy and zero mining waste policies, giving the mining industry an important seal of sustainability, not only in the case of copper, but also for other critical elements for world industry, such as REE, cobalt, nickel, magnesium, and others. Another bioleaching area of current interest is the treatment of complex concentrates that contain appreciable amounts of toxic elements, such as arsenic, which currently either cannot be processed by smelters or are sold to markets with significant economical penalties. Bioleaching could make feasible the economical production of copper-enriched liquors, leaving behind the contaminants, with the additional potential to use the idle capacity of many Chilean SX-EW plants. Finally, the search for synergies between bacterial activity and alternative chemical leaching technologies, such as chloride leaching, has become of great interest to develop more effective and efficient solutions for particularly challenging ores.

Acknowledgements We are grateful for the assistance given by Dr. Enrique Román (Biosigma Technological Development Director 2006-2011, and currently an independent mining consultant) with data in Table 9.1 and various personal communications. Mario Vera and Alvaro Videla want to acknowledge funding support of ANID-PCI international collaboration program, through the ERAMIN Microbial Consortia for enhanced Copper Recovery (MiCCuR) project.

References

Bobadilla-Fazzini RA, Cortés MP, Maass A, Parada P (2014) *Sulfobacillus thermosulfidooxidans* strain Cutipay enhances chalcopyrite bioleaching under moderate thermophilic conditions in the presence of chloride ion. AMB Exp 4:84. https://doi.org/10.1186/s13568-014-0084-1

Bobadilla-Fazzini RA, Pérez A, Gautier V, Jordan H, Parada P (2017) Primary copper sulfides bioleaching vs. chloride leaching: advantages and drawbacks. Hydrometallurgy 168:26–31. https://doi.org/10.1016/j.hydromet.2016.08.008

COCHILCO (2017) Sulfuros primarios: desafíos y oportunidades. Available from: https://www.cochilco.cl/

Codelco (2007) Press release. https://www.codelco.com/codelco-logro-cosechar-su-primer-catodo-biolixiviado-en-division-andina/prontus_codelco/2011-02-21/123513.html

Codelco (2009) Press release. https://www.codelco.com/codelco-invertira-us-5-5-millones-en-planta-industrial-que-producira/prontus_codelco/2011-04-02/221522.html

Codelco (2014) Press release. https://www.codelco.com/biosigma-crea-tecnologia-industrial-para-producir-cobre-gracias-a-bacterias/prontus_codelco/2014-08-26/133613.html

Domic EM (2001) Hidrometalurgia. Fundamentos, procesos y aplicaciones. Santiago de Chile. ISBN 978-956-291-083-5

Domic EM (2007) A review of the development and current status of copper bioleaching operations in Chile: 25 years of successful commercial implementation. In: Rawlings DE, Johnson DB (eds) Biomining. Springer, Berlin, Heidelberg. https://doi.org/10.1007/978-3-540-34911-2_4

Gentina JC, Acevedo F (2016) Copper bioleaching in Chile. Minerals 6:23. https://doi.org/10.3390/min6010023

Lagos G, Peters D, Videla A, Jara JJ (2018) The effect of mine aging on the evolution of environmental footprint indicators in the Chilean copper mining industry 2001–2015. J Clean Prod 174:389–400. https://doi.org/10.1016/j.jclepro.2017.10.290

Martínez P, Gálvez S, Ohtsuka N, Budinich M, Cortés MP, Serpell C, Nakahigashi K, Hirayama A, Tomita M, Soga T, Martínez S, Maass A, Parada P (2013) Metabolomic study of Chilean biomining bacteria *Acidithiobacillus ferrooxidans* strain Wenelen and *Acidithiobacillus thiooxidans* strain Licanantay. Metabolomics 9:247–257. https://doi.org/10.1007/s11306-012-0443-3

Martínez P, Parada P (2013) BioSigma bioleaching seeds (BBS): a new technology for managing bioleaching microorganisms. Adv Mat Res 825:305–308. https://doi.org/10.4028/www.scientific.net/AMR.825.305

Martinez P, Vera M, Bobadilla-Fazzini RA (2015) Omics on bioleaching: current and future impacts. Appl Microbiol Biotech 99:8337–8350. https://doi.org/10.1007/s00253-015-6903-8

Schwarz A, Suárez JI, Aybar M, Nancucheo I, Martínez P, Rittmann BE (2020) A membrane-biofilm system for sulfate conversion to elemental sulfur in mining-influenced waters. Sci Total Environ 740:140088. https://doi.org/10.1016/j.scitotenv.2020.140088

Suárez JI, Aybar M, Nancucheo I, Poch B, Martínez P, Rittmann BE, Schwarz A (2020) Influence of operating conditions on sulfate reduction from real mining process water by membrane biofilm reactors. Chemosphere 244:125508. https://doi.org/10.1016/j.chemosphere.2019.125508

Chapter 10
Heap Bioleaching of an Enargite-Dominant Ore Body: Minera Yanacocha, Perú

Francisco Figueroa Roberto and Hamer Arévalo Lara

Abstract It is unquestionable that natural resources including metals and minerals are being depleted globally, with remaining metal resources being increasingly difficult to extract and recover. Base metals, such as copper that were once plentiful near-surface as easily extracted oxides and sometimes as native metals, are now present at greater depths and distributed in more difficult to extract primary, refractory, and complex sulfide ores. For copper, substantial deposits remain of primary sulfides like chalcopyrite, enargite, and bornite; while chalcopyrite and bornite can be processed by conventional methods, enargite requires alternative mineral processing strategies including flotation concentration, roasting, pressure oxidation, and bioleaching at elevated temperatures. This chapter will describe the commercial-scale demonstration of enargite bioleaching at the Yanacocha mine in northern Perú.

Keywords Enargite · Bioleach · Primary sulfide · Copper

10.1 Enargite as a Copper Resource

The primary copper sulfide enargite (Cu_3AsS_4) is emerging as a major form of copper mineralisation as mining operations advance to greater depths in South America, the Philippines, Papua New Guinea, Bulgaria, and other mining areas of the world. Like the more abundant primary copper sulfide chalcopyrite ($CuFeS_2$), process options for enargite include production of a flotation concentrate followed by other pyrometallurgical or hydrometallurgical steps to extract and recover copper (discussed in the next section). Enargite poses an additional complication due to the presence of arsenic, which reduces the value of enargite concentrate destined for

F. F. Roberto (✉)
Technical Services Processing and Metallurgy, Newmont USA Limited, Denver, CO, USA
e-mail: frank.roberto@newmont.com

H. Arévalo Lara
Minera Peñasquito, Mazapil, Zacatecas, Mexico
e-mail: hamer.arevalo@newmont.com

D. B. Johnson et al. (eds.), *Biomining Technologies*,
https://doi.org/10.1007/978-3-031-05382-5_10

smelting (arsenic is a so-called penalty element), and the decision by China in 2006 to bar importation of concentrates containing >0.5% As effectively created an international limit for saleable arsenic-containing concentrates. Liberation of arsenic during hydrometallurgical processing also requires additional measures to recover, stabilise, and sequester arsenic in tailings impoundments, often as a ferric arsenate compound such as scorodite. Capture of arsenic from pyrometallurgical operations like roasting, pressure oxidation, or smelting may produce arsenic trioxide, a compound once widely used in the preparation of chromated copper arsenate (CCA) to pressure treat and protect wood products from insects and fungal decomposition. Use of CCA was voluntarily discontinued in the United States in products for sale to the general public in 2004 and consequently the global market for arsenic trioxide has contracted more than 50 percent.

10.2 Alternatives for Copper Recovery from Enargite

Several pyrometallurgical processes exist for recovery of copper from enargite (Conner and Anderson 2013; Safarzadeh et al. 2014). These process options include partial roasting (developed by Outotec) where roasting at 600–750 °C produces a low-arsenic copper calcine that can be smelted to form arsenic trioxide as a by-product.

The Namibia Custom Smelter (Dundee Precious Metals—Tsumeb) is a custom smelter that processes high-arsenic gold and copper concentrates from around the world. A subsidiary of Dundee Corporation, Dundee Sustainable Technologies (DST) developed and demonstrated a trademarked vitrification process, GlassLock™, to sequester arsenic in a stable waste form containing up to 20% As.

Atmospheric leaching of enargite can be accomplished with sodium hypochlorite under alkaline conditions releasing arsenic as anionic arsenate and copper as copper oxide. Copper and arsenic are dissolved under acidic conditions and slightly elevated temperature (60–95 °C) by ferric sulfate. At higher temperatures (170 °C) in acidic sulfate and chloride, enargite is completely dissolved, while lower temperatures result in the passivation of enargite particles with elemental sulfur. Enargite can also dissolve slowly in ammonia solutions.

High temperature (220 °C) pressure oxidation in an oxygen-enriched atmosphere oxidises sulfides to sulfates producing sulfuric acid and liberating copper and arsenic. The latter can be removed by precipitation as scorodite. This process has been considered to be sub-economic for copper concentrates unless there is appreciable gold content which can be recovered by subsequent cyanidation of the washed and neutralised residue, despite reported high copper recovery either through production of a low-arsenic concentrate or addition of pyrite to the enargite. However, it should be noted that Freeport McMoran's Morenci mine in Arizona has found pressure oxidation to be profitable through the production of sulfuric acid as a by-product.

10.3 Bioleaching of Enargite

Bioleaching of enargite, like chalcopyrite, is thermodynamically less favourable than bioleaching of chalcocite or pyrite. As a consequence, even the suite of microorganisms suited to bioleaching of these primary copper sulfides is different, with higher growth temperatures and preference for using reduced sulfur rather than iron as electron donor seemingly being superior in bioleaching enargite. Early studies showed the greater efficiency of the thermo-acidophile, *Sulfuracidifex metallicus* (formerly *Sulfolobus acidocaldarius*) strain BC, to extract copper from enargite. Contradicting these results, some researchers had success using an arsenic-tolerant strain of the mesophile *Acidithiobacillus ferrooxidans*. Nevertheless, it is generally accepted that enargite (like chalcopyrite) bioleaching under thermophilic growth conditions proceeds at higher rates (Muñoz et al. 2006; Takatsugi et al. 2011).

Newmont and other mining companies have examined the potential for bioleaching enargite ores and concentrates to reduce the high capital and energy intensity of alternative processing approaches such as roasting, smelting, pressure oxidation, and alternative lixiviants as briefly described in the preceding section. Bioleaching, even under thermophilic growth conditions, has the added advantage of retaining arsenic in solution rather than generating an off-gas containing arsenic trioxide. Following on from earlier column leach work with copper ores containing secondary copper sulfide minerals and lesser quantities of enargite (Acar et al. 2005), Lee et al. (2011) compared the temperature dependency of copper bioleaching of chalcocite (Cu_2S), covellite (CuS), and enargite-dominated ore composites using consortia of mesophilic bacteria and thermophilic archaea. Bioleaching was performed at 20–22 °C and 65 °C with two composites of each copper mineralogy and monitored for 300 days. All composites leached effectively (>80% Cu extraction) at 65 °C, but little copper was extracted from the covellite or enargite ore composites at ambient temperatures. Copper extraction ranged from 81 to 98%, with an average of 92%, average iron extraction of 37%, and average sulfide-S oxidation of 47%. Copper minerals in the bioleached column residues were examined using Mineral Liberation Analysis (MLA) and showed that chalcocite and pyrite were oxidised under mesophilic conditions, but covellite and enargite were not altered unless thermophilic conditions were imposed. An extended 160 day lag period was observed for both covellite and enargite leaching which was relieved by multiple water washes of the ore within the columns and reinoculation with the thermophilic consortium. This preliminary work informed an extensive column bioleach campaign which included an additional 24 master composites to examine the influence of temperature, copper, and gangue mineralogy (primarily pyrophyllite, a phyllosilicate clay mineral) on copper extraction from enargite, as well as to gain information on projected effluent water quality (pH, dissolved metals, and sulfate).

10.4 The Verde Bioleach Demonstration Facility

The Yanacocha mine (Minera Yanacocha S.R.L., now solely-owned by Newmont Corporation) is located approximately 600 km northwest of Lima and 125 km east of the Pacific Ocean in the Cajamarca region of northern Perú. The mine is one of the richest high-sulfidation epithermal gold deposits in the world (Teal and Benavides 2010), producing more than 37 million ounces (1150 tonnes) of gold since commercial operation began in 1993 (with a peak production of 3.3 million ounces—93.5 tonnes—in 2005). Yanacocha can no longer claim to be the largest gold mine (by production) in South America and declining gold grades and increasing copper mineralisation with depth have led to a focus on the Yanacocha Sulfides (Yanacocha Sulfuros[1]) project for the future. Enargite and pyrite dominate the mineralogy of deeper regions of the mine, resembling that of other enargite-rich deposits such as Cerro de Pasco in Perú, Zijinshan in China, Motomboto in Indonesia, El Indio-Tambo and the copper porphyry orebodies of Collahausi, Escondida, and Chuquicamata in Chile.

Sustained testing from the early 2000s to 2010 reinforced the feasibility of bioleaching enargite-dominant sulfides from the Yanacocha Verde deposit. Concurrent studies concluded that it would be difficult to produce a saleable concentrate from this ore due to the high probable concentration of arsenic originating in enargite which is difficult to separate from other copper sulfide minerals through flotation. Early work indicated the potential for a significant deposit with over 2 billion lb of copper that would transition the mine from primarily a gold and silver producer to a copper mine with appreciable gold and silver by-products.

Planning for the pilot leach facility included a small SX-EW plant to confirm copper recovery subsequent to bioleaching as copper cathode. An existing waste dump site at Yanacocha with ready access to power, water, and acid water treatment was selected for construction of the demonstration facility.

10.4.1 Process Flowsheet and Design Criteria

As originally envisioned, the bioleach demonstration facility would comprise an ore crushing circuit, microbial inoculum production (bioreactor) plant, agglomeration drum, lined leach pads, air blowers, and distribution network, drip irrigation circuit flowing from the raffinate pond to the heap and back to the PLS, or rich solution pond, and solvent extraction-electrowinning (SX-EW) plant. Due to high annual precipitation (2550 mm) and regular rainy season, it was anticipated that a solution bleed would be necessary to account for the positive water balance. The crushing circuit was designed to provide run-of-mine (ROM), and $P_{80} = 37$ mm and $P_{80} = 12$ mm particle sizes produced by sequential jaw crushing followed by

[1] bnamericas.com accessed November 7, 2021.

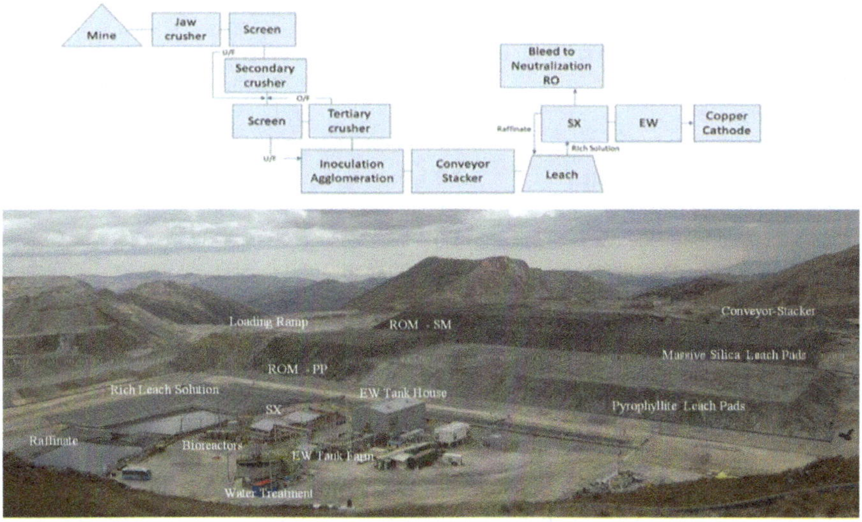

Fig. 10.1. Flowsheet for the enargite bioleach and actual layout of the bioleach demonstration facility at YanocOcha

gyratory crushing and screening to generate the smaller particle sizes. The ROM was truck-dumped and the smaller particle sizes were conveyor-stacked after agglomeration with the microbial inoculum. The ore was stacked in 2 lifts of 7.5 m in height each. Two parallel heaps were stacked to examine the influence of pyrophyllite ("PP" in this case study) content on leaching efficiency and geotechnical stability compared to ore considered to be of massive silica ("SM") alteration. Pyrophyllite (geologic abbreviation "Prl") is an aluminosilicate mineral sometimes associated with hydrothermal deposits and minerals like andalusite. The clay-like properties of pyrophyllite were a concern given the acidic conditions of the bioleach operation and the potential for swelling and other geotechnical issues. A schematic representation of the flowsheet and image of the completed demonstration facility are shown in Fig. 10.1.

The design criteria for the demonstration plant, originally planned to operate for 26 months are listed in Table 10.1

The heap leach pads were irrigated using a drip emitter array laid out over the top of each pad. Metal-rich solution was collected from the combined drainage of all PP and SM leach cells (each cell's drainage was sampled independently prior to collection in the rich solution pond for metal accounting) in a 10,000-m^3 lined pond prior to feeding into the small pilot-scale SX-EW plant for copper recovery. Spent solution was discharged into the 3000 m^3 raffinate pond after solvent extraction (Fig. 10.1).

Table 10.1. Design criteria for verde bioleach demonstration facility

Enargite bioleach key design criteria	
Total ore under leach	1 million tonnes (670 kt first lift, 330 kt second lift)
Copper ore grade	~ 0.5%
Copper recovery (projected)	40% ROM, 60% 37 mm, 70% 12 mm
Leach solution application rate	7.3 L hr^{-1} m^{-2}
Process plant flow rate (SX)	190 m^{-3} hr^{-1}
Rich leach solution grade (Cu)	1.0 g L^{-1}
Copper production (26 months)	3.8 million lb (1700 tonnes)

10.4.2 Dump and Heap Construction

The base of each parallel leach pad was 92 m × 276 m (each ore "cell" was 92 m long) with 7.5 m lift height (total height of 15 m). The underlying impermeable geotextile membrane was sloped to permit gravity collection of solution from the base of each cell. The crushed ore cells were stacked by conveyor, while the ROM ore was dumped via a truck ramp (the location and relative positions of the cells are indicated in Fig. 10.1).

While the original mine plan and heap design accommodated two full-height SM-ROM and PP-ROM cells, and two sequentially stacked lifts each of the SM-37 mm, SM-12 mm, PP-37 mm, and PP-12 mm crushed ore stockpiles, geotechnical instability of the first stacked 12 mm ore lift led to the decision to only stack SM- 37 mm and PP-37 mm ore after the first lift of PP-12 mm ore. Therefore, a total of 208.5 kt of SM-ROM, 202.3 kt SM-37 mm, 120 kt SM-12 mm, 219.7 kt PP-ROM, 194.2kt PP-37 mm, and 142.1kt PP-12 mm ore were stacked. Altogether, 1.09 million tonnes of enargite-dominant copper sulfide ore were placed in the two parallel leach pads in 6 distinct cells.

Air distribution for the crushed ore cells was provided by two continuously operated blowers through large-diameter manifold piping from which regularly spaced 76 mm diameter distribution pipes perforated approximately every 1 m (3 mm diameter perforations) were laid within a gravel bed at the base of each crushed ore cell (Fig. 10.2). The SM-ROM pad can be seen directly to the left, illustrating the heterogeneity in particle size and potential for large boulder-sized "particles" in the SM-ROM ore.

10.4.3 Inoculation and Commissioning of Bioleach Demonstration Plant

The bioleach inoculum (a mixture of mesophilic and moderately thermophilic bacteria and thermophilic archaea) was cultivated in a bioreactor circuit comprising four 15 m^3 stainless steel tanks that could be operated independently at ambient or

Fig. 10.2 Air distribution pipe at the base of a crushed ore cell during installation (left) and the drip emitter array at Yanacocha (right)

elevated temperature to produce the blend of microorganisms required. Bioreactor solutions from intermediate laboratory-scale bioreactors operated in the metallurgical services laboratory at Yanacocha were stockpiled for initial inoculation of the 15-m^3 bioreactors at the demonstration facility. Agitation was provided only by air bubble turbulence associated with large Wilfley diffusers attached to ring aerators installed near the bottom of the tanks. Modified Kelly Medium (MKM; 0.4 g L^{-1} $(NH_4)_2SO_4$, 0.4 g L^{-1} $MgSO_4 \cdot 7H_2O$, 0.04 g L^{-1} K_2HPO_4, pH 2.0; Lee et al. 2011) containing 6 g L^{-1} $FeSO_4$ as a reduced energy source, was used throughout the demonstration, supplemented with 0.02% yeast extract to promote growth of moderately thermophilic and thermophilic species in the bioreactors dedicated to their cultivation.

The bioreactor plant was commissioned in early 2013 using electricity produced by a diesel genset. The poor reliability of the genset led to several power outages that resulted in loss of heating and aeration sometimes for several days early in production. Inferior quality materials of construction and assembly of heater units also resulted in unexpected failures which further delayed production of the thermophilic inoculum and an extended period where the ambient temperatures ranged from 2 to 7 °C daily. As a consequence, microbial analysis of stockpiled inoculum from the first few months of operation showed *Leptospirillum ferriphilum* to be dominant. Inoculum from the bioreactor plant was used to agglomerate crushed ore prior to delivery to the leach pads by conveyor (see Fig. 10.1). Once stable operation of the bioreactor plant resumed, most production focused on moderate thermophiles and archaeal species as primarily mesophilic inoculum had been used for several months and mesophilic iron-oxidising bacteria were quantified in high numbers in the leach solutions. Over the duration of the demonstration, the bioreactor plant produced more than 6 million litres of microbial inoculum used initially for agglomeration of the stacked 37 mm and 12 mm ore and subsequently for continuous inoculation of the leach cells (including ROM pads). Bioreactor solution was discharged to the spent solution (raffinate) pond for distribution onto the leach pads.

10.4.4 Operational Performance

A preliminary description of the project design and commissioning was presented by Roberto and Arévalo (2015). Irrigation of the SM-ROM pad commenced in March 2013 and the first copper cathode was produced in September 2013, but other leach pads were not placed under leach until November 2013 due to delays in construction and commissioning of the SX-EW plant. Each leach pad was planned to be leached initially for 270 days, followed by secondary leach cycles for the bottom lifts of the crushed ore heaps as additional ore was stacked. By the end of 2014, 909 tonnes (2.0 million pounds) of copper cathode had been produced. An additional 650 tonnes (1.43 million pounds) had been produced by November 2015. A total of 2670 tonnes (5.87 million pounds) of copper cathode was produced by the conclusion of the demonstration in 2017. It should be noted that, accounting for copper in solution (both circulating leach/process solutions and bleed to water treatment), an additional 742 tonnes of copper (1.63 million pounds) was estimated to have been leached from all leach pads for a total of 3412 tonnes (7.51 million pounds) of copper extracted from the SM and PP ores. Irrigation of the heaps was terminated in January 2017, and production of cathode copper ceased in June 2017.

A key performance requirement for enargite leaching that was established during the laboratory column leach studies was heap leach temperatures above 50 °C. This heat was generated by the oxidation of pyrite and other sulfide minerals. Average leach cell temperatures at the end of 2014 ranged from 20 °C for the leach pad with shortest time under leach (60 days) to 41 °C for the SM-37 mm leach pad under leach for 286 days. It should be noted that all pads achieved at least 50 °C at some point while under leach, and that average leach cell temperatures represent the average of all Resistance Temperature Detector (RTD) temperature measurements reported from each heap. As built, each lift was constructed with 15 RTDs distributed in 5 PVC temperature wells buried within the ore in a 5-spot "star" pattern, each containing a sensor near the bottom, middle, and top of the heap). Unfortunately, data were not consistently or reliably obtained due to frequent lightning strikes which damaged the entire measurement network. A hand-held IR temperature gun was used to confirm continued high temperatures above 50 °C near the surface of the heaps, but since this was limited to line of sight down the temperature wells, this did not extend more than about 3 m below the leach pad surface. As second lifts were stacked, the lower lift temperatures were observed to cool down (through the RTD network) with respect to temperatures measured in the upper lifts, probably as a result of additional compaction and reduced aeration, in addition to depletion of the sulfide minerals.

Average heap internal temperatures had increased to between 31 and 49 °C by November 2015, several months after a supplemental neutralisation plant had been commissioned. Initial plans to incorporate a neutralisation plant in the process design were abandoned for financial reasons, but a consequence of this decision was that a bleed of solution was insufficient to prevent extremes of pH (~ 0.6), free acid (>15 g L^{-1}) and concentrations of total soluble iron exceeding 60 g L^{-1} to build

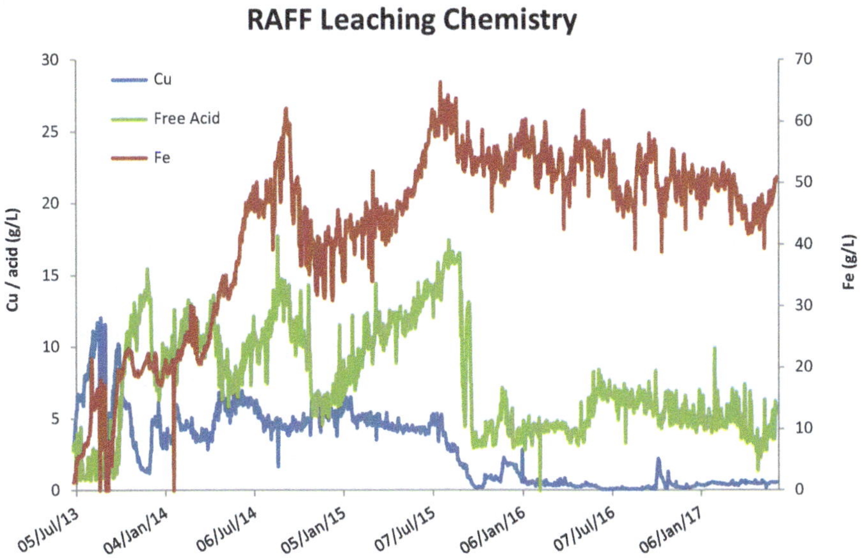

Fig. 10.3 Changes in raffinate chemistry during the operation of the pilot-scale test at Yanococha. The introduction of the neutralisation plant in September 2015 restored target copper concentrations in the rich solution to ~1 g L^{-1}

up in the leach circuit (Fig. 10.3), and the neutralisation plant was unable to bring total dissolved iron concentrations to below ~50 g L^{-1}. Continuous addition of aqueous calcium hydroxide ("milk of lime") in the neutralisation plant was able to raise the pH and lower the free acid concentration but large quantities of sludge (largely gypsum and ferric iron salts) were produced despite the pH remaining below 2.0. It was noted that ferrous iron concentrations in the leach solutions never exceeded 5 g L^{-1} and were rarely >2 g L^{-1}, so virtually all of the soluble iron was present in the oxidised ferric state.

These extreme conditions were noted to negatively impact total cell numbers in the leach solution (Fig. 10.4), and also alter the microbial composition substantially as described below. These compounded effects on microbial numbers and composition were ultimately deemed to have delayed and inhibited the heating of the leach pads through sulfide oxidation.

Installation of the neutralisation plant and raw water additions to the heaps to further dilute the high ionic strength of the irrigation solution took place in late 2015. These corrective measures designed to improve the solution chemistry, coupled with halting the leaching of the 12-mm SM and PP heaps, also resulted in bringing the Cu concentration in solution within the original design criterion of 1 g L^{-1}. Ultimately only the SM-12 mm first lift was placed. Average heap temperatures were also observed to increase with the improved solution chemistry, leading to increased enargite leaching. The increased heap temperatures also aided in controlling soluble arsenic concentrations and increased sequestration of arsenic in iron–arsenic precipitates such as scorodite.

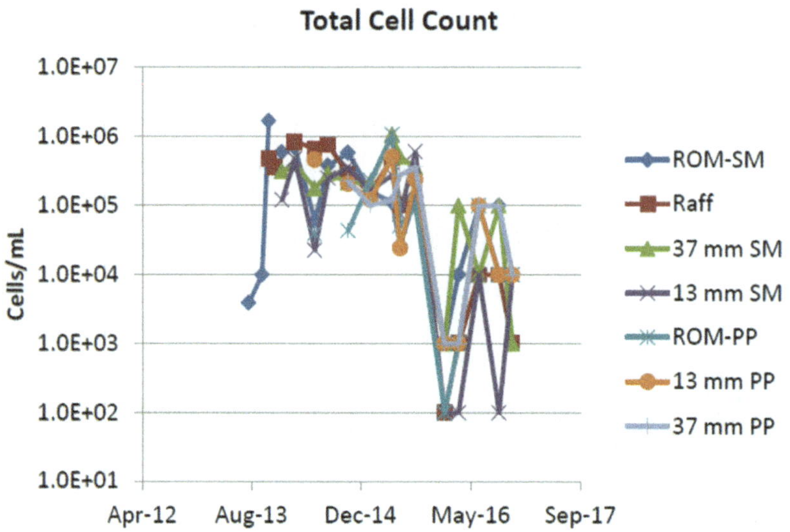

Fig. 10.4 Viable mesophilic iron-oxidising bacteria cells in solution from different leach cells and raffinate ponds

Measurements of copper, arsenic, and sulfide-sulfur in the ore found that, contrary to the initial project resource model and mine plan, the SM leach pad contained 14% more copper, 11% more sulfide-sulfur, and 27% more arsenic than the initial design. Head assays, post-leach drilling, and Cu/As mass balances estimated a copper content of 0.77% with 59% of the copper mineralisation as enargite. The SM-ROM heap actually had an estimated grade of 0.87% Cu, with 81% copper mineralisation as enargite, and the PP leach pad contained 12% more Cu, 14% more As, and 1% more sulfide-sulfur than had been anticipated (overall Cu grade of 0.52%, 31% from enargite). The overall copper grade of the combined heaps was 0.64%, which was close to the original design criteria. Secondary copper sulfides accounted for more copper than originally considered (present in chalcocite and covellite) which leached quickly and led to much higher rich solution copper tenors and the need to expand the electrowinning circuit from an original design capacity of 3.2 tonnes d^{-1} (7040 pounds d^{-1}) to 5.3 tonnes d^{-1} (11,660 pounds d^{-1}) to accommodate the additional copper in solution.

Final copper extraction calculated from head assays, post-heap residue mineralogy, and solution mass balances, was estimated to be 52–95% for the secondary copper sulfides chalcocite and covellite and 30–60% from enargite, depending on which leach cell and alteration type was considered. No size dependence was evident. While laboratory test work demonstrated strong temperature dependence for enargite bioleaching, the lack of reliable temperature measurement throughout the life of the project precluded making any strong conclusions about the influence of internal heap temperatures on final copper extraction efficiency from enargite in each leach cell, although in general lower observed temperatures, when these occurred, were linked to lower extraction of copper from enargite. Initial pyrite concentrations ranged from 8.4 to 11.7% and post-leach drilling indicated pyrite oxidation ranged from 37 to 56%.

First drilling of leach pads in October 2016 permitted examination of ore and residues and microbiological analysis. Final drilling in 2017 indicated additional copper leaching continued as solution flow was gradually reduced and copper tenors reduced to permit a safe shutdown of the EW plant.

Over the 34 months of active leaching, solution samples were collected to allow monitoring of the microbial populations. It is generally accepted that planktonic cells may only account for a small proportion of the overall bioleaching communities in the heaps, but sampling actively leaching heaps poses operational and safety concerns that cannot be ignored. Ore samples were not collected until October 2016 just prior to the end of active leaching. Total viable counts of mesophilic iron oxidisers declined as the solution chemistry became more extreme, with some evidence of rebound when neutralisation and raw water irrigation were commenced towards the end of 2015. However, as shown in Fig. 10.4, there was nearly a 4-order of magnitude decline in viable cell numbers prior to improvements in leach solution chemistry, and an overall ~3-order magnitude decline in those organisms over the course of the project. Attempts were made to quantify viable moderately thermophilic iron-oxidising bacteria and thermophilic archaea in the same samples, but plate counts rarely returned more than 10–100 cells mL^{-1}. 16S rRNA gene library analysis of solution samples (Roberto 2018) showed initial dominance of *Acidithiobacillus* sequences which were replaced by *Leptospirillum* within the first 6 months of leaching, after which *Leptospirillum* was the only mesophilic sequence detected above 1% of the identified sequence. It is likely that *L. ferriphilum* was present, as some strains are considered to be thermotolerant up to ~48 °C. *Sulfobacillus* sequences were identified in all samples but less than 1% of the total sequences recovered. In parallel, archaeal libraries showed that less than 1% of the total sequences identified were associated with *Acidianus* and *Thermogymnomonas* spp. while over 98% of the archaeal sequences recovered through the life of the project were identified as *Ferroplasma* spp.

Ore residue samples were similarly analysed and showed quite different results. Among bacterial sequences recovered, *Acidithiobacillus* sequences dominated (up to 94.5% of the identified sequences) the various drill holes sampled near the surface, in the middle of the lift, and at the bottom while *Leptospirillum* represented no more than 1.4%. Sequences of the moderately thermophilic bacteria *Sulfobacillus* and *Ferrimicrobium* represented as much as 65.9% of the identified sequences in some samples. In parallel archaeal libraries, *Acidiplasma*, *Thermogymnomonas*, and *Thermoplasma* sequences represented small (less than 1%) proportions of the identified sequences which were dominated by *Ferroplasma* spp., similar to the leachate samples.

While these results are limited, they do suggest that despite the inability to detect common iron-oxidising bacteria that would be expected to be active during sulfide mineral oxidation at low temperatures using solution samples, those microorganisms are present on the ore itself throughout the life of the project. The dominance of *Leptospirillum* in leachates appears not to be the case on the surface of the ore particles, and this apparent discrepancy is probably explained by that organism's much higher tolerance for ferric iron (and greater affinity for ferrous iron) with the possibility of different local chemistry at the ore surface. Likewise, analysis of the ore residues revealed that moderately thermophilic species were present on the ore despite the inability to detect them in solution. Finally, analysis of the archaeal communities in leachates and on solid samples agrees with the conclusion that the heap leach environment of this demonstration project favoured the growth of *Ferroplasma*. Since it was difficult to demonstrate that internal heap temperatures achieved the high temperatures initially considered desirable to enargite leaching (i.e., >50 °C) it is not unreasonable that more active archaeal iron- and sulfur-oxidising microorganisms like *Sulfolobus, Sulfuracidifex, Acidianus*, and *Metallosphaera* were not detected. It should be noted that seeps occasionally noted with spontaneous ore heating on cyanide heap leach pads at Yanacocha (visible as localised steam venting from the heaps on cold days) have previously been the source of *Ferroplasma* at the mine, so it was not surprising to identify the microbe during microscopic examinations of the enargite bioleach solutions and in the 16S rRNA gene libraries. It was, however, surprising that it emerged to dominate the archaeal sequences detected from the very beginning of the project, despite efforts to inoculate the heaps with strains of the more thermophilic archaea *Acidianus, Sulfolobus*, and *Metallosphaera* as was done in laboratory column test work. However, early problems with reliability of the bioreactor heaters (requiring shut down and replacement of all the heaters that were initially installed, which sometimes took several weeks) also likely influenced the populations of thermo-acidophilic archaea inoculated onto the heaps, as it was not always possible to restart the bioreactors with fresh inoculum from the laboratory at Yanacocha.

10.4.5 Lessons Learned

The Verde Bioleach Demonstration Facility (VBDF) has taken small-scale column leach test work to a small industrial scale, 1 million tonne demonstration, of the feasibility of bioleaching enargite. Copper recovery as copper cathode was proved and exceeded the original design production for the plant due to amounts of secondary copper sulfides exceeding the original design criterion.

A decision made during the engineering design of the demonstration to eliminate a neutralisation circuit for cost reasons was likely to have had a major impact on the overall success of the project, as extreme leach chemistry ultimately led to construction of the neutralisation plant within a year of full operation. Discrepancies in ore characterisation also resulted in some leach pads containing much higher copper grades than initially planned, and more secondary copper sulfides which also contributed to solution copper concentrations exceeding design by fivefold, requiring additional electrowinning capacity to be added mid-project. Despite the expensive modifications made to the facility after commissioning, the solution chemistry remained extreme with respect to pH (lime supply to adjust the pH higher was limited by the capacity of the local Newmont-owned slaker) and total dissolved iron concentration and inhibitory to the desired microbial populations which had been previously demonstrated to be effective in the laboratory.

It is difficult to assess how the dominance of *Ferroplasma* impacted the overall rates of enargite bioleaching. *Ferroplasma* spp. have temperature optima between 42–50 °C and may tolerate temperatures as high as 60 °C. In addition, these archaeal species have pH optima between 0 and 1.7 (Golyshina and Timmis 2005). The closest industrial analogue to VBDF at this time is the Zijinshan copper mine, where similar extremes of pH and total iron have been observed (Ruan et al. 2011). Similar dominance of the solution microbial populations by *Leptospirillum* and *Ferroplasma* have also been observed there. Emergence of sulfur-oxidising, moderately thermophilic bacteria was not observed at VBDF as was seen at Zijinshan, though leach solution temperatures in the VBDF project never achieved the high temperatures (>50 °C) observed regularly at Zijinshan. It is also possible that heterotrophic species of *Ferroplasma* were able to out-compete the sulfur-oxidisers by utilising carbon derived from autotrophic bacteria or organics from the SX-EW circuit. Other notable differences between the two operations include lower copper grade (0.4% vs 0.64%), lower pyrite content (5.8% vs 8.4–11.7%), lack of active aeration, and truck dumping of the ores at Zijinshan. Nevertheless, similar extremes of pH and dissolved iron concentrations have seen the introduction of a neutralisation plant at Zijinshan as well to control the total iron in solution.

Unlike Zijinshan, where secondary copper sulfides represent the bulk of the copper mineralisation, VBDF was specifically designed to demonstrate bioleaching of enargite at commercial scale. Despite the challenges encountered during operation of the demonstration, it was considered a success in that regard, and generated information that will be essential to full-scale commercial operation of copper bioleaching at Yanacocha, and other sites where enargite is a significant copper mineral, in the future.

Acknowledgements The authors gratefully acknowledge the many talented Newmont technical staff and operators in the USA and Perú who contributed to the success of this project between 2011 and 2017. The insights and access to detailed heap leach analyses provided by Dr. Matt Jeffrey have been particularly helpful to the compilation of this case study.

Author Note Newmont Corporation strives to demonstrate its commitment to diversity and inclusion in many different ways. This chapter has been written to reflect a recent initiative to remove several historic mining terms in English that have been identified as insensitive to inclusion, namely replacing the term "PLS, or pregnant (leach) solution" with "rich (leach) solution", and "barren solution" with "spent solution". Newmont is pleased that its business partners are embracing and adopting this more inclusive language.

References

Acar S, Brierley JA, Wan R-Y (2005) Conditions for bioleaching a covellite-bearing ore. Hydrometallurgy 77:239–246. https://doi.org/10.1016/j.hydromet.2004.05.004

Conner K, Anderson C (2013) Enargite treatments and pressure oxidation of concentrates. J Met Eng (ME) 2:115–123

Golyshina OV, Timmis KN (2005) *Ferroplasma* and relatives, recently discovered cell wall-lacking archaea making a living in extremely acid, heavy metal-rich environments. Environ Microbiol 7:1277–1288. https://doi.org/10.1111/j.1462-2920.2005.00861.x

Lee J, Acar S, Doerr DL, Brierley JA (2011) Comparative bioleaching and mineralogy of composited sulfide ores containing enargite, covellite and chalcocite by mesophilic and thermophilic microorganisms. Hydrometallurgy 105:213–221. https://doi.org/10.1016/j.hydromet.2010.10.001

Muñoz JA, Blazquez ML, Gonzalez F, Ballester A, Acevedo F, Gentina JC, Gonzalez P (2006) Electrochemical study of enargite bioleaching by mesophilic and thermophilic microorganisms. Hydrometallurgy 84:175–186. https://doi.org/10.1016/j.hydromet.2006.05.012

Roberto FF, Arevalo H (2015) The Verde bioleach demonstration facility: a 1million tonne test of enargite bioleaching in northern Peru. In: Mubarok MZ, Chaerun SK, Minwal WP, Muhammad F, Pratama K (eds) Proceedings of the 21st international biohydrometallurgy symposium. Institut Teknologi Bandung, Indonesia

Roberto FF (2018) 16S-rRNA gene-targeted amplicon sequence analysis of an enargite-dominant bioleach demonstration in Peru. Hydrometallurgy 180:271–276. https://doi.org/10.1016/j.hydromet.2018.08.005

Ruan R, Liu X, Zou G, Chen J, Wen J, Wang D (2011) Industrial practice of a distinct bioleaching system operated at low pH, high ferric concentration, elevated temperature, and low redox potential for secondary copper sulfide. Hydrometallurgy 108:130–135. https://doi.org/10.1016/j.hydromet.2011.03.008

Safarzadeh MS, Moats MS, Miller JD (2014) Recent trends in the processing of enargite concentrates. Miner Process Extract Metall Rev 35:283–367. https://doi.org/10.1080/08827508.2012.723651

Takatsugi K, Sasaki K, Hirajima T (2011) Mechanism of the enhancement of bioleaching of copper from enargite by thermophilic iron-oxidizing archaea with the concomitant precipitation of arsenic. Hydrometallurgy 109:90–96. https://doi.org/10.1016/j.hydromet.2011.05.013

Teal L, Benavides A (2010) History and geologic overview of the Yanacocha mining district, Cajamarca, Peru. Econ Geol 105:1173–1190. https://doi.org/10.2113/econgeo.105.7.1173

Chapter 11
Biooxidation of Gold Ores in Russia and Kazakhstan

Aleksandr Vasilyevich Belyi and Olga Vladimirovna Tupikina

Abstract Bioprocessing of refractory gold-bearing sulfides is represented in the Commonwealth of Independent States (CIS) by two technologies: BIOX® and BIONORD®. This chapter focuses on the Suzdal mine in Kazakhstan owned by Nordgold and the Olimpiada mine in Russia owned by Polyus. The Suzdal mining operation is amongst the most technologically advanced of the Nordgold operations. It currently utilises three Outotec-licensed processes: the Bacterial Oxidation (BIOX™) process for the liberation of sulfide-entrained refractory gold, the Activated Sludge Tailings Effluent Remediation (ASTER™) process for the destruction of thiocyanate, and the High Temperature Caustic Conditioning (HiTeCC™) process for the reduction of gold losses due to preg-robbing. Suzdal is the first BIOX plant that operates at sub-zero ambient temperatures and has the second ASTER and HiTeCC implementations worldwide. Polyus is the largest gold producer in Russia, while the first industrial process for the biooxidation of refractory gold-bearing ores in Russia, BIONORD, was launched at the Olimpiada Gold mine in 2001. In subsequent years, three more plants were put into operation at the site: BIO-2, BIO-3, BIO-4. After commissioning the BIO-4 in October 2017, the total throughput capacity of the BIO plants grew to 1500 tonnes per day (t d^{-1}). The oxidation of sulfide flotation concentrates reached 90% and increased gold recovery to 94%. This helped the production of over 30 t of gold in 2019 using the BIONORD process.

Keywords Refractory sulfide ores · Gold · BIOX® · BIONORD® · Nordgold · Suzdal · Polyus · Olimpiada · ASTER™ · HiTeCC™

A. V. Belyi
JSC "Polyus Krasnoyarsk", Krasnoyarsk, Russia
e-mail: BelyiAV@polyus.com

O. V. Tupikina (✉)
Nordgold, Moscow, Russia
e-mail: olga.tupikina@nordgold.com

D. B. Johnson et al. (eds.), *Biomining Technologies*,
https://doi.org/10.1007/978-3-031-05382-5_11

11.1 Introduction

The recovery of noble metals (gold in particular) from technologically refractory ores that are not treatable by conventional methods, is of particular concern in the Commonwealth of Independent States (CIS) as alluvial gold and oxidised ore reserves are being depleted and a strategic necessity for increasing the treatment of refractory gold-containing ores arises. Lode gold-containing ores represent about 40–60% of proven gold reserves.

Refractory gold ores contain fine gold particles intimately associated with sulfide minerals, and harmful/inhibitory impurities, such as arsenic, antimony, and organic carbon, which result in low gold recoveries when using conventional cyanidation technologies (10–60% gold recovery depending on concentrate composition). These impurities make such concentrates unsuitable for direct concentrate smelting and require special treatment methods for recovering noble metals, such as autoclave (pressure oxidation) or bacterial oxidation. All three processes are currently used worldwide. However, in the context of the worsening global environmental situation and increasing legislative and societal pressure, when selecting and developing effective technological processes for extracting gold from refractory materials, preference has been given to biohydrometallurgical methods: BIOX® in Kazakhstan and BIONORD® in Russia.

11.2 The Nordgold Suzdal Mine, Kazakhstan

The Suzdal mine site is located in east Kazakhstan, 55 km from Semey city (Semipalatinsk). The climate is classified as “warm summer continental” (Dfb; Köppen Climate Classification) with temperatures reaching minima of −50°C in winter, and maxima of +42°C in summer. The altitude varies between 320 and 345 metres above sea level.

The Suzdal gold deposit was discovered in 1983 by a Soviet geological exploration expedition. Suzdal is typically a shear-hosted mesothermal gold deposit of the Palaeozoic age, in which gold is hosted in sediments along fractures, and comprises both oxide and sulfide ores (Kovalev et al. 2012). Sulfide ores occur in massive terrigenous-carbonate rocks with carbonaceous matter and quartz-carbonate rocks. Siltstones, aleuropelites, and limestones contain laths, concretions, spots, and short veins of carbonaceous matter. Gold is locked within the sulfides (pyrite, arsenopyrite, and pyrrhotite) found as inclusions within the host rock, and measures 0.5–5.0 microns. The difference in the nature of the oxidised and sulfide ores requires the utilisation of different gold recovery processes. During 1999–2005, the oxide ore was mined by open-pit and then processed using the gold cyanidation heap leach method.

Nordgold acquired Suzdal in 2009 and it is now the most technologically advanced complex of the Nordgold operations. The processing plant includes crushing, grinding, flotation, BIOX, Counter Current Decantation (CCD), Carbon in Leach (CIL), ASTER, and HiTeCC units. The Suzdal BIOX processing circuit was the first of this type to be launched in Eurasia. The plant currently processes concentrates that can be classified as double refractory (refractory and preg-robbing).

11.2.1 The Suzdal BIOX Installation

Detailed engineering of the Suzdal BIOX plant commenced in early 2004 with Bateman Engineering, South Africa, as the contractor. The first module (BIOX Module I) was successfully commissioned in May 2005 and was designed to treat flotation concentrate at a nominal rate of 173 t d^{-1}, a sulfide-S grade of 12%, pulp density of 20% with $P_{80} = 75$ μm (and < 1% greater than 150 μm). A second BIOX module (Module II) was added to the operation in 2009. Both modules were designed and operated in the first years as a classical BIOX generation III circuit (Chap. 4) consisting of three primary reactors in parallel and three secondary reactors in series. Each reactor is 650 m^3. Due to the prevalence of pyrrhotite in the ore, the BIOX circuit configuration was later changed to four primary reactors and two secondary reactors. All of the BIOX reactors are located outside the main processing building and the internal temperature of the reactors is maintained between 39 and 41°C. During winter, the temperature in the final secondary reactor can decrease to 35°C, indicating complete sulfide oxidation.

At Suzdal, the biooxidation process takes around 5 days (120 h), the sulfide-S oxidation level is high, varying between 93 and 97% and yielding a cyanide leach gold recovery of 83–86%, compared to the 20–40% recovery if the concentrate were leached before BIOX treatment. Sulfide-S oxidation in the primary reactors reaches up to 80% resulting in 75–80% leach recovery. The operation and performance of the primary reactors are the key focus of the Suzdal BIOX operation. The Suzdal BIOX plant consumes 180–225 kg t^{-1} of concentrated sulfuric acid owing to a fairly high carbonate content in the ore and the prevalence of pyrrhotite. The pH in the primary reactors is controlled within the range of 1.4–1.6 through the addition of sulfuric acid. This systematically results in a decrease of the pH to 1.2 in the secondary reactors. Suzdal has successfully processed concentrates from other sites with different mineralogical compositions. Technical grade diammonium hydrogen phosphate, ammonium sulfate, and potassium sulfate are used as nutrients, with total additions of 10.3 kg t^{-1} of concentrate. This was decreased during later years of the operation, to 4.6 kg t^{-1} of concentrate.

11.2.1.1 The Suzdal BIOX Microbial Culture

In 2019, the Suzdal BIOX® microbial culture was analysed using two methods: real-time PCR (qPCR; Smart et al. 2017) and metagenomic analysis (Bulaev et al. 2017). Both methods showed similar results with respect to the proportion of key microorganisms in the population. However, the measured or inferred 16S rRNA gene copy numbers differed by several orders of magnitude. This could be due to different transport times and conditions of the samples during shipment to the Centre for Bioprocess Engineering Research, University of Cape Town in South Africa, and to the Winogradsky Institute of Microbiology, Research Center of Biotechnology of the Russian Academy of Sciences in Moscow, Russia. *Ferroplasma acidiphilum* was found to be the dominant species in all the Suzdal plant reactors, accounting for up to 95% of the total prokaryotic community. Species of the Genera *Acidithiobacillus* and *Leptospirillum* represented 0.5%–16% of the reactor microbial communities, during different time periods in 2019.

11.2.2 Suzdal ASTER Installation

The ASTER technology was implemented at the Suzdal mine in 2013 to remove cyanide (CN^-) and thiocyanate (SCN^-), which are both toxic for BIOX microorganisms, from decant water of historical cyanide tailings storage facilities (TSF) and to recycle the water for grinding, flotation, and BIOX processes. The circuit includes two aerated and heated 300 m^3 reactors, and one 51 m^3 settler. The design capacity of the circuit is 500 m^3 per day at 1200 ppm SCN^- in the feed. The circuit reached a stable operation of 500 m^3 d^{-1} with 1400 ppm SCN^- in the feed. The addition of a stock tank, a secondary reactor, and a secondary settler allowed for an increase in processing volume to 1380 m^3 d^{-1} with 3400 ppm SCN in the feed during the summer months of 2017.

The carbon source and nutrients used included 0.1–0.2 kg m^{-3} molasses and 0.1–0.2 kg m^{-3} diammonium hydrogen phosphate. The pH was maintained above pH 7 by the addition of sodium hydroxide, and 99% degradation of CN^- and SCN^- was generally achieved within 16–21 h. ASTER microorganisms form flocs that allow them to sink in a settler from where they can be recirculated back to the bioreactors. This allows reduction in the retention time that is limited only by degradation reactions. The reaction pathways of CN^- and SCN^- biodegradation and species of the ASTER microbial community are described in Chap. 4. While the usual optimal temperature range for the ASTER operation is 18–28 °C, the Suzdal ASTER has operated at a temperature range of 11–30 °C. The implementation of the ASTER™ process has resulted in an improved water balance, increased life of the cyanide TSF, and cost-effectiveness at the Suzdal mine operation.

11.2.3 Suzdal HiTeCC Installation

Preg-robbing is commonly described as the process whereby dissolved gold species (typically as the complex $Au(CN)^{2-}$) are adsorbed onto naturally occurring finely disseminated organic carbon particles contained in the ore. In most cases, the degree

Table 11.1 Basic Suzdal HiTeCC process design criteria (van Buuren et al. 2018)

Description	Value
Plant productivity	93.4%
Treatment rate	16 t $hr.^{-1}$
Liquid solid ratio	3
Ratio of CIL product to TSF solids	1
NaOH addition	30 kg t^{-1}
NaCN in feed	1–6 kg t^{-1} feed
Desorption reactors	3
Desorption temperature	80–85 °C
Adsorption reactors	3
Adsorption temperature	< 30 °C
Adsorption carbon	40 g L^{-1}
Desorption carbon	40 g L^{-1}
Reactor volume	260 m^3
HiTeCC feed	7.5 g t^{-1}
Gold recovery	40–70%

of adsorption is weak and preg-robbing is reversible in the presence of highly active adsorbents, such as activated carbon or resins. Gold is efficiently desorbed from loaded activated carbon compounds in the presence of an ionic solution (generally sodium hydroxide). High temperatures dramatically increase the rate of desorption as does the rate of desorbant (eluant) flow, and therefore sodium hydroxide was added and the pulp contained preg-robbed gold heated in order to facilitate release of the gold–cyanide complexes. This was the preferred treatment route for the BIOX process at Fosterville Gold Mine in Victoria which installed the first commercial circuit for the recovery of the preg-robbed gold in 2010 (Binks and Wemyss 2012).

In June 2016, Suzdal became the second mine worldwide to commission the innovative Outotec HiTeCC process to recover gold from both historical and current carbon in leach (CIL) tailings (van Buuren et al. 2018). In the tailings at the plant, most of the gold (50–75%) is occluded by organic carbon present in the ore. In collaboration with Outotec, all HiTeCC test work was conducted on-site. This resulted in a design to recover between 40 and 70% of gold from the CIL tailings. The Suzdal HiTeCC circuit operates on a 24-hour cycle. The desorption stage at 80–85 °C in the presence of sodium hydroxide occurs in the first 12 h followed by the 12 h adsorption stage at ambient temperature. Design parameters are shown in Table 11.1.

The plant reached design recovery in the first year of the operation The HiTeCC™ project increased the Suzdal production output by 9–14 kOz (255–397 kg) of gold doré per year.

11.3 The Polyus Olimpiada Mine, Russia

The industrial processing of refractory gold–arsenic ores is one of the immediate challenges for the gold extraction industry in the Russian Federation. Such ores contain up to 40–60% of current gold reserves (Borovkov et al. 1985; Serdyuk 1997). The issue is complicated by the fact that large gold–arsenic ore deposits are located in the hardest-to-reach areas of the far north of the country. In Russia, the leading extractor of gold from refractory ores is the ore production company at the Olimpiada deposit in Severo-Eniseysky District, Krasnoyarsk Krai. The Olimpiada deposit, located 600 km to the north of Krasnoyarsk, in Severo-Eniseysky District, was discovered in 1975 by Li (2003), and the field's geochemical parameters have been described by Kuzmin et al. (2000) and Genkin et al. (1994). Exploration of the field was completed in 1992 by the North Geological Survey Expedition of the Production Geological Association "Krasnoyarskgeologiya." Total C1 + C2 (estimated + inferred) category gold reserves amount to 417 t.

The primary ores of the Olimpiadinsky deposit are represented by massive and shale rocks of mica–quartz–carbonate composition with increased (up to 1%) carbonaceous matter content, hydrothermally modified to varying degrees (chloritisation, muscovitisation, calcification) and containing sulfides: pyrrhotite, arsenopyrite, antimonite (stibnite), and pyrite. According to the content of sulfides, there are two main types of primary ores: the main pyrrhotite-arsenopyrite with an admixture of antimonite and pyrite, and the secondary—essentially antimonite with a subordinate content of other minerals. Gold in the ore is mainly represented by fine and submicroscopic forms in close association with sulfide minerals—pyrrhotite (FeS), arsenopyrite (FeAsS), pyrite (FeS_2), and antimonite (Sb_2S_3). The average mineral content in the ore is 4% FeS, 1.45% FeAsS, 0.7% FeS_2, and 0.3% Sb_2S_3. The average gold grade is 3.2–3.5 g t^{-1}. The close association of gold with minerals is the reason for the persistence of the ores of the Olimpiadinsky deposit during their processing and cyanidation. The flotation concentrate produced has a sulfide content of up to 58%, and in the concentrate contains on average 23.1% FeS, 13.4% FeAsS, 10.7% FeS_2, and 4.5% Sb_2S_3.

11.3.1 History of Field Development

The field contains two ore types: oxide and primary ores. Primary ores are refractory gold–arsenic sulfide ores. The Polyus CJSC Olimpiada Mine commenced development of the Olimpiada gold ore deposit in 1996. Ore recovery from oxide ore was set up at Mill 1. The ore processing technology was developed and plants were constructed by Polyus CJSC with assistance from the Central Research Institute of Geological Prospecting for Base and Precious Metals (TsNIGRI) and Irkutsk research institute of precious and rare metals and diamonds (Irgiredmet) institutes.

The achieved oxide ore processing throughput capacity reached 1.5 Mt. y^{-1} using a direct cyanidation process.

In late 2001, Mill 2 was commissioned to process refractory primary ores from the Olimpiada deposit. The design capacity of the mill was 3 Mt. y^{-1}. However, oxide ore reserves of the Olimpiada deposit were coming to an end but the company's strategic development plan did not provide for a reduction in gold production at the Olimpiada Mine. The company's management resolved to maintain gold production at the previous level by processing higher volumes of primary ore. For this purpose, construction, and commissioning of Mill 3 with a design capacity of 5 Mt. y^{-1} was completed in 2007. In 2016 due to the depletion of all oxide ores, Mill 1 was transferred to processing primary ores. The total primary ore processing throughput of Mill 1, Mill 2, and Mill 3 reached 11.0 Mt. y^{-1}.

11.3.2 Technological Scheme of BIONORD® Processing

Primary ore is processed by gravity separation followed by flotation using the BIONORD® concentrate biological oxidation technology (Fig. 11.1), which is unique for high latitude climatic conditions.

The full process cycle consists of the following operations:

- Primary ore is ground to 100% passing 74 μm.
- Through flotation and slurry thickening, gold-containing sulfide concentrate is obtained from the ore.
- Flotation concentrate is reground to 100% <40 μm.
- In bioreactors, the concentrate is subject to biological oxidation using BIONORD® technology, containing a consortium of microorganisms isolated at the Olimpiada deposit.
- The concentrate biooxidation product is subject to sorption cyanidation (CIP) at the hydrometallurgical plant.
- Metal from the gold-containing solution obtained as a result of cyanidation is precipitated by electrolysis.
- The obtained gold is smelted into bullion.

11.3.3 Industrial Pilot Research on Biooxidation of Concentrates

11.3.3.1 The Pilot Biooxidation Plant Research

The key provisions for the biological oxidation technology for the ore concentrate used at the Olimpiadinskoe deposit were developed in the USSR in the 1970–1980s at a pilot plant with a concentrate throughput of 60 kg d^{-1} made by the Tula Branch

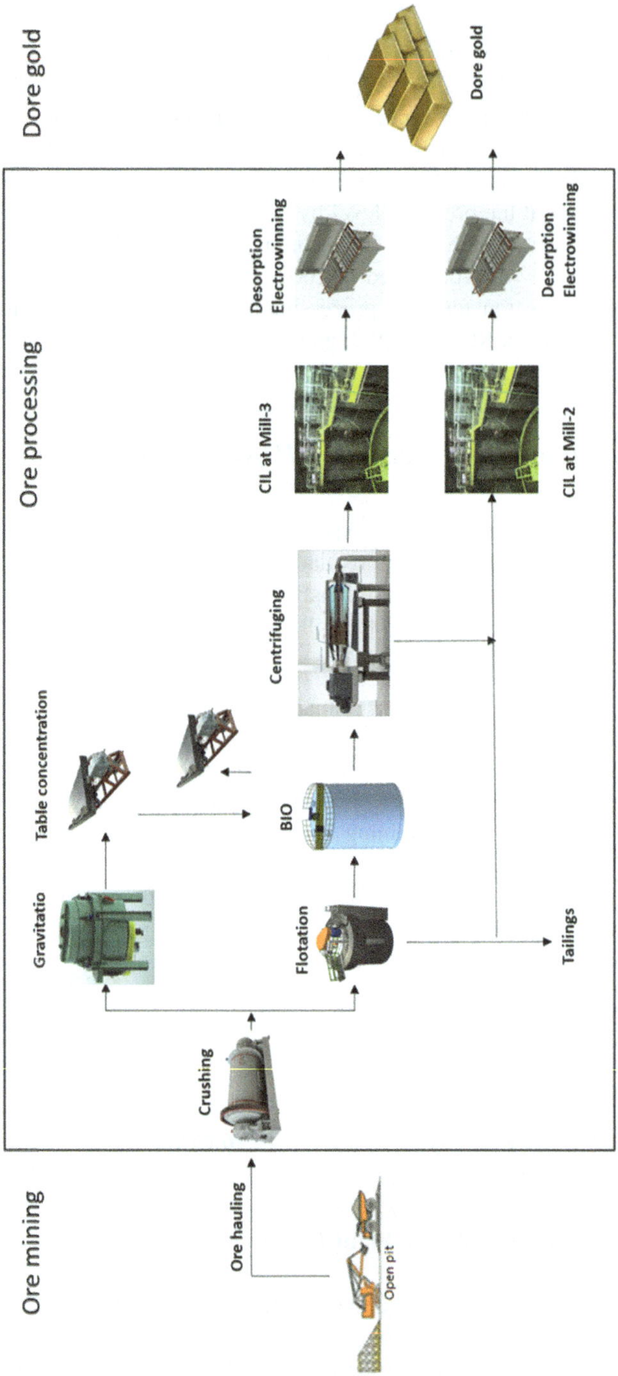

Fig. 11.1 BIONORD® Process flowsheet at the Olimpiada Mill

of the TsNIGRI institute. However, the technology was not tested in pilot conditions or run to ensure readiness for industrial use. Therefore, in 1997 a pilot plant was established at Mill 1 to develop and test the technology for gold recovery from refractory ores using microbial oxidation of concentrates.

The pilot biooxidation plant included 6 reactors connected sequentially, each with a volume of 2.5 m^3. This plant was used to study the impact on the process of the parameters such as pH, redox potential, temperature, slurry density (solid content), the necessary grade of solid particles, blending, and aeration. The suitability was determined of pit waters and low-end mineral salts for preparation of BIO feedstock. For the industrial concentrate biooxidation process, a proprietary medium was developed with replacement of pure nitrogen and phosphorus salts with "ammophos" fertiliser.[1] Comparison with the classical 9 K medium during iron (II) oxidation demonstrated that iron oxidation rates were virtually identical. During cultivation on the concentrate, the biomass grew and the cell concentration increased up to 9–10 $\times 10^9$ cells mL^{-1}. These studies demonstrated that the necessary nitrogen concentration in the medium was 0.24–0.32 g L^{-1} with ammonium sulfate consumption of 0.7–1 g L^{-1}, which ensured biomass growth and efficient sulfide oxidation. The long-term operation of the plant using media with ammophos (0.5 g L^{-1}) with ammonium sulfate added at 0.7–1 g L^{-1} during 2000 confirmed the results of the laboratory tests regarding the efficiency of bacteria cultivation on these media. During the biooxidation testing period, ammonium sulfate consumption was 5 kg t^{-1}, ammophos was 2.5 kg t^{-1} of the concentrate or 0.15–0.09 kg t^{-1} of the ore. The temperature in the reactors was kept at 34°C.

The biooxidation process was run with a solid-to-liquid ratio in the plant feedstock S:L = 1:5–6, on different grade concentrates with a grinding grade of 90–96% <74 μm. The liquid slurry phase accumulated up to 5 g L^{-1} of arsenic (V), up to 25–30 g L^{-1} of ferric iron with the slurry pH decreasing from 1.9–2.1 to 1.5–1.7. The slurry liquid phase contained 8.2–9.8 $\times$ 10^9 cells mL^{-1}, and oxidation of arsenic (III) to arsenic (V) reached 93.8–97.0%. The content of sulfide-S in the biooxidation products decreased to 0.7–0.9% with 93–95% oxidation of sulfide-S. When the reactors were operating in the most efficient mode, addition of sulfuric acid (2.9 kg t^{-1}) was only required to prepare the medium (pH 1.7), which confirmed the high activity of concentrate oxidation with an industrial culture.

11.3.3.2 Composition of the Microbial Consortium during Industrial Pilot Testing

Microscopic studies of pulps from different reactors of the pilot plant showed the presence and active oxidation of sulfide minerals, Fe^{2+} and sulfur compounds. In the

[1] Nitrogen–phosphorus concentrated soluble fertiliser, containing approximately 10–12% N and 52% P_2O_5 mainly composed of monoammonium phosphate ($NH_4H_2PO_4$) and partly of diammonium phosphate ($(NH_4)_2HPO_4$).

concentrate of the Olimpiada mine, several species of microorganisms belonging to different phylogenetic groups were identified (*Acidithiobacillus ferrooxidans*, *Acidithiobacillus thiooxidans*, *Leptospirillum ferrooxidans*, *Sulfobacillus thermosulfidooxidans*) as well as *Archaea* and acidophilic heterotrophs. Each microorganism present in the mixed population was isolated and studied in pure culture. Analysis of the chromosomal DNA structure by restriction profile pulse-field gel electrophoresis (PFGE) of the *At. ferrooxidans* cultures isolated from the slurries of the first and fifth reactors demonstrated the presence of the same strain in the sequential reactors 1 and 5; this strain was absent from the inoculum used to inoculate the slurry when the plant was launched. The chromosomal DNA structures of the strains *At. ferrooxidans* and *At. thiooxidans* were analysed using PFGE, along with *Sulfobacillus*, isolated as pure cultures from the reactors during trials in February 1998. Chromosomal DNA restriction digests of *At. ferrooxidans* isolated from reactors 1 and 3 were similar to the restriction samples of the chromosomal DNA of the strain of *At. ferrooxidans* isolated from the reactors in 1997, and the restriction pattern of the wild strain of *At. ferrooxidans* isolated from a non-inoculated medium with a concentrate. This result showed that the bacterial process of the Olimpiada concentrate was dominated by a wild-type strain of *At. ferrooxidans* and not by the strains of *At. ferrooxidans* that had been inoculated at the plant's launch and indicated that the indigenous strains present in the concentrate were more active at oxidising the concentrate. The number of *Leptospirillum* spp. increased in the final reactors, which was thought to be related to their lower sensitivity to low pH, higher resistance to ferric ions, and higher affinity for ferrous iron.

During the concentrate biooxidation tests, highly active microbial biomass was obtained that was adapted to the concentrate biooxidation conditions. This was used to achieve a high degree of oxidation of sulfide minerals and extraction of the associated gold with further sorption extraction of up to 97–98% with a residual gold content of 1–2 g t^{-1} in the tailings.

11.3.4 Industrial Development of Technology for Processing Refractory Gold–Arsenic Ores Using Bacterial Oxidation

11.3.4.1 BIONORD® Generation I: BIO-1 Plant (2001)

Based on the results of the technology tests and trials at the pilot plant, a process procedure for processing by-product refractory ores was developed and authorised by the appropriate supervisory and environmental entities of the Krasnoyarsk Krai in 1997–2000. Based on this, and for the first time in Russia, Mill 2 for processing refractory gold–arsenic ores of the Olimpiadinskoe deposit was developed and commissioned for industrial operation on October 1, 2001. The first industrial BIO plant was equipped with 5 lines, each with 6 reactors, and the capacity of each bioreactor was 450 m^3. The mill was designed to process 3 × 10^6 t of primary sulfide ore y^{-1} from the Vostochnyi pit of the Olimpiada deposit.

Since this was the first time that such reactors had been constructed in Russia, and therefore, along with production objectives, process tasks were resolved and settled. These included biooxidation process control and management, determination of the thresholds of the process parameters such as slurry density, optimum flotation particle size, flow rates/retention times, temperature, pH, agitator speed, and air consumption. Various designs of agitator impellers, aeration systems, and feedstock dosing systems for the charging reactors and process schemes for reactor connections in lines were investigated.

During the operation of the BIO-1 plant, several deficiencies were identified. The plant was equipped with piston compressors and some of the oil, together with the air, entered the slurry of the bioreactors. Feedstock supply and slurry density were regulated manually, which did not allow for high-precision compliance with the process parameters. As a result, regulation precision was determined by the operator's experience and professionalism, so that the human factor played a significant role. The gentle slope of the charging tubes from the slurry splitters to charging reactors, together with small ducts, did not enable the uninterrupted delivery of feedstock. The high pyrrhotite content in the flotation concentrate (up to 40%) resulted in overheated reactors and to remove the thermal load, a scheme was used with five charging reactors in parallel and a single subsequent stage reactor in rotation. Despite these deficiencies, the extent of oxidation for the sulfide flotation concentrates reached 88% and 92% of the gold present in the concentrate was recovered. The slurry solid content was (in 2021) 182 g L^{-1} (16.11%), the process time for the entire reactor chain was 110 h, and the feedstock supply speed was up to 5.0 m^3 h^{-1}. At that time, the plant was charged with 450 t d^{-1} (100% <40 μm).

In accordance with the company's development plans, the throughput of the Olimpiada Mine was increased incrementally, new BIO production capacities were introduced, with the objective of meeting and improving on results. Each new BIO plant was another step forward, both in terms of technical and technological improvement of the biohydrometallurgical process.

11.3.4.2 BIONORD® Generation II: BIO-2 Plant (2007)

Based on the experience gained from the operation of the BIO-1 plant, BIO-2 plant was launched in 2007 with a design capacity of 5 × 10^6 t of primary sulfide ore. The total processing throughput at the two BIO plants reached 8 × 10^6 t of sulfide ore per year. The plant was equipped with 3 lines, each with 6 reactors each, and the capacity of each reactor tank was 1000 m^3. The reactors in the line were twinned, and slurry splitters were installed so that the feedstock supply tubes of the reactors had a sufficient slope and were not too long, in order to prevent blockage with the solid slurry phase. Turbine compressors were installed instead of piston compressors, and a separate pump was used to supply feedstock slurry to each of the three lines. The slurry solid content (in 2021) was 182 g L^{-1} (16.11%), the process time 110 h, and the feedstock supply speed was up to 10.0 $m^3 h^{-1}$. The plant was charged with 510 t concentrate d^{-1}, and therefore the annual throughput of concentrate was about 186,000 t, produced from 5 × 10^6 t of sulfidic ore.

11.3.4.3 BIONORD® Generation III: BIO-3 Plant (2014)

Based on industrial operation and analysis of the operation of the BIO-1 and BIO-2 plants, an improved concept BIO-3 plant was developed. The plant was put into operation in 2013 and equipped with one line with six 1000 m^3 reactors. The basis of the concept was automation of the biooxidation process and the display of the control and management process parameters on monitors in the control room. The following elements were changed:

(i) Two homogenisation reactors with a capacity of 1000 m^3 each were installed to ensure stability of the composition of the orthophosphoric acid delivered to the BIO as a feedstock.
(ii) A system for the automatic regulation of the set slurry density supplied as feedstock of the BIO charging reactors.
(iii) A new scheme of slurry splitters was installed which helped to regulate the feedstock supply speed with precision of up to 0.1 $m^3 h^{-1}$.

These changes helped to stabilise considerably the biooxidation process and enhance the plant's throughput. This was related to the fact that the regulation precision of the slurry flow speed was high and enabled the plant to operate at nearly maximum flow rates, in stable high-performance mode, preventing process failure. Along with improvement of the process engineering support, the system was also automated, which helped to significantly improve the engineering support of the biooxidation process. The slurry solid content (in 2021) was 250 g L^{-1} (21.36%), the process time 110 h, and the rate of nutrient feedstock supply was 8.0–10.0 m^3/h. The plant was charged with 250 t of concentrate d^{-1}.

11.3.4.4 BIONORD® Generation IV: BIO-4 Plant (2017)

The maximum improvement of unit performance was always the key issue for the economy of the production process, as this impacted the capital expenditures and the marginal cost of the manufactured product. This was the focus during the development of the concept of Generation IV of the BIONORD process and its further development, considering the experience of the previous development stages (Sovmen et al. 2007a, b; Belyi et al. 2017; Belyi et al. 2018). As a result of a detailed analysis of the latest modification of BIONORD (III generation), decisions were made on modifying the design of the mixing and aeration systems, the redistribution of energy consumption between these systems, and the concept of reactor cooling was revised. The main focus was a reduction in the total power consumption for agitation, aeration, and cooling. In this context, great attention was

Fig. 11.2 History of BIONORD technology development at the Olimpiada Mine

paid to improving the mass-exchange performance of the aeration and agitation systems.

The BIO-4 plant was designed and constructed based on the conducted studies and calculations. The plant was commissioned in October 2017. Figure 11.2 shows the key milestones of BIONORD technology development at the Olimpiada Mine. Each BIO plant is actually a new plant for processing gold-containing raw materials using biohydrometallurgical principles.

The upgrade was based on Generation III 1000 m^3 bioreactors from the BIO-3 plant. The plant was a module of 8 reactors and composed of six primary and two secondary reactors. For BIO-4 plant bioreactors, 110 kW EKATO motor gearboxes were purchased with a Combijet+ agitator (Jung and Keller 2014; Keller et al. 2019). This agitator was selected, as preliminary studies conducted by the Polyus JSC Research Center under laboratory conditions had proven their high potential for reducing costs for aeration and increasing the slurry operating density. The impeller's design made it possible to move away from standard circular manifold air dispersal systems, as air is dispersed through the impeller blades.

The higher efficiency of the biooxidation process is directly related to two process parameters—the slurry flow rate and the slurry solid content. The flow rate is determined by the rate of reproduction of microorganisms in the community and cannot exceed a certain limit, otherwise, the microorganisms are washed out. The slurry solid content is determined by the specifications of the bioreactor's agitation and aeration systems, which create the most favourable conditions for the physical, chemical, and biological processes necessary for the stable high-performance functioning of a community of microorganisms.

A major characteristic of the mineralogy of the flotation concentrates of the Olimpiada deposit ores is their high content of pyrrhotite, which produces a large amount of heat during the biooxidation process. Bioslurry can be overheated above the maximum growth temperature of the consortium of moderately thermophilic microorganisms, which results in the failure of the biooxidation process. Therefore, the inefficient cooling system is a constraining factor for improvement of the bioplant's performance. To increase the efficiency of agitation and aeration systems

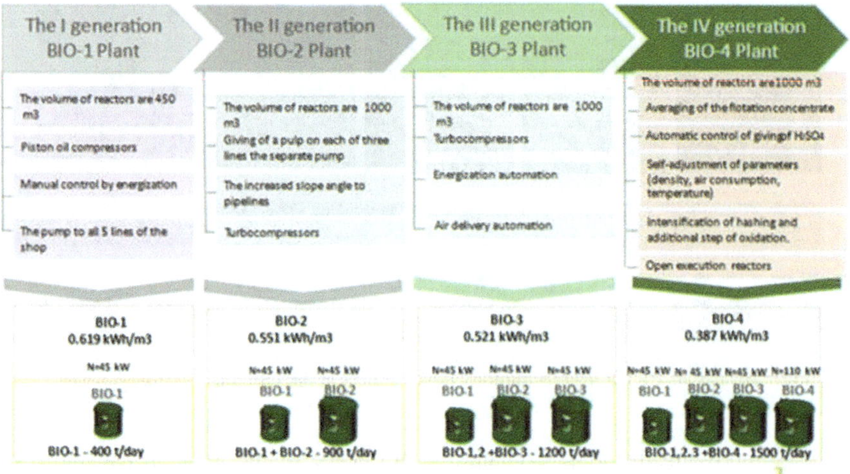

Fig. 11.3 Growth of technical and process improvements of the BIO plants in terms of biooxidation of flotation concentrate of the Olimpiada deposit ores (N is the power of the electric motor in kW)

and improve their mass exchange parameters, a cooling system is required that is equally as efficient as these. The first step towards resolving this issue was to reduce the thermal load on the primary bioreactors by increasing their number. For this purpose, at the BIO-4 plant, concentrate feed with nutrient salts was supplied to each of the six primary bioreactors. The second step in reducing costs for the cooling system and to improve its efficiency was the "open-type" design of the reactors, where most of the reactor surface was located outside the plant premises and was cooled by the ambient air temperature, especially during colder periods. In contrast, the BIO-1, BIO-2, and BIO-3 plants were equipped with older agitation and aeration systems and the reactors were located inside the plant building ("closed type").

Most of the power consumption was used by the aeration systems. The BIO-1, BIO-2, and BIO-3 plants used 0.426 kW h m^{-3} to 0.405 kW h m^{-3}. The switch to the EKATO COMBIJET+ impeller for the BIO-4 plant made it possible to reduce aeration power consumption to 0.205 kW h m^{-3} while the agitation power consumption grew to 0.131 kW h m^{-3}. On the whole, power consumption redistribution for aeration and agitation devices helped to reduce the power consumption from 0.521–0.619 kW h m^{-3} to 0.387 kW h m^{-3} (Fig. 11.3). Cooling consumption was reduced from 0.01–0.018 kW h m^{-3} to 0.007 kW h m^{-3}.

The plant is charged with 300–360 t concentrate d^{-1}. The slurry density is 250–260 g L^{-1} (21.36–22.0%), the process time is 110–120 h. The feedstock supply speed is up to 10 m^3 h^{-1}. The oxidation degree of sulfide flotation concentrates reached 90% and the gold recovery up to 94%.

The use of new agitators resulted in a sufficient improvement in the mass exchange parameters of the bioreactors, which impacted the efficiency of the biooxidation process. Biomass concentration in the BIO-4 reactors reached values higher than 3 g L^{-1}, while for BIO-1, BIO-2, and BIO-3 it did not exceed 1.5 g L^{-1}. As feedstock with the same mineralogic composition was supplied to all BIO plants, higher biomass concentration in the BIO-4 plant reactors implied a more efficient biooxidation process in the BIO-4 plant reactors. Sulfuric acid consumption in BIO-4 reactors was 24–45 kg t^{-1} of flotation concentrate, while at other plants it is 122 to 164 kg t^{-1} of flotation concentrate. Due to better mass exchange parameters of the COMBIJET+ agitators, the biooxidation process was more intensive, which was expressed in a greater oxidation of sulfides of flotation concentrates, higher biomass concentration, and lower sulfuric acid consumption.

11.3.5 Composition of Microbial Consortia of Microorganisms During Industrial Biooxidation of Concentrates

The technology of bacterial leaching of gold–arsenic concentrates from the Olimpiada deposit has been developed using the iron/sulfur-oxidiser *At. ferrooxidans*, which successfully oxidised sulfide minerals at a temperature of 28–32°C. However, after transition to large-volume equipment, the process operating temperature grew from 34°C to 38–40°C and the microflora composition changed. The results of molecular biological analyses of consortia of indigenous acidophilic microorganisms in the BIO plant reactors conducted in 2019 confirmed the monitoring data obtained in 2015–2018 (Kondrat'eva et al. 2015; Bulaev et al. 2017), which highlighted the stability of the composition of the microbial consortium driving the sulfide concentrate biooxidation process at the BIO plants at the Olimpiada Mill. However, it was noted that relative abundances of the dominating groups of microorganisms were constantly changing, probably due to fluctuations in the slurry's physical and chemical parameters as impacted by changes in the composition of the mineral feedstock. The dominating microorganisms in 2019, according to the averaged data of the metabarcoding 16S rRNA gene analysis, were *Acidithiobacillus* spp. (13%) and *Acidiphilum* spp. (13%), *Acidiferrobacter* spp. (29%), *Leptospirillum* spp. (20%), and *Ferroplasma* spp. (26%).

Identification of indigenous microorganisms from 16S rRNA gene sequencing showed that there were no differences in their gene sequences with those of the previously isolated strains obtained from the slurry of reactors at the BIO-1, 2, 3, and 4 plants and deposited in the Core Facility "*Collection of Unique and Extremophilic Microorganisms*" on the basis of the Federal State Institution Federal Research Centre "*Fundamentals of Biotechnology*" of the Russian Academy of Sciences. The degree of sulfide mineral oxidation in the reactor slurry was evaluated based on differences in the content of sulfide-As, sulfide-Fe, sulfide-S, and sulfide-Sb in microbial oxidation cakes.

11.3.6 BIO Process Management

The key criteria for evaluating the BIO plant's operation is throughput capacity and the quality of the obtained "biocake" (a biooxidation product of the flotation concentrate) for the further hydrometallurgical process stage. The plant's throughput is based on the supplied slurry density in the BIO feedstock and its flow rate through the biooxidation reactors. The slurry density is determined by the specifications of bioreactors, which help to maintain the uniform distribution of the solid phase and the air oxygen throughout the reactor volume, and maintenance of the set temperature and slurry acidity (Chap. 3).

The studies conducted by the Polyus CJSC Research Center in the laboratory and at an industrial scale determined the limit and optimum values of all the important biooxidation process parameters. These included: chemical and mineralogical composition of feedstock, feedstock solid grade, solid content in the BIO feedstock, feedstock supply rate, agitator rotations, air consumption for aeration, slurry temperature, and slurry redox potential. Understanding the effect of these parameters on the biooxidation operation resulted in the creation of the BIO plant management algorithm. In terms of process management in a stable high-performance mode, the key significant parameters include the total content of sulfur in the flotation concentrate and the pyrrhotite/pyrite ratio. The process is implemented focused on redox potential, and the optimum parameters for all the key process metrics are maintained. Automation of the key process parameters helped to improve the precision of parameter regulation and the time of their processing, which resulted in a throughput increase at the BIO plants to 1510 t d^{-1}.

Most of the sulfide minerals (up to 80%) are oxidised in the charging reactors (first oxidation stage). Their further oxidation takes place in the secondary reactor. As a result, the oxidation of sulfide minerals reaches: arsenopyrite, 92–96%; pyrite and pyrrhotite, 76–91%; sulfide-S, 84–89%; stibium (antimony sulfide minerals) 47–72%. This degree of oxidation of sulfide minerals facilitates gold extraction of up to 94%.

The key conditions for the highly efficient and stable operation of BIO plants are described below:

(i) The homogenisation reservoirs (surge tanks), tank equipment with an agitator, help to level abrupt jumps, both for sulfide-S and for acute fluctuations in the ratio of sulfide minerals in the flotation concentrate. These considerably stabilise the biooxidation process and help improve the throughput of the BIO plants by 15–20%.
(ii) The use of the set slurry density automatic regulation system removes density pulsation of the BIO feedstock and ensures stable conditions for active oxidation activity of microorganisms. This enabled a 10–15% efficiency increase of the BIO plants.
(iii) The new design of slurry splitters can regulate the feedstock supply speed with precision of up to 0.1 m^3 h^{-1}. The slurry flow speed regulation precision was consequently high and enabled the plant to operate at next to maximum flow

rates, preventing process failure. This helped to increase process efficiency by 10–15%.

(iv) Upgrading the BIO-1 and BIO-2 plants in 2015–2016 with an automation system, which justified itself during operation at the BIO-3 plant, helped to improve the throughput capacity of all flotation concentrate processing from 850–900 t d^{-1} to 1180–1250 t d^{-1}. After commissioning the BIO-4 plant, the total throughput capacity of the BIO plants grew to 1500 t d^{-1}. This helped to produce over 30 t of gold in 2019 using the gold-containing flotation concentrate biooxidation technology.

(v) Capacity redistribution between the agitation and aeration system helped to reduce consumption by 110 kW (32%) per bioreactor, and aeration consumption fell by 50–60%. The "open-type" design of the reactors helped to reduce cooling power consumption by up to 30%. Due to the upgrade of agitators, slurry solid content increased from 17.2% to 21.7%. By using efficient agitation and aeration systems, the degree of oxidation of sulfide minerals of the flotation concentrate grew by 17% and reached 98%, whereas acid consumption decreased two to threefold. The amount of gold recovered from the biocake increased by 2.0–2.5% and reached 94%.

11.4 Conclusions

The Suzdal project is an example of the successful implementation of progressive new technologies that deal with challenges of a changing ore deposit (preg-robbing and sulfide mineral composition) and water supply and balance limitations. The adoption of these technologies has provided an advantageous and stable, cost-effective processing operation.

11.5 Technical Note

Polyus JSC currently holds the rights to patented technologies of bacterial processing of concentrates of refractory gold-containing sulfide ores, a patented consortium of microorganisms engaged in the process, and the process technology is protected with the "BIONORD" trademark.

References

Belyi A, Chernov D, Solopova N (2017) Production development of Olimpiadinskoe gold processing plant through BIONORD® technology processing. Proceedings 22nd International Biohydrometallurgy Symposium, Trans Tech Publications, p 12–15. https://doi.org/10.4028/www.scientific.net/SSP.262.12

Belyi AV, Chernov DV, Solopova NV (2018) Development of BIONORD® technology on Olimpiada deposit refractory arsenic-gold ores treatment in conditions of extreme north. Hydrometallurgy 179:188–191. https://doi.org/10.1016/j.hydromet.2018.04.010

Binks M, Wemyss P (2012) Fosterville gold mine heated leach process. Mineral Processing and Extractive Metallurgy 121(4):224–227. https://doi.org/10.1179/1743285512Y.0000000023

Borovkov GN, Li LV, Sherman ML (eds) (1985) Geology and metallogeny of the Yenisei ore belt. KNIIGIMS, Krasnoyarsk, p 291

Bulaev A, Belyi A, Panyushkina A, Solopova N, Pivovarova T (2017) Microbial population of industrial biooxidation reactors. Solid State Phenom 262:48–52. https://doi.org/10.4028/SSP.262.48

Genkin AD, Lopatin VA, Savelyev RA, Safonov YG, Sergeev NB, Kerzin AL, Tsepin AI, Amstutz H, Afanasyeva ZB, Wagner F, Ivanova GF (1994) Gold ores of the Olimpiada deposit (Yenisei ridge, Siberia). Geol Ore Deposits 36:111–136

Jung J, Keller W (2014) Process and cost optimized solutions for gassed agitated CSTR's. ALTA Gold Precious Metals Proceedings:37–53

Keller W, Rümmele A, Last W (2019) On the plus side of mixing. PROCESS Worldwide 21:38–40

Kondrat'eva TF, Bulaev AG, Muravyov MI (2015) Microorganisms in biogeotechnologies of sulfide ores processing. Nauka, Moscow

Kovalev KR, Kalinin YA, Borisenko AS, Naumov EA, Kolesnikova MK, Polynov VI, Kydyrbekov EL, Netesov MI, Klimenko AG (2012) The Suzdal gold-sulfide deposit in the black shale of eastern Kazakhstan. Geol Ore Deposits 54:254–275. https://doi.org/10.1134/S1075701512040034

Kuzmin MI, Zorina LD, Spiridonov AM, Mitrofanov GL (2000) The main types of gold deposits in Siberia (composition, genesis, problems of development). Non-ferrous metals 8:4–8

Li LV (2003) Olimpiadinskoe Deposit, disseminated gold-sulfide ores. Krasnoyarsk, KNIIGIMS, p 120

Serdyuk SS (1997) Gold-bearing and gold-platinum-bearing provinces of Central Siberia: Geological and metallogenic structure and prospects for the development of the raw material base. In: Proceedings of the Geology and minerals of Central Siberia. Krasnoyarsk: KNIIGIMS 1997. p 89–183

Smart M, Huddy RJ, Edward CJ, Fourie C, Shumba T, Iron J, Harrison STL (2017) Linking microbial community dynamics in BIOX® leaching tanks to process conditions: integrating lab and commercial experience. Solid State Phenom 262:38–42. https://doi.org/10.4028/SSP.262.38

van Buuren C, Yakovenko G, Krutskikh S, Beisebayeva G, Tupikina O (2018) Application of the HiTeCC technology at NORDGOLD's Suzdal operation to combat preg-robbing and enhance gold recovery. In: Proceedings of the ALTA 2018 Gold-PM. Perth, Australia

Sovmen VK, Guskov VN, Belyi AV, Kuzina ZP, Drozdov SV, Savushkina SI, Maiorov AM, Zakrevsky MP (2007a) Processing of gold-bearing ores using bioleaching under far north conditions. Nauka, Novosibirsk, p 144

Sovmen VK, Guskov VN, Belyi AV, Lipatova TV, Kondrat'eva TF, Gish AA (2007b) RU Patent:2346063

Chapter 12
Biomining in Finland: Commercial Application of Heap and Tank Bioleaching Technologies for Nickel Recovery

Mariekie Gericke, John William Neale, and Pasi Määttä

Abstract The use of tank and heap bioleaching technologies for the extraction of nickel and cobalt from sulfide ores and concentrates has found commercial application in Finland. At the Terrafame mine (previously Talvivaara), the complexity of the black schist orebody precluded the production of a suitable concentrate and a decision was made to bioleach the crushed ore in heaps. To date, it is the only industrial-scale mine to utilise heap bioleaching for nickel production. In the case of Mondo Minerals, a sulfide concentrate, generated as a by-product from talc mining operations, contains significant quantities of nickel and cobalt, but also a small but significant quantity of arsenic. Previously the concentrates had been sold to toll smelters, but the arsenic content has made this option less attractive. Tank bioleaching technology was identified as the most suitable option for the recovery of nickel and cobalt from this sidestream and led to the first commercial implementation of nickel sulfide concentrate bioleaching. This chapter will briefly cover developments at the Terrafame mine since the previous edition of this book, while the main emphasis will be on the technical development of the Mondo Mineral tank bioleaching process from bench- to commercial-scale application.

Keywords Bioleaching · Nickel · Cobalt · Tank · Heap

12.1 Introduction

Bioleaching is a well-established process used in the extraction of base metals and refractory gold from sulfide ores and concentrates. The process has been commercially applied in a wide variety of locations for the pre-oxidation of refractory gold

M. Gericke (✉) · J. W. Neale
Biometallurgy Division, Mintek, Randburg, South Africa
e-mail: mariekieg@mintek.co.za; Johnn@mintek.co.za

P. Määttä
Elementis Minerals B.V. Suomen Sivuliike, PL, Kajaani, Finland
e-mail: Pasi.Maatta@elementis.com

D. B. Johnson et al. (eds.), *Biomining Technologies*,
https://doi.org/10.1007/978-3-031-05382-5_12

concentrates and for the recovery of copper from secondary sulfide ores (Gericke et al. 2009). Current estimates are that biomining accounts for around 10–20% of global copper production, ~1% of gold production, and smaller percentages of other metals such as zinc, nickel, and uranium (Chap. 1).

Nevertheless, the application of bioleaching has remained a niche, rather than a mainstream, technology in the mining sector. Some unique benefit is typically required for bioleaching to be selected as the process option of choice. Examples, where bioleaching processes do show significant advantages over conventional mineral extraction processes, include: (i) treatment of refractory gold ores, (ii) recovery of metals from low-grade ores and tailings, (iii) treatment of sulfide concentrates containing significant quantities of impurities that would incur smelter penalties, and (iv) under conditions where production of a suitable flotation concentrate is problematic (Gericke et al. 2009; Johnson 2018).

In this chapter, two Finnish case studies are described where bioleaching has been selected as the technology of choice for the treatment of nickel-sulfide-containing ores and concentrates. At the Terrafame mine (previously Talvivaara), the complexity of the polymetallic black schist ore body precluded the production of a suitable concentrate and the decision was made to bioleach the crushed ore in heaps. Nickel production commenced in 2008 and it is still the only industrial-scale mine to utilise heap bioleaching for nickel production (Riekkola-Vanhanen 2007). The bench-scale development of the heap leach process applied at the Terrafame mine was covered in a previous edition (Puhakka et al. 2007). The developments at Terrafame since then are described in this chapter.

In the case of Mondo Minerals, a sulfide concentrate, produced as a by-product from their talc mining operations, contains valuable quantities of nickel and cobalt, but also a small but significant amount of arsenic. While sales to smelters had been the long-established method of commercially dealing with the concentrate, Mondo chose to create a value-added nickel product to enhance its revenue and profitability streams and avoid environmental liabilities. Tank bioleaching technology was identified as the most suitable option for the recovery of nickel and cobalt from this side stream. The commercial plant was commissioned in 2015 and it is the first commercial implementation of nickel sulfide concentrate bioleaching (Neale et al. 2015). The technical development of the process from bench-scale test work to piloting and commercial application are highlighted in this chapter.

12.2 The Talvivaara/Terrafame Project

The Talvivaara deposit, located in Sotkamo, Finland, is the largest known nickel sulfide deposit in Europe (Riekkola-Vanhanen 2007). The deposit comprises two polymetallic orebodies hosted by a black schist, Kuusilampi and Kolmisoppi, and the size of the resource is estimated to be 1550 million tonnes (Heikkinen and Korte 2019). The complex ore has average grades of 0.27% nickel, 0.56% zinc, 0.14% copper, 0.02% cobalt, 10.3% iron, 8.4% sulfur, and 7.2% carbon. The main minerals present are pyrrhotite, pyrite, pentlandite, sphalerite, violarite, chalcopyrite, and graphite (Riekkola-Vanhanen 2007).

Owing to the complex and low-grade nature of the Talvivaara ores, exploitation of the deposits was not economically viable using conventional processes. The application of bioleaching for the treatment of the Talvivaara ores was studied extensively in laboratory-scale and column leaching test work over a period of two decades, leading to the construction of a 17,000-tonne demonstration heap in 2005. Over a period of 500 days, 92% nickel, 82% zinc, 14% cobalt, and 2.5% copper were recovered. Copper is present as chalcopyrite, which explains the low copper recoveries achieved (Riekkola-Vanhanen 2007).

In February 2007, the primary demonstration heap was reclaimed and restacked to a secondary heap with the aim of enhancing copper and cobalt recoveries. Transferring the material to a secondary leaching stage has the added advantage of enhancing the recovery of metals from those parts of the primary heaps where the leaching solution has poor contact with the ore particles. Such areas include the slopes of the heaps, and areas within the bulk of the heap where solution flow was not ideal (Riekkola-Vanhanen 2011). The acid consumption was 15 kg t^{-1} in the primary leaching stage and 2 kg t^{-1} in the secondary leaching stage (Riekkola-Vanhanen 2011). The final recoveries after an additional 21 months of secondary leaching were 99% nickel, 99% zinc, 35% cobalt, and 22% copper (Riekkola-Vanhanen 2011).

12.2.1 Commercial Implementation

Based on the successful operation of the demonstration heap, commercial implementation of the bioheap leaching process was initiated in July 2008 and the first metal sulfides were produced in October 2008. The mining method is large-scale open-pit mining. The heap leaching process is divided into two steps, a dynamic primary leaching and multi-lift secondary leaching. The process flowsheet and the challenges experienced during commercial implementation of the heap leach operation have been described in detail in literature (Riekkola-Vanhanen 2007, 2011, 2013; Saari and Riekkola-Vanhanen 2012; Ahoranta et al. 2018).

The ore is crushed and screened to $P_{80} = 8$ mm (80% of particles <8 mm) followed by agglomeration with pregnant leach solution (PLS). The ore is then stacked on the primary heaps which are irrigated with an acidic solution (pH around 2) and aerated. Most of the acid required is generated in the heaps. The residence time on the primary leaching pads is approximately 13–14 months, after which the ore is reclaimed and restacked to permanent secondary heaps to continue the bioleaching process. After secondary leaching for a further three and a half years, the barren ore remains permanently in the secondary heaps. The secondary leaching pads are constructed on top of waste rock dumps which reduces earthwork quantities, the final footprint of the operation, and the rehabilitation costs. The secondary pads are planned to be stacked with four 15 m lifts, and the 60 m high heap to be eventually covered and revegetated.

In the metals recovery process, the metals are precipitated in stages from the PLS using gaseous hydrogen sulfide. The resulting intermediate products are transported to various refineries for further processing (Riekkola-Vanhanen 2007, 2011, 2013). After the target metals have been recovered, the solution is further purified to remove unwanted metals and returned to irrigate the heaps. During removal of residual metals, the pH of the PLS is raised to 9–10 with lime slurry, leading to the precipitation of residual metals (mostly manganese, magnesium, and iron) as hydroxides, together with gypsum and calcium carbonate. The resulting slurry is thickened and the thickener underflow is directed to gypsum waste ponds (Ahoranta et al. 2018; Tuovinen et al. 2018).

Challenges with crushing and the aeration systems at the beginning of industrial-scale leaching delayed the increase in metal recovery, but after the first two operational years the leaching results have improved significantly. The planned nickel production of 50,000 t y^{-1} was anticipated to be reached in 2012, with additional production targets of 90,000 t y^{-1} of zinc, 15,000 t y^{-1} of copper, and 1800 t y^{-1} of cobalt. Due to several challenges that were experienced, considerably lower metal production was achieved. By 2011, the production volumes achieved were 16,087 t y^{-1} of nickel and 31,815 t y^{-1} of zinc (Riekkola-Vanhanen 2013).

12.2.2 Talvivaara Becomes Terrafame

By 2014, the Talvivaara operation was facing serious operational and environmental challenges. The production delays combined with several leaks of metal-contaminated tailings which threatened local waterways, drove the company to bankruptcy. The asset was acquired by Terrafame in 2015, after which it went into a new commissioning phase (Arpalahti 2017; Ahoranta et al. 2018).

The fortunes of the operation have improved substantially in the years after the acquisition by Terrafame. Several practical factors have been addressed to improve the operability of the plant. One of the major complications was that the material hardens significantly during primary leaching, so much so that it fuses together and loses its granularity entirely, making re-mining of the ore after primary leaching difficult. To solve this issue, Terrafame developed mobile surface mining as a more feasible solution for the second round of reclamation. Several other practical aspects of the operation were also optimised and improved. These included repositioning the aeration and drainage pipes, and optimising the movement and stacking of the ore after re-mining of the primary leach pad, to minimise downtime (Arpalahti 2017).

In the period between September 2015 and early 2017, the third complete dynamic cycle of re-mining and re-stacking of the leach pads was completed (Arpalahti 2017). As a result, in 2017 20,864 t of nickel and 47,205 t of zinc were produced. Annual increases in production were recorded and nickel production of 27,468 t and zinc production of 55,222 t were achieved during 2019 (Neale 2020).

12.2.3 Future Developments

There are currently two strategic developments underway at the Terrafame operation, namely uranium production and the establishment of a battery chemicals plant (Ahoranta et al. 2018). The Finnish government granted a uranium extraction permit to Terrafame in February 2020, but the decision has been appealed, and it is expected that it will take a further 2 years before a final decision is made (Neale 2020).

Terrafame also announced a plan to invest in a battery chemicals plant to produce nickel and cobalt chemicals to be used in the electric vehicle (EV) industry. The decision was based on forecasts of growing demand for EV batteries, coupled with an indication that the share of nickel in battery applications is also increasing. The plant is intended to have an annual production capacity of around 150,000 tonnes of nickel sulfate and 5000 t of cobalt sulfate, making Terrafame one of the largest nickel sulfate producers in the world (Ahoranta et al. 2018). The plant is currently under construction, and commercial production is planned to commence in 2021 (Neale 2020).

12.3 The Mondo Minerals Tank Bioleaching Project

Finland hosts some of the world's largest talc deposits and is a significant global producer of this mineral ($Mg_3Si_4O_{10}(OH)_2$). Mondo Minerals (acquired by Elementis plc in 2018) is the world's second-largest talc producer with mining operations at two sites in Finland, Sotkamo, and Vuonos. At both sites, a nickel-rich sulfide concentrate, containing pyrrhotite ($Fe_{(1-x)}S$ (x = 0 to 0.2)), pentlandite ($(Fe,Ni)_9S_8$), pyrite (FeS_2), gersdorffite (NiAsS), magnesite ($MgCO_3$), and talc, is produced as a by-product of the flotation process. Nickel is the main metal of interest, but the concentrates also contain a small amount of cobalt and a small but significant quantity of arsenic (Table 12.1).

Previously, the concentrates had been sold to toll smelters but the arsenic contents have made this option less attractive. Mondo, therefore, elected to produce a value-added nickel product to enhance its revenue and profitability streams and avoid environmental liabilities. Mondo tested and evaluated several processing options before selecting Mintek's proprietary bioleaching technology as the most suitable for the recovery of nickel and cobalt from these side streams (Neale et al. 2016).

It is recognised that bioleaching is particularly suited to the treatment of concentrates containing problematic elements such as arsenic. In this project, those circumstances are present—the concentrate contains arsenic, which makes it increasingly more expensive to treat the material via smelting, and stockpiling of the material is undesirable as it would create an unacceptable environmental liability.

A 2-year-long test work programme was undertaken which developed and successfully demonstrated the application of Mintek's technology to treat the by-products from Mondo's talc production process. The study formed the basis for

Table 12.1 Composition of the Sotkamo and Vuonos concentrates

	Before pyrrhotite rejection		After pyrrhotite rejection	
Component	Sotkamo (%)	Vuonos (%)	Sotkamo (%)	Vuonos (%)
Ni	10.10	7.62	15.2	14.0
Co	0.44	0.32	0.71	0.56
Fe	46.6	40.5	44.2	47.6
Si	0.70	0.38	0.55	0.73
As	2.11	0.98	2.18	0.92
Sulfide S	32.7	24.3	33.9	32.9
Elemental S	<0.10	1.50	0.16	0.21
Total S	32.8	26.7	35.6	34.7
Carbonate	3.47	12.80	1.53	1.44
Mineral				
Pyrrhotite	63.3	56.5	47.8	57.7
Pentlandite	20.9	18.4	37.8	36.7
Pyrite	6.0	1.0	8.8	1.3
Gersdorffite	3.3	1.5	4.6	1.7
Magnesite	3.7	18.1	0.7	0.8
Talc	1.6	1.1	0.6	1.2

a feasibility study that showed that bioleaching, followed by a nickel- and cobalt-precipitation process, was an economically viable option for Mondo Minerals to derive value from the by-product. An important aspect of the process is that it includes the production of a stable arsenic-bearing waste, suitable for impoundment.

12.3.1 *Metallurgical and Pilot-Scale Studies*

The main objectives of the laboratory test work conducted by Mintek were to confirm the technical suitability of bioleaching to recover nickel and cobalt from the Sotkamo and Vuonos concentrates and to define the optimum bioleach operating parameters. The programme conducted incorporated regrinding of the concentrates, bioleaching, iron and arsenic precipitation, stability testing of the product, and nickel–cobalt (or mixed) hydroxide (MHP) precipitation. The bench-scale facilities used in the test work programme provided for close control and monitoring of the bioleaching process, allowing simulation of near-commercial scale conditions and control.

The bioleach test work was conducted on 1:1 blends of the Sotkamo and Vuonos concentrates. The typical chemical and bulk modal analyses of the concentrates used in the test work program are summarised in Table 12.1.

A moderately thermophilic culture from Mintek's culture collection was used in the bioleach test work programme. The culture has been adapted over a number of years to tolerate high soluble nickel and iron concentrations and has an optimum growth temperature of 45 °C. The dominant organisms present in the microbial consortium include *Acidithiobacillus caldus, Leptospirillum ferriphilum, Sulfobacillus benefaciens* and *Sulfobacillus thermosulfidooxidans.* Three phases of laboratory test work were performed, as described below.

12.3.1.1 Phase 1: Laboratory Amenability Test Work

Bioleach amenability and optimisation tests were carried out on the blended concentrate (without pyrrhotite rejection) in multi-stage laboratory-scale, continuously operated reactors with a total working volume of 7 L.

The effect of process parameters such as residence time, feed solids concentration, and grind size of the concentrates on metal extractions was evaluated. The initial results of this phase of the test work programme showed that nickel and cobalt extractions of 95% could be obtained in a five-stage continuous plant, at an overall residence time of 7 days, a feed grind size of $P_{80} < 30$ μm, and a feed solids concentration of 15%.

Optimisation of the feed solids concentration to 17.5% resulted in a marginal decrease in the nickel and cobalt extractions, to 93.5% and 94.0%, respectively. Further reduction of the feed grind size to $P_{80} = 20$ μm improved the nickel and cobalt recoveries to 97% and 95%, respectively. Under these operating conditions, the pH level in the first stage reactor was not controlled and could be maintained at pH 1.5 without acid addition. Stable operation was demonstrated in the presence of around 45 g L^{-1} total soluble metals and 120 g L^{-1} sulfate in the first stage growth reactor. There was a linear relationship between sulfide oxidation and nickel dissolution obtained in the bioleach process, indicating that high nickel recoveries were dependent on achieving a high degree of sulfide oxidation (Gericke et al. 2014).

12.3.1.2 Phase 2: Mini-Pilot Plant Scale Test Work

The results of the Phase 1 tests were used to define the bioleach operating parameters for Phase 2, which was performed in a continuously operated bioleach mini-plant with a total operating volume of 120 L. In addition, the product from the mini-plant was collected and used to assess the operating parameters for iron and arsenic removal from the bioleach slurry with minimal nickel loss, to confirm the stability of the neutralised iron- and arsenic-bearing precipitates and to demonstrate the production of a high-grade mixed hydroxide precipitate with a combined nickel and cobalt content of more than 40%.

The results from this phase of the test work demonstrated that nickel and cobalt extractions of 94% could be obtained in a four-stage continuous plant, at an overall residence time of 7.2 days, a feed grind size of $P_{80} = 20$ μm, and a feed solids concentration of 17.5%. The slightly lower extractions in this test were ascribed to the mini-plant system having four and not five stages, and a greater potential for some short-circuiting of solids in the gravity overflow system used in the mini-plant.

(The laboratory-scale reactor system used in Phase 1 employed pumps to transfer the pulp between the reactors).

A continuous mini-plant test was undertaken for the iron/arsenic precipitation step, employing a recycle for seeding and limestone ($CaCO_3$) as the neutralising agent. Iron and arsenic removal of over 95% could be achieved in a six-stage neutralisation plant, operated at a temperature of 35 °C, with the pH level being controlled at between 3.0 and 3.5. No nickel and cobalt losses were observed. The precipitated product was tested for stability using the European Standard EN 12457–3 procedure (EN 12457–3, 2002) and since no nickel, iron, cobalt, or arsenic were released, the precipitate could be classified as regular waste (Gericke et al. 2014).

MHP production was evaluated in a batch test using magnesia (MgO) as the neutralising agent. A MHP containing around 42% nickel, 2.4% cobalt, and between 1.8 and 2.0% magnesium could be produced at a precipitation pH of between 7.0 and 7.8. A two-stage approach for MHP production was recommended, in which between 80 and 90% of the nickel and cobalt would be precipitated in the first stage, to minimise contamination of the MHP precipitate with unreacted MgO, and therefore to maximise the nickel grade. Following a solid/liquid separation step, the nickel and cobalt remaining in the barren liquor would be recovered by precipitation with lime (CaO), and the resultant solids recycled to the iron/arsenic precipitation process, where the target metals would be re-solubilised (Neale et al. 2015).

12.3.1.3 Phase 3: Test Work on the Upgraded Concentrate Blend

It was subsequently determined that rejection of the pyrrhotite contained in the concentrates would result in a significant reduction in the size and therefore the cost of the bioleach plant. Pyrrhotite rejection was undertaken by magnetic separation, with minimal loss of nickel. Additional upgrading of the concentrate by flotation, to reject talc and magnesite from the concentrate, was also undertaken. The rejected pyrrhotite had a low nickel content, and was a potential saleable product.

The suitability of bioleaching to recover nickel and cobalt from the upgraded concentrate after pyrrhotite rejection was demonstrated in laboratory-scale, continuously operated reactors, and a continuously operated bioleach mini-plant. Additional test work was also conducted on a synthetic solution to confirm the design of the iron/arsenic precipitation process, and the solid–liquid separation process that followed. This was deemed necessary because the upgraded concentrate had a significantly higher nickel and iron content, and so the concentrations of these metals in the bioleach product were also higher.

Nickel and cobalt extractions of 97% and 98%, respectively, were obtained at an overall residence time of 7 days, a feed grind size of 80% < 20 μm, and a feed solids concentration of 15%. Very high soluble metal concentrations were measured and the total metal concentration (of iron, nickel, cobalt, and arsenic) approached 65 g L^{-1} in the first-stage growth reactor. Under these conditions, the redox potential in the first-stage bioleach reactor did not exceed +550 mV (vs Ag/AgCl), and so the maximum recommended feed solids concentration for the process was set at 15%. Under these operating conditions, the pH level in the first-stage bioleach reactor was controlled at 1.6, resulting in an acid consumption of 120 kg t^{-1} (Gericke et al. 2014).

At very high iron (60 g L^{-1}) and nickel (30 g L^{-1}) tenors in the feed to the iron/arsenic precipitation unit, efficient agitation of the slurry was not possible for pH values greater than 1.5, since the slurry viscosity increased significantly. The introduction of a seed recycle was able to decrease the viscosity of the slurry, owing to the dilution of the iron tenor in the feed solution. Iron removal of greater than 99% could be achieved in a five-stage neutralisation plant, with minimal nickel and cobalt losses. By implementing a seed recycling, viscosity effects were eliminated, the slurry could be agitated efficiently, and the product could be thickened and filtered (Neale et al. 2015).

The overall conclusion reached at the end of the metallurgical test work programme was that a process consisting of concentrate regrinding, magnetic separation, flotation, bioleaching, iron/arsenic removal by lime precipitation, metal precipitation to produce a MHP, and tailings neutralisation, had been successfully demonstrated and these results provided the process design specifications for the commercial-scale plant design.

12.3.2 Process Design Criteria

Based on the outcomes of the metallurgical test work programme, a set of process design criteria, a process flowsheet, and a mass balance were developed; these formed the basis for a feasibility study that was commissioned by Mondo Minerals and executed by Tenova Mining & Minerals (Neale et al. 2015).

A simplified block diagram of the process flowsheet is shown in Fig. 12.1.

The sulfide treatment plant is modest in size and is required to treat 35 t d^{-1} of sulfide concentrate at a nickel production rate of 1000 t y^{-1}.

The concentrate preparation section includes the following stages:

- A regrinding circuit to grind the concentrate to a P_{80} of 20 μm.
- Magnetic separation to remove some of the pyrrhotite from the concentrates.
- A flotation circuit to upgrade the nonmagnetic fraction further by removal of some of the remaining gangue materials, predominantly magnesite and talc, and some additional pyrrhotite and pyrite. The flotation circuit design includes:
  - 1 Flotation train
  - 4 Flotation cells per train
  - 3-stage rougher flotation configuration
  - Target of 99.9% pentlandite and 65.8% gersdorffite recovery

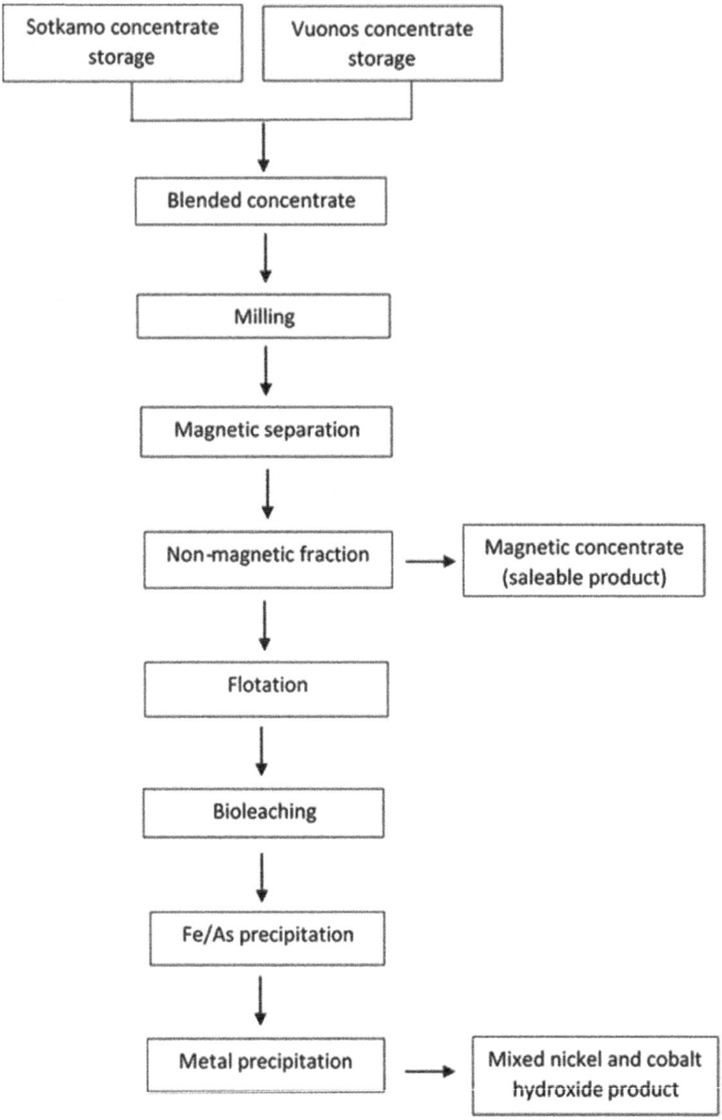

Fig. 12.1 General process flowsheet of the Mondo Minerals plant

The magnetic separation and flotation processes achieved an upgrading of almost 50%, reducing the quantity of concentrate requiring bioleaching to 18 t d^{-1} compared with 35 t d^{-1} being fed to the mill (Neale et al. 2015).

A summary of the process design criteria that were developed for the bioleach, iron/arsenic precipitation, and metal recovery sections is presented in Table 12.2. This is by no means an exhaustive list of the design criteria, but merely highlights some of the key features of the design.

Table 12.2 Target bioleach and metal recovery design criteria (Neale et al. 2015)

Criterion	Value	Unit
Concentrate treatment rate	18.0	$t\ d^{-1}$
Bioleach feed solids concentration	15.0	% (kg/kg)
Bioleach reactor configuration		
• Number of primary reactors.	3	–
• Number of secondary reactors.	4	–
Overall bioleach reactor residence time	7.0	d
• Primary bioleach reactor residence time	3.0	d
• Secondary bioleach reactor overall residence time	4.0	d
• Secondary bioleach reactor individual residence time	1.0	d
Design operating temperature in all bioleach reactors	45.0	°C
• Maximum operating temperature in all bioleach reactors	49.0	°C
• Minimum operating temperature in all bioleach reactors	41.0	°C
pH levels		
• Target pH level in primary bioleach reactors	1.6	–
• Anticipated pH ranges in secondary bioleach reactors	1.3–1.4	–
• Sulfuric acid addition to primary bioleach reactors	125	$kg\ t^{-1}$
Overall dissolution/oxidation levels in bioleach reactors		
• Ni	95.0	%
• Co	99.5	%
• Fe (based on soluble Fe concentrations)	73.8	%
• As	95.0	%
• Sulfide	97.6	%
Inorganic nutrient requirements		
• $(NH_4)_2SO_4$	12.60	$kg\ t^{-1}$
• K_2SO_4	2.10	$kg\ t^{-1}$
• K_2HPO_4	2.25	$kg\ t^{-1}$
Iron/arsenic removal		
• Number of tanks	5	–
• Overall residence time	6	h
• Target final pH level	2.5	–
• Operating temperature	35	°C
• Neutralising agent	$CaCO_3$	–
Metal precipitation		
• Number of tanks	5	–
• Overall residence time	5	h
• Target final pH level	7.0–7.2	–
• Operating temperature	30–34	°C
• Neutralising agent	MgO	–

12.3.3 Process Economics

The feasibility study included an estimation of the capital and operating costs of the plant, which were used to derive measures of profitability (Neale et al. 2015). Costs and prices prevailing in October 2013 were used.

The capital cost was developed as a Class 2 estimate, with approximately 10–15% accuracy. The estimated capital cost for the engineering development, procurement, and construction of the processing facility was in the range of €13–15 million, comprising 80% direct and 20% indirect costs. With the addition of a contingency and escalation, the capital cost estimated was in the range of €15–16 million.

The operating cost estimate included elements of varying accuracy, and was considered accurate to −12/+20%. The annual operating cost was estimated to be approximately €2.5–4.0 million, based on various operating parameters.

A detailed financial evaluation was undertaken, taking into account metal price forecasts, market potential, and market competition. A nickel price of US$20,000 t^{-1} and a discount rate of 8% were used for the base case, and sensitivity analyses were based on that rate.

The project was evaluated by considering various inflation scenarios for the prices of energy, consumables and metals, and labour rates. For the base case—with no inflation—the net present value at a discount rate of 8%, or NPV(8), was approximately €36 million, and the internal rate of return (IRR) was 19.8%. For the various inflation scenarios that were investigated, the NPV(8) ranged between €32 million and €41 million, and the IRR varied between 19% and 22%.

Sensitivity analyses showed that the nickel price was the most significant determinant of project economics. For a lower case nickel price of ±US$14,000/t, the NPV(8) was around €15 million and the IRR was about 9%. For an upper case nickel price of ±US$30,000/t, the NPV(8) rose to approximately €77 million and the IRR to about 35%.

The project economics determined in the financial evaluation were deemed acceptable by Mondo, and the project was approved.

12.3.4 Process Description

The nickel sulfide treatment plant was constructed on the site of the existing Vuonos talc concentrator plant. The plant was fully integrated with the existing plant in terms of labour, services, and utilities. The location of the plant in eastern Finland meant that it would experience very low temperatures in winter. The minimum ambient external temperature specified in the design criteria was −15 °C, with the capacity to operate at short-term temperatures as low as −30 °C. For this reason, the existing concentrator and most of the new treatment plant are located indoors, in order to shield the plant and its operators from the harsh winter conditions. Only the bioleach tanks, which are heat generating, are located outside of the existing buildings (Neale et al. 2015).

12.3.4.1 Concentrate Preparation

The sulfide treatment plant was designed with a feed rate of 35 t d^{-1}, typically comprising a 50:50 blend of concentrates derived from Mondo's two talc production plants at Sotkamo and Vuonos. The concentrate preparation circuit comprises stockpiling of concentrate, repulping, milling, magnetic separation, and flotation.

The stockpiled Sotkamo concentrate is transferred by a pipe conveyor into a repulp tank. The repulp tank also receives fresh concentrate as slurry directly from the Vuonos plant. The combined feed slurry is then pumped to the mill where the concentrate is ground to $P_{80} = 20$ μm. The regrind mill is an ultrafine vertical grinding mill operating in an open circuit. The mill discharge is then pumped to the magnetic separator, where a concentrate that contains 3 to 5% nickel is produced. The concentrate is filtered on a disc filter and the nonmagnetic fraction is pumped to flotation for further upgrading.

Flotation is conducted in three rougher cells and one scavenger cell, with a conditioner ahead of the first stage. Flotation tailings are returned to the existing Vuonos flotation circuit tails disposal tank and pumped to the tailings facility. The flotation concentrate is thickened to reduce the level of flotation reagents in the bioleach feed slurry. The thickener underflow is pumped to the bioleach feed tank where the density is reduced from 65% to 50% solids using recycled water, and nutrients to support microbial growth are added (Neale et al. 2015).

12.3.4.2 Bioleaching

The slurry discharged from the bioleach feed tank is diluted in-line to 15% solids, from where it is distributed to the three primary bioleach reactors. The bioleach circuit originally consisted of seven 112 m^3 tanks, with an overall residence time of 7 days at the design flow rate (Fig. 12.2). Three of the tanks were configured as primary oxidation reactors, followed by four tanks configured as secondary

Fig. 12.2 The bioleach reactors at the Mondo Minerals plant

Fig. 12.3 The dual P4/P3 impeller system in the three primary bioleach reactors at the Mondo Minerals nickel sulfide plant

oxidation reactors. However, a fourth primary reactor was added at a later stage to provide additional capacity in the plant, if and when required. Each tank is fitted with an agitator for dispersing air, suspending solids and maintaining homogeneity, an air sparge ring for injecting air supplied by a blower, a separate pipe for adding gaseous carbon dioxide (CO_2), and a number of vertical cooling coils which perform a dual function as baffles (Fig. 12.3).

The concentrate is fed into the primary reactors operating in parallel. The secondary oxidation stage comprises four tanks in series, which provides sufficient residence time for the target level of nickel recovery to be achieved and sufficient reactor units to minimise short-circuiting of the slurry (Neale et al. 2016).

The main services for the bioleach plant include concentrated sulfuric acid (H_2SO_4), blower air, and CO_2, which is stored in a pressurised tank. Carbon dioxide is required in the bioleach process for microbial growth and is supplied by adding CO_2 gas directly to the primary reactors.

The pH level in the reactor bank is allowed to vary as the reaction proceeds and typically remains within a range of 1.2 to 1.6, with the higher pH values in the primary oxidation reactors, and declining as the level of sulfide oxidation increases in the secondary oxidation stages. Provision is made for pH control, if needed, by the addition of either H_2SO_4 or $CaCO_3$.

The oxidation reactions are exothermic and the slurry temperature inside the bioleach reactors tanks is maintained at around 46 °C by banks of cooling coils located in each tank. It has been determined that the operating temperature should not exceed 49 °C, as this will have a deleterious effect on the performance of the moderately thermophilic bioleaching consortium. The cooling coils also act as baffles in each tank, and are supported by brackets mounted on the tank walls.

The agitators in the bioleach reactors were supplied by Afromix. Each agitator in the four primary bioleach reactors is fitted with a dual impeller system, similar to that shown in Fig. 12.3, which is a relatively new innovation in bioleach reactor design.

The lower impeller, known as the P4, is a downward-pumping, four-blade, high-solidity-ratio hydrofoil impeller of the type that has become the standard in bioleach reactors. It is designed to disperse high volumes of air while also maintaining solids in suspension and promoting heat transfer. Such impellers are characterised by their ability to operate at high gas volumes without flooding. The upper impeller is the innovative aspect: it is an upward-pumping, three-blade medium-solidity-ratio impeller, known as the P3, which was originally designed for high-viscosity applications, but has now found application in three-phase (gas-liquid-solid) mixing systems. Improved mass-transfer performance is achieved through surface air induction created by the top impeller, and enhanced gas hold-up from the specific mixing pattern that is created by the dual-impeller configuration.

The four secondary bioleach reactors, which have considerably lower gas dispersion duties than the primary reactors, are fitted with single P4 impeller systems (Neale et al. 2015). The slurry from the final oxidation tank is pumped to the iron/arsenic removal circuit.

12.3.4.3 Iron and Arsenic Precipitation

In the iron and arsenic removal circuit, iron and arsenic are eliminated from the pregnant solution in a six-stage precipitation circuit, comprising one conditioning and five precipitation tanks, a thickener, and a horizontal filter press. Iron and arsenic are precipitated in aerated tanks with $CaCO_3$ to form a stable ferric arsenate. Slurry from the final iron/arsenic removal tank is pumped to a thickener, and the thickener overflow is pumped to a cartridge filter, which acts as a clarifier and the clarified solution proceeds to the metal precipitation circuit for nickel and cobalt recovery. The filtered ferric arsenate cake is re-pulped with water and pumped via the neutralisation circuit to the tailings pond (Laukka et al. 2018).

12.3.4.4 Metal Precipitation

The objective of the metal precipitation circuit is to precipitate a mixed nickel and cobalt hydroxide product with a nickel content of typically 42%, bagged for sale. The metal precipitation circuit is a five-stage precipitation circuit followed by thickening and precipitation. In the primary stage of precipitation, the pH is raised to a level of 8.5 by addition of MgO. The precipitated product is thickened and the thickener underflow is filtered in two horizontal filter presses. The filter cake from the filter presses is bagged and sold. The thickener overflow and the filtrate from the filter presses are directed to a secondary precipitation and filtration stage, where slaked lime ($Ca(OH)_2$) and MgO are used to further increase the pH to a level of 10. The filtrate from the second-stage filter can either be used as recycled water (for example, as wash water for the iron/arsenic filter or as repulp water for the iron/arsenic filter cake), or directed to the tailings stream (Laukka et al. 2018).

12.3.4.5 Recycle Water Treatment

Neutralised and clarified water from the metal recovery circuit is used as a process water source for the plant. However, it has too high a magnesium sulfate level for reuse in the bioleaching process or for safe environmental discharge and is therefore processed in the recycle water treatment circuit. The water is fed to a neutralisation tank, where CaO is used to raise the pH to a level of around 9.5. The resulting precipitated hydroxides and gypsum are then removed from the solution via a thickener. The thickener underflow is pumped to tailings, while the water, now containing low levels of sulfates and magnesium, is reused in the process (Neale et al. 2016).

12.3.4.6 Tailings Neutralisation

The tailings from the nickel sulfide treatment plant consist of repulped iron/arsenic filter cake and tailings water from the metal recovery circuit, which usually contains in the range of 0.01–0.1 g nickel L^{-1}. These streams are neutralised in a three-stage neutralisation circuit (comprising two neutralization tanks and a pumping tank with $Ca(OH)_2$ addition), where the pH target is 10.3–10.5 for optimal nickel precipitation.

The neutralised tailings stream is pumped to a separate area in the plant's tailings storage facility in order to minimize the impact of the discharge of sulfates to the existing tailings dam, from which the talc concentrator and talc refinery derive their process water. Clarified water from this area is combined with the water stream from the talc operation's tailings (Laukka et al. 2018).

12.3.5 Inoculation and Commissioning of the Commercial Plant

The build-up of the microbial inoculum for the bioleach plant was conducted in several phases. Initially, an inoculum was prepared at Mintek's facilities in South Africa and air-freighted in the form of a filter cake to the plant site. The on-site inoculum buildup comprised a number of successive stages, starting with a small-scale five-stage continuous mini-plant and progressing through batch operation in 1, 3, and 6 m^3 reactors.

The inoculation and start-up of the production plant commenced in late September 2015. The first two attempts failed, and this was attributed to the highly reactive pyrrhotite in the Sotkamo concentrate (which was used in the commissioning phase) reacting with acid in the absence of aeration and producing hydrogen sulfide (H_2S) gas, which poisoned the inoculum as soon as it was introduced into the vessels. This behaviour was not encountered during the metallurgical test work programme, for several reasons: the scale of operation in the laboratory- and pilot-scale test work prevented reducing conditions from forming, and pre-acidification was conducted under aerated (and therefore oxidising) conditions. The inoculation method was adjusted, and on the third attempt, successful inoculation of one of the primary bioleach reactors was achieved. From there, the other bioleach reactors were filled, allowing feeding of the bioleach plant (Neale et al. 2016).

During commissioning, several practical challenges, mainly equipment related, were encountered that needed to be addressed. These included a build-up of agglomerated lumps and small pebbles in the conically bottomed concentrate repulp tank, which then passed into the suction of the peristaltic pump that fed the regrind mill feed tank, causing regular blockages of the pump. This was fixed by the installation of a makeshift filter basket in the repulp tank, and lifting the withdrawal point (which was close to the bottom of the tank) above the conical section of the tank (Neale et al. 2016). A screen was also subsequently installed in the regrind mill feed tank (Laukka et al. 2018).

Initially, the upgrading circuit—comprising magnetic separation and flotation—was not commissioned, which resulted in the appearance of a persistent foam in the bioleach reactors, considered to be caused by the talc that occurs in the non-upgraded concentrates. To counter this problem, the flotation section of the upgrading circuit was designed to remove almost all of the talc in the concentrates, and foaming was combated by the use of an antifoaming agent.

The greatest challenge in the iron/arsenic precipitation circuit has been to find the correct parameters and operational methods for iron–arsenic filtration. This included issues such as the selection and blinding of the filter cloth material, finding the correct flocculant type and dosage, and the correct handling of the non-flocculated fines from the thickener. The commissioning of the hydroxide precipitation circuit was straightforward without major issues (Neale et al. 2016).

One of the main features of the design of this bioleach plant is that it needs to withstand the harsh Finnish winter. This aspect of the design was fully tested in January 2016, when the temperatures were particularly low for a sustained period. Temperatures as low as −30 °C were experienced and the impact on the plant was quite severe, with all of the pipework and instrumentation above the bioleach reactors becoming enveloped in ice. During the coldest week, the regrind mill feed tank pump failed, which in turn resulted in the bioleach feed pump being stopped and caused the bioleach feed line to freeze. This caused the exothermic sulfide oxidation reactions in the bioleach reactors to slow down, and the pulp temperatures to drop. This then caused the cooling water flow rates to decrease, resulting in the cooling water supply lines also freezing. Steam generators were brought to site, the lines were defrosted, and normal operation resumed. The bioleach pulp temperatures began rising once feeding was re-instituted, and the cooling water flow rates returned to the expected levels with no long-term impact on the process or microbial performance (Neale et al. 2016).

12.3.6 Operational Performance

During April and May 2016, a mass balance sampling campaign was instituted over the bioleach plant. At the time, the design throughput had not been reached yet and the bioleach plant was operating at a feed solids concentration of 11% and a grind size of 80% $<$ 50 μm. The average nickel and cobalt extractions attained in the bioleach plant over this period were 97.4% and 98.4%, respectively. Although the plant was not operating under full design load, these results provided confidence in the robustness, stability, and efficiency of the bioleaching section of the plant.

The residue from the iron–arsenic precipitation circuit was subjected to environmental stability testing, which showed that the neutralised product met the requirements for classification as a regular waste. The metal precipitation plant was also successfully commissioned and at the time a product containing around 42 to 43% nickel hydroxide and about 1% cobalt hydroxide was produced (Neale et al. 2016).

Over the following 2 years, the complete circuit was commissioned and production was ramped up, although a few bottlenecks remained which prevented the attainment of the design throughput. A mass balance campaign conducted during 2018 at a feed solids concentration of 17% and an overall residence time of 9 days, indicated nickel and cobalt recoveries of 87.8% and 90.7%, respectively, in the bioleaching circuit. An overall sulfide oxidation level of 94.9% was estimated. The P_{80} of the feed material was 58.7 μm and it is anticipated that metal extraction would improve with further optimisation of the regrind mill circuit.

The overall nickel recovery achieved in the treatment plant was 80% of the MHP product from the bioleach feed. It was thought that further improvement to a desired range of 85–90% could be possible, particularly with optimisation of pH control in the precipitation circuits and of the iron/arsenic filter wash sequences.

The quantity of the mixed hydroxide precipitate being produced was of excellent quality, typically containing 47% nickel and 2% cobalt. The MHP produced was sold and used for the production of battery-grade nickel and cobalt sulfates (Laukka et al. 2018).

Further developments during this time included investigation of the possibility to derive additional value from the gold and platinum group metals occurring in low concentrations in some of the nickel concentrates that form the feed to the nickel bioleaching plant. This led to the development and piloting of a process to recover gold and platinum group metals from the bioleach residue. These metals can successfully be recovered through the production of an upgraded gold-containing concentrate derived from the bioleach residue which is suitable for sale to refiners. The process, developed with the Geological Survey of Finland (GTK), utilises Knelson concentrators and a strong acid wash (Laukka et al. 2018).

In the last quarter of 2018, Mondo Minerals was acquired from US private equity firm Advent International by the British speciality chemicals company Elementis plc. In view of the decline in the nickel price during the second half of 2018, the new owners took the decision to suspend operations, and the plant is currently being kept in "care and maintenance" mode, though this halt in production is considered to be temporary. Prior to the cessation of the operation, good progress had been made in ramping up the production, and the throughput was just below the design value (Neale 2020).

12.4 Conclusions

The suitability of bioleaching as an extraction process for nickel and cobalt sulfides has been successfully demonstrated at commercial scale. However, the application of mineral processing biotechnologies remains relatively limited and some unique feature of the feed material is typically required for bioleaching to be selected as the technology of choice. In the case of the Talvivaara/Terrafame project, the complexity of the orebody was the overriding factor that led to the selection of a heap leaching process, while at Mondo Minerals, the presence of arsenic in the sulfide material was the reason for the selection of a tank bioleaching treatment process.

Although the demand for nickel is driven mainly by the world stainless steel market, it is anticipated that the demand for high-quality nickel and cobalt sulfate will surge as the trend towards electric vehicles and the development of battery-based technologies for large-scale energy storage increases. It is assumed that, if demand grows to the extent that is forecast, exploration may lead to the discovery of new nickel- and cobalt-containing sulfide deposits, which may increase the potential for future applications of bioleaching technologies, particularly since they are ideally placed to enable integration with downstream value addition.

References

Ahoranta S, Natunen H, Arpalahti A (2018) Process update on dynamic heap bioleaching of a black schist ore. In: Proceedings of the ALTA 2018 Nickel-Cobalt-Copper Conference. ALTA Metallurgical Services, Melbourne, pp 499–505

Arpalahti A (2017) Experiences on dynamic heap leaching of a black schist ore. In: Proceedings of the ALTA 2017 Nickel-Cobalt-Copper Conference. ALTA Metallurgical Services, Melbourne, pp 202–209

Gericke M, Neale JW, van Staden PJ (2009) A Mintek perspective of the past 25 years in minerals bioleaching. J S Afr Inst Min Metall 109:567–585

Gericke M, Pawlik C, Dew DW, van Aswegen P, Barnett SCC, Mikkola T (2014) Bioleaching of nickel-cobalt-arsenic containing flotation by-products of the talc industry. In: Proceedings of the Biohydrometallurgy'14 Conference, Falmouth, Minerals Engineering International

Heikkinen V, Korte M (2019) Bioheapleaching in boreal conditions–temperature profile inside the heaps and microbiology in elevated temperatures. In: Proceedings of the ALTA 2019 Nickel-Cobalt-Copper Conference. ALTA Metallurgical Services, Melbourne, pp 404–414

Johnson DB (2018) The evolution, current status, and future prospects of using biotechnologies in the mineral extraction and metal recovery sectors. Fortschr Mineral 8:343. https://doi.org/10.3390/min8080343

Laukka A, Seppälä J, Neale J, van Aswegen P, Barnett S, Taskinen A (2018) Mondo minerals nickel sulfide bioleach project: maximising value through gold recovery. In: Proceedings of the ALTA 2018 Nickel-Cobalt-Copper Conference. ALTA Metallurgical Services, Melbourne, pp 507–524

Neale J, Gericke M, Pawlik C, van Aswegen P, Barnett S, Seppälä J (2015) The Mondo minerals nickel sulfide bioleach project: from test work to design. In: Proceedings of the ALTA 2015 Nickel-Cobalt-Copper Conference. ALTA Metallurgical Services, Melbourne, pp 373–396

Neale J, Seppälä J, Laukka A, van Aswegen P, Barnett S (2016) The Mondo Minerals nickel sulfide bioleach project: Construction and commissioning. In: Proceedings of the ALTA 2016 Nickel-Cobalt-Copper Conference. ALTA Metallurgical Services, Melbourne

Neale J (2020) Bioleaching of nickel and cobalt–the progress and the potential. In: Proceedings of the ALTA 2020 Nickel-Cobalt-Copper Conference. ALTA Metallurgical Services, Melbourne

Puhakka JA, Kaksonen AH, Riekkola-Vanhanen M (2007) Heap leaching of black schist. In: Rawlings DE, Johnson DB (eds) Biomining. Springer-Verlag, Berlin, pp 139–151. https://doi.org/10.1007/978-3-540-34911-2_7

Riekkola-Vanhanen M (2007) Talvivaara project–Bioheapleaching of a complex sulfide ore in boreal conditions. In: Proceedings of the ALTA 2007 Nickel-cobalt conference. ALTA Metallurgical Services, Melbourne

Riekkola-Vanhanen M (2011) Talvivaara Sotkamo mine–bioheapleaching of a polymetallic nickel ore in subarctic conditions. In: Proceedings of the ALTA 2011 Nickel-Cobalt-Copper Conference. ALTA Metallurgical Services 150–163

Riekkola-Vanhanen M (2013) Talvivaara mining company–from a project to a mine. Miner Eng 48: 2–9. https://doi.org/10.1016/j.mineng.2013.04.018

Saari P, Riekkola-Vanhanen M (2012) Talvivaara bioheapleaching process. J South Afr Inst Min Metall 112:013–1020

Tuovinen H, Pelkonen M, Lempinen J, Pohjolainen E, Read D, Solatie D, Lehto J (2018) Behaviour of metals during bioheap leaching at the Talvivaara mine, Finland. Geosci 8:66. https://doi.org/10.3390/geosciences8020066

Chapter 13
Mineral Bioleaching in Brackish and Saline Environments

Miao Chen and Michael Schlömann

Abstract In addition to the limited availability of freshwater in arid regions, the presence of chloride in gangue or high-value minerals, possibly increased by evaporation processes, have the consequence and challenge, of carrying out biomining operations in the presence of elevated concentrations of chloride and other inorganic solutes. Examples of chemical leaching of minerals such as chalcopyrite in brackish and saline environments illustrate the potential of chloride as lixiviant and raise the question whether it is possible to combine bioleaching with chloride leaching. However, chloride inhibits bioleaching microorganisms to different degrees, while some possess physiological responses that allow them to tolerate relatively high levels of chloride. Advantages of NaCl-containing solutions for bioleaching have been published and a number of corresponding processes have been patented, but are not, as yet, used on an industrial scale. The use of chloride-tolerant strains of mineral-oxidising acidophiles together with measures to control chloride concentrations to below inhibitory levels could facilitate the development of industrial-scale biomining with brackish or saline waters.

Keywords Chloride leaching · Chloride inhibition · Halotolerant acidophiles · Mineral bioleaching

M. Chen (✉)
CSIRO Mineral Resources, Clayton South, VIC, Australia
e-mail: miao.chen@csiro.au

M. Schlömann
Technische Universität Bergakademie Freiberg, Institute of Biosciences, Freiberg, Germany
e-mail: michael.schloemann@ioez.tu-freiberg.de

D. B. Johnson et al. (eds.), *Biomining Technologies*,
https://doi.org/10.1007/978-3-031-05382-5_13

13.1 Introduction

13.1.1 Demand for Bioleaching in Saline/Brackish Waters to Meet Challenges

Over the past decades, bioleaching has been successfully applied to process a variety of sulfide ores that cannot be economically utilised by traditional methods. One of the challenges that biomining operations face is the need for water to run the process. In the larger context, efficiency of water usage has become a priority for global minerals industries, especially in regions with limited freshwater supplies, such as northern Chile, southern Peru, Australia, northern China, and some regions of North and South Africa, and Asia. Due to the high cost of sourcing freshwater, mining companies are searching for alternatives to freshwater in these arid regions, which often means using saline and brackish water supplies. In general, (i) waters with total dissolved salts or mineral constituents (TDS) concentration less than 3000 mg L^{-1} are considered freshwater; (ii) waters with 3000–10,000 mg L^{-1} TDS are considered to be brackish; (iii) waters containing 10,000 mg L^{-1} or more TDS are considered saline; (iv) groundwaters with salinity greater than seawater (about 33,000–48,000 mg L^{-1} TDS) are usually referred to as brines.

Even where saline or brackish waters are not used for irrigating bioheaps or dumps, biomining operations may have to deal with high TDS-containing lixiviants, since gangue minerals may dissolve and release ions, including chloride. Especially in arid regions, dissolved salts may reach very elevated concentrations, due to the evaporation of water from the heaps or dumps and due to the recycling of the raffinate or of water from tailings storage facilities. One specific source of chloride in heaps containing mixed oxidic and sulfidic copper ores may be the occurrence of atacamite ($Cu_2Cl(OH_3)$), which readily dissolves in acid and releases chloride.

13.1.2 Additional Opportunities of Bioleaching in Saline/Brackish Waters

Biomining with saline or brackish waters is not only an issue to meet the challenges described above, but also provides additional opportunities. This results from the fact that chloride has been demonstrated to promote mineral leaching under abiotic conditions by chemical means (Watling 2014). Metal chlorides are generally more soluble than corresponding sulfates. Furthermore, chloride salts appear to facilitate the diffusion of leaching agents via the formation of a more porous sulfur layer, thus improving the kinetics of the system. Consequently, there has been a number of industrial applications and smaller-scale investigations of purely chemical chloride leaching of minerals and other materials (Table 13.1).

One of the first processes using seawater for leaching was the Cuprochlor® process developed by Minera Michilla (patents CL 40891, EP 1559799B1) and

Table 13.1 Examples of nonbiological mineral processes using brackish/saline water

Leaching process/plant	Product	Chloride concentration ($g\ L^{-1}$) (chloride salt used)	pH	Temp °C	Other parameters	Reference/ patent number
Cuprochlor® Minera Michilla, Antofagasta minerals	Cu	30–130 (seawater or other saline water, $CaCl_2$)	<2.5	30–60	Agglomerated with $CaCl_2$ and H_2SO_4	CL 40891 EP 1424403 B1 EP 3246420 B1
BHP	Cu, Ni, Zn	130–230 (NaCl, $MgCl_2$, KCl or $AlCl_3$)	< 2.5		$E_H > 700$ mV	EP 2888380 B1
Platsol™	Au, Cu, Ni, PGM from complex concentrate	3–12 (NaCl)	–	200–225	O_2 over-pressure 7 bar	Aylmore (2016)
Brenda mine	Purification of Mo from Cu and Pb	84 ($FeCl_2$) 45 (NaCl) 29 (HCl)	–	100	Cl_2	US4500496
Intec/Nikko	Cu, Au, Ag	170 (NaCl)	<2	85–95	28 $g\ L^{-1}$ NaBr	Aylmore (2016)
Nichromet	Au	170–180 (NaCl)	1–2	35–45	Roasting as pretreatment 2.7–3 $g\ L^{-1}$ NaBr Cl_2	Aylmore (2016)

then by Antofagasta Minerals (patent EP 3246420 B1). It involves agglomeration with $CuCl_2$ and H_2SO_4, curing, leaching, and washing. More recently the "heap leaching method" (EP 2888380 B1) making use of chloride concentrations of 130–230 g L^{-1} was patented by BHP Billiton. It has been claimed that the Cuprochlor® process can be carried out with seawater directly and offers a number of advantages over bioleaching including higher recoveries and lower leaching times for secondary copper sulfides.

In the context of the most relevant applications, copper leaching from sulfidic ores, the leaching power is related to the fact that chloride ligand stabilises both cuprous and cupric ions (Watling 2014), thus avoiding cuprous ion disproportionation to cupric and elemental copper, and making it available for a cupric/cuprous redox cycle. The cupric ion in turn plays a crucial role, for example, in the mechanism of chalcopyrite leaching by attacking the H_2S probably released from chalcopyrite ($CuFeS_2$) under acidic conditions. Chalcopyrite is the most abundant copper mineral in the lithosphere, comprising roughly 70% of the world's accessible copper reserve. Chalcopyrite leaches much more slowly than copper oxides and other copper sulfides such as chalcocite (Cu_2S) and bornite (Cu_5FeS_4), and therefore improvement of chalcopyrite leaching is of high relevance.

Besides the effect on cupric and cuprous ion stability, the higher copper extraction from chalcopyrite in a chloride medium has also been associated with the lower activation energy for chalcopyrite dissolution. Chloride leaching of chalcopyrite is especially effective at elevated temperatures. The positive effect of chloride could be attributed to the formation of a more porous sulfur product thus allowing continued diffusion. A decreased passivation by sulfur layers obviously also plays a role in bioleaching of chalcopyrite.

While biomining is a proven technology to process low-grade ores, as described in sect. 13.1.1, the presence of elevated chloride concentrations can be problematic for various reasons (scarcity of freshwater, dissolution of minerals, recycling). The purely chemical processes mentioned above demonstrate the leaching power of chloride, and this raises the question whether the leaching power of chloride and microorganisms can be combined. Therefore, the application of biohydrometallurgy using seawater or saline water represents an attractive alternative for the mining industry, in particular in areas where freshwater sources are scarce. However, the efficacy of bioleaching may be compromised in the presence of saline and brackish waters as the majority of known acidophilic microorganisms are relatively intolerant of elevated concentrations of chloride (sect. 13.2.).

13.2 Inhibition of Bioleaching Microorganisms by Chloride

Given the potential advantages of bioleaching in the presence of chloride, growth, and iron or mineral oxidation of the more well-known mineral bioleaching microorganisms have been studied in media containing elevated concentrations of chloride (Zammit and Watkin 2016). It has been reported that iron oxidation activity of *Acidithiobacillus ferrooxidans* can be significantly (by ~50%) inhibited at concentrations of 100 mM chloride (NaCl or KCl) or even less, and more than 90% inhibited at concentrations below 200 mM (Rivera-Araya et al. 2020). Interestingly, sulfur oxidation, usually represented by oxidation of elementary sulfur, thiosulfate, or tetrathionate seems to be less affected than ferrous iron oxidation, though it should be considered that in some cases the starting pH was higher for growth with the sulfur compound than for that with iron (e.g., Zammit et al. 2012), a fact that will have influenced the inhibitory effect, as detailed below. Similar values as for *At. ferrooxidans* were also observed for iron oxidation of non-defined mixed cultures. Interestingly, over the years several acidophilic and halophilic or halotolerant iron-oxidising strains belonging to or related to the genus *Acidithiobacillus* have been isolated, e.g., from geothermal sites on islands in the Mediterranean Sea (Norris et al. 2020). Also, some *Leptospirillum* strains have been shown to be less sensitive to chloride than *At. ferrooxidans* (Zammit et al. 2012; Rivera-Araya et al. 2020), and adaptation to increased chloride levels has been reported. Thus, by *L. ferriphilum* Sp-Cl after prolonged incubation a redox potential of 600 mV (vs. Ag/AgCl) was still reached at ca. 600 mM chloride (WO 2010/009481A2). Acidophilic bacteria that can not only oxidise iron or sulfur, but also grow mixotrophically or heterotrophically (like *Sulfobacillus thermosulfidooxidans* or *Ferrimicrobium acidiphilum*) seem to tolerate higher chloride concentrations even without prolonged adaptation (Huynh 2021; Rivera-Araya et al. 2020). Iron oxidation by *Sb. thermosulfidooxidans* has been observed to be possible up to 600 mM NaCl and bioleaching of pyrite up to 400 mM NaCl (Huynh 2021). As with *At. ferrooxidans*, cells grown on a sulfur compound (tetrathionate) were reported to be more tolerant than ferrous iron-grown cells (Huynh 2021).

The extremely thermophilic archaea *Sulfuracidifex* (*Sulfolobus*) *metallicus* and *Acidianus brierleyi* have been reported to operate efficiently at chloride concentrations up to 1.1 M and therefore have been proposed to be "halophilic in character" (Deveci et al. 2008). *Sulfolobus acidocaldarius* has successfully been used for chalcopyrite leaching at 1 M NaCl, though no growth was observed under these conditions (Martins et al. 2019). On the other hand, *Sa. metallicus,* requiring media with less than 30 mM chloride, has been reported to be "particularly sensitive" to chloride in solution. Also, the review by Watling et al. (2016) on representatives of the genera *Sulfolobus/Sulfuracidifex, Acidianus* and *Metallosphaera* did not support activity at 1 M chloride or above, and included reports of lack of growth for two *Acidianus* species in the presence of 170 mM, growth "up to" 170 mM for *Metallosphaera cuprina*, minimum inhibitory concentrations of 320 mM chloride for several *Sulfolobus* species and tolerance of up to 513 mM for *Sa. metallicus*.

One reason for the relatively high sensitivity of acidophiles to chloride is considered to be the inside-positive membrane potential (Alexander et al. 1987). While sulfate can still be excluded from entering the cells, the inside-positive potential attracts chloride ions, thus reducing the membrane potential and favouring subsequent (or simultaneous) entry of protons, resulting in an acidification of the cytoplasm. The membrane potential is greatly affected by the pH of the external (bathing) liquor and becomes increasingly less positive and eventually negative) as the latter increases. This helps explain why the inhibitory effect of chloride is more pronounced at lower pH than at higher pH, while the inhibitory effect of cationic transition metals such as copper shows the opposite behaviour (Falagán and Johnson 2018). The pH dependency and the simultaneous effect of copper ions may explain some of the differences in the inhibitory effects referred to above.

Although bioleaching organisms are considered to be relatively tolerant to concentrations of sulfate as high as 1 M and sometimes greater, in addition to damaging the membrane potential and acidifying the cytoplasm, chloride will also contribute to the osmotic stress of the cells (Rivera-Araya et al. 2019). In addition, it has been reported that sodium chloride exposure induces oxidative stress in the cells (Rivera-Araya et al. 2019).

13.3 Responses and Adaptations to Chloride

As has been observed in neutrophilic microorganisms, acidophiles may respond to NaCl exposure by accumulating potassium and/or compatible solutes in the cells (Zammit and Watkin 2016). Genome analyses suggest that corresponding transport proteins or enzymes for the synthesis of, e.g., (hydroxy)ectoine or trehalose are present in acidophiles (Rivera-Araya et al. 2020). The formation of trehalose in response to NaCl has been reported for *L. ferriphilum*, *Sb. thermosufidooxidans*, and *Acidiphilium cryptum*, with *L. ferriphilum* additionally producing glucose, and *At. thiooxidans* glucose and proline, but not trehalose (Galleguillos et al. 2018). Proteomic analyses of *At. caldus* and *Acidimicrobium ferrooxidans* revealed up-regulation of proteins involved in membrane biosynthesis, in the synthesis of various amino acids possibly as a form of osmoprotectant and in CO_2 fixation (Zammit et al. 2012). Other compatible solutes that have been reported for acidophiles are taurine and glycine betaine (Zammit et al. 2012).

Of particular interest are the species of *Acidihalobacter*, since these are not only moderately halotolerant iron- and sulfur-oxidising acidophiles, but also appear to require some chloride (Khaleque et al. 2020b) and may be designated as moderate halophiles. The first strains of *Acidihalobacter prosperus* (formerly *Thiobacillus prosperus*) were isolated from the Mediterranean island, Vulcano. Recently, *Acidihalobacter aeolianus, Ah. ferrooxydans*, and *Ah. yilgarnensis* have been described as additional species based on strains isolated from similar locations (Khaleque et al. 2020b). Based on the genome sequence of *Ah. prosperus,* rusticyanin I has been suggested to have a more negative surface potential possibly explaining activity at elevated chloride concentration (Dopson et al. 2017). Proteomic and transcriptomic investigations of different *Acidihalobacter* representatives have revealed numerous chloride effects, among them increases in the expression of genes involved in ectoine biosynthesis, ectoine uptake, and redox balance (Khaleque et al. 2020a).

Halotolerant microorganisms of the genera *Acidithiobacillus, Alicyclobacillus*, and *Sulfobacillus* were detected (by 16S rRNA gene analyses) in a mine tailings-contaminated beach in Chile, and iron-oxidising enrichment cultures were found to be active in media containing up to 1 M NaCl (Korehi et al. 2013). *Alicyclobacillus* sp. S09 has recently been isolated from a tailing-contaminated beach in Spain and has been shown to oxidise iron in the presence of 1.5 M NaCl (Huynh 2021). Genome analysis suggested that it is a novel species of this genus, and numerous genes predicted to be involved in iron oxidation, heavy metal and arsenic tolerance, oxidative stress response, and osmoadaptation were identified.

13.4 Bioleaching Processes in the Presence of Chloride

Based on the work on *Acidihalobacter,* BHP Billiton filed a patent in 2001 (AU 2002254782B2) describing a process to leach sulfide concentrates using *Ah. ferrooxydans* V8 or *Ah. aeolianus* V6 (both formerly *T. prosperus*) at a pH below 3 and a chloride concentration in excess of ca. 280 mM. Leaching of copper from Escondida ore by *Ah. aeolianus* V6 has later been claimed to be possible at 430–860 mM NaCl (Davis-Belmar et al. 2008). While most *Acidihalobacter* strains tend to be relatively sensitive to copper, which lowers their potential for use in biomining operations, the recently described species *Ah. yilgarnensis* appears to be more tolerant (Khaleque et al. 2020a).

Two patents of JX Nippon Mining & Metals, filed in 2006, focussed on an *Acidithiobacillus* strain able to oxidise sulfur at high chloride concentration and thereby leach copper sulfide ores (US 8497113 B2) or more specifically chalcopyrite containing copper sulfide ores using chloride concentrations between 170 and 510 mM, and concentrations of copper (II) of 8–80 mM and iron (II) of 9–90 mM, respectively (AU 002007203317B2).

In 2008, BHP Billiton filed a patent that invoked using a mixed culture of a halophilic sulfur-oxidiser and *L. ferriphilum* Sp-Cl to leach sulfide minerals at chloride concentrations between 40 mM and 850 mM (US 8597933). Column experiments on a chalcocite-rich ore in a corresponding system with *L. ferriphilum* Sp-Cl and an *At. thiooxidans*-like strain with chloride concentrations up to 170 mM gave extraction yields of around 80% (Davis-Belmar et al. 2014). The authors concluded that "biologically assisted leaching ... in the presence of moderate chloride is achievable", that seawater may be incorporated and the recirculation of process waters may be increased. In 2010, BHP Chile filed a patent for a two-step process in which the minerals are first treated by chemical leaching using chloride concentrations between 0.2 and 2.3 M followed by a bioleach cycle using chloride concentrations below 0.17 M (WO 2012001501A1).

Other patents in this area include one filed by Biosigma (in 2013) using *Sulfobacillus thermosulfidooxidans* strain Cutipay, which the company claimed allowed "up to 73.7% (increase) in copper recovery from primary sulfides mainly chalcopyrite" (AU 2013405779 B2), and another by Compañía Minera Zaldivar (2014) which did not give specific strains, but referred to *Acidithiobacillus*, *Leptospirillum*, and *Sulfolobus* strains adapted to high chloride concentrations in an "annexed plant of bioreactors". According to the latter patent, the lixiviant uses Fe (II) and Fe (III) at redox potentials <550 mV (vs. Ag/AgCl) and chloride concentrations up to 5.6 M (WO 2016/26062 A1). A further patent application, filed by Universidad de Antofagasta in 2017, claimed to leach chalcopyrite by a consortium of *Acidithiobacillus* and *Acidiphilium* in the presence of 25%–100% seawater (WO 2019/126891A1).

In addition to patented developments in the minerals industry, numerous experiments on bioleaching in presence of chloride have been performed at research institutions, in many cases focussing on chalcopyrite. While relatively good leaching of chalcopyrite by *At. ferrooxidans* and *At. thiooxidans* in the presence of 100 mM NaCl had previously been described, Noguchi and Okibe (2020) reported a synergy for chalcopyrite leaching involving chemical chloride leaching and the "low-E_H-bioleaching" effect at 140 mM chloride; only chemical chloride leaching was observed in the presence of 560 mM chloride. With the moderately halophilic strain *Alicyclobacillus* sp. S09 chalcopyrite could be leached using artificial seawater, at rates considerably faster than under abiotic conditions (Huynh 2021).

While most bioleaching work has focussed on copper, the effect of chloride has also been investigated for some other metals, including zinc. While 50 mM chloride does not appear to impede the bioleaching of sphalerite (ZnS), inhibition in the presence of 140 mM chloride has been reported. Leaching of chalcopyrite by *Sb. thermosulfidooxidans* is not affected by the presence of 200 mM chloride, sphalerite leaching seemed to be inhibited under these conditions (Huynh 2021). Sadeghieh et al. (2020) have reported optimal leaching of copper, nickel, and cobalt from tailings by a moderately thermophilic consortium at 45 °C in the presence of 170 mM NaCl.

13.5 Summary and Future Prospects

In arid regions, where freshwater is not readily available, it would be highly advantageous if mineral processing in general, and bioleaching in particular, could be performed using brackish water or seawater. This would not only conserve water for domestic purposes but also avoid the significant capital and operating costs involved in desalination.

Bioleaching in the presence of elevated chloride concentrations may also be necessary where there are significant amounts of gangue or valuable minerals (such as atacamite) that contain chloride, especially if the raffinate is recycled. In addition, as shown by various large-scale processes, chloride itself has a strong leaching potential and it would be tempting to combine the leaching power of microorganisms that generate acidic ferric iron leachates with the leaching power of chloride. While this has resulted in a number of patents being filed and promising investigations published that advocate bioleaching in the presence of chloride, to date no full-scale industrial operation using this strategy has been established. This raises the question as to the obstacles to bioleaching in presence of chloride.

Compared to "normal" bioleaching, an obvious problem associated with using chloride is its corrosive nature. This necessitates the use of more costly corrosion-resistant materials, a fact which would be anticipated to play a bigger role in tank bioleaching than in heap bioleaching. On the other hand, the industrial processes with chemical chloride leaching show that such materials are available at a price, that under certain conditions may be competitive. In addition, combining bioleaching with abiotic chloride leaching holds the promise of possibly using lower chloride concentrations than for a purely chemical process, thereby probably lowering somewhat the costs incurred in utilising corrosion-resistant equipment.

Another factor to consider is that seawater has a pH of ~8.4 and contains large amounts of bicarbonate alkalinity, both of which would require greater acid demand (compared to freshwater) to generate the extremely acidic liquors required for oxidative bioprocessing of sulfidic ores. Brackish waters would have similar disadvantages. In addition, these waters contain other dissolved ions which may interfere with the leaching process or the winning of the target metal(s) from the pregnant leach solution.

Chloride inhibition of typical bioleaching microorganisms, combined with simultaneous effects of metal ions and low pH, remains the major barrier to making use of bioleaching in presence of chloride on a larger scale. However, the identification of chloride-tolerant acidophiles that can also tolerate elevated concentrations of dissolved metals continues to provide new candidates for such processes. Even with such acidophiles, it will be necessary to control the chloride concentration at an appropriate level in order to avoid toxic concentrations. This may be relatively easy in tank operations, but more of a challenge in bioheaps and dumps, especially in arid regions, where evaporation will tend to increase the salt concentration in a temperature-dependent way and thus with different rates depending on time and location in the heap. (Dilution by rain water would have the opposite effect.) An aspect related to counteracting evaporation is that any excess salt separated from the raffinate in the form of brines needs to be disposed of in a sustainable way.

Despite the obstacles mentioned, there is confidence among scientists researching this area that at least for specific ores or concentrates the combination of bioleaching and abiotic chloride leaching in the near future could become a robust and economically viable strategy to win metals.

References

Alexander B, Leach S, Ingledew WJ (1987) The relationship between chemiosmotic parameters and sensitivity to anions and organic acids in the acidophile *Thiobacillus ferrooxidans*. J Gen Microbiol 133:1171–1179. https://doi.org/10.1099/00221287-133-5-1171

Aylmore MG (2016) Alternative lixiviants to cyanide for leaching gold ores. In: Adams MD (ed) Gold ore processing, 2nd edn. Elsevier, pp 447–484. https://doi.org/10.1016/B978-0-444-63658-4.00027-X

Davis-Belmar CS, Nicolle JLC, Norris PR (2008) Ferrous iron oxidation and leaching of copper ore with halotolerant bacteria in ore columns. Hydrometallurgy 94:144–147. https://doi.org/10.1016/j.hydromet.2008.05.030

Davis-Belmar CS, Cautivo D, Demergasso C, Rautenbach G (2014) Bioleaching of copper secondary sulfide ore in the presence of chloride by means of inoculation with chloride-tolerant microbial culture. Hydrometallurgy 150:308–312. https://doi.org/10.1016/j.hydromet.2014.09.013

Deveci H, Jordan MA, Powell N, Alp I (2008) Effect of salinity and acidity on bioleaching activity of mesophilic and extremely thermophilic bacteria. Trans Nonferrous Met Soc China 18:714–721. https://doi.org/10.1016/S1003-6326(08)60123-5

Dopson M, Holmes DS, Lazcano M, McCredden TJ, Bryan CG, Mulroney KT, Steuart R, Jackaman C, Watkin ELJ (2017) Multiple osmotic stress responses in *Acidihalobacter*

prosperus result in tolerance to chloride ions. Front Microbiol 7:2132. https://doi.org/10.3389/fmicb.2016.02132
Falagán C, Johnson DB (2018) The significance of pH in dictating the relative toxicities of chloride and copper to acidophilic bacteria. Res Microbiol 169:552–557. https://doi.org/10.1016/j.resmic.2018.07.004
Galleguillos PA, Grail BM, Hallberg KB, Demergasso CS, Johnson DB (2018) Identification of trehalose as a compatible solute in different species of acidophilic bacteria. J Microbiol 56:727–733. https://doi.org/10.1007/s12275-018-8176-2
Huynh ND (2021) Effect of sodium chloride on acidophilic iron- and sulfur-oxidizing microorganisms and their bioleaching activity. Doctoral dissertation. TU Bergakademie Freiberg, Germany
Khaleque HM, Fathollazadeh H, González C, Shafique R, Kaksonen AH, Holmes DS, Watkin ELJ (2020a) Unlocking survival mechanisms for metal and oxidative stress in the extremely acidophilic halotolerant *Acidihalobacter* genus. Genes 11:1392. https://doi.org/10.3390/genes11121392
Khaleque HN, González C, Johnson DB, Kaksonen AH, Holmes DS, Watkin ELJ (2020b) Genome-based classification of *Acidihalobacter prosperus* F5 (=DSM 105917=JCM 32255) as *Acidihalobacter yilgarnensis* sp. nov. Int J Syst Evol Microbiol 70:6226–6234. https://doi.org/10.1099/ijsem.0.004519
Korehi H, Blöthe M, Sitnikova MA, Dold B, Schippers A (2013) Metal mobilization by iron- and sulfur-oxidizing bacteria in a multiple extreme mine tailings in the Atacama Desert, Chile. Environ Sci Technol 47:2189–2196. https://doi.org/10.1021/es304056n
Martins FL, Patto GB, Leão VA (2019) Chalcopyrite bioleaching in the presence of high chloride concentrations. J Chem Technol Biotechnol 94:2333–2344. https://doi.org/10.1002/jctb.6028
Noguchi H, Okibe N (2020) The role of bioleaching microorganisms in saline water leaching of chalcopyrite concentrate. Hydrometallurgy 195:105397. https://doi.org/10.1016/j.hydromet.2020.105397
Norris PR, Davis-Belmar CS, Calvo-Bado LA, Ogden TJ (2020) Salt-tolerant *Acidihalobacter* and *Acidithiobacillus* species from Vulcano (Italy) and Milos (Greece). Extremophiles 24:593–602. https://doi.org/10.1007/s00792-020-01178-w
Rivera-Araya J, Huynh ND, Kaszuba M, Chávez R, Schlömann M, Levicán G (2020) Mechanisms of NaCl-tolerance in acidophilic iron-oxidizing bacteria and archaea: comparative genomic predictions and insights. Hydrometallurgy 194:105334. https://doi.org/10.1016/j.hydromet.2020.105334
Rivera-Araya J, Pollender A, Huynh D, Schlömann M, Chávez R, Levicán GP (2019) Osmotic imbalance, cytoplasm acidification and oxidative stress induction support the high toxicity of chloride in acidophilic bacteria. Front Microbiol 10:2455. https://doi.org/10.3389/fmicb.2019.02455
Sadeghieh SM, Ahmadi A, Hosseini MR (2020) Effect of water salinity on the bioleaching of copper, nickel and cobalt from the sulphidic tailing of Golgohar iron mine. Iran Hydrometallurgy 198:105503. https://doi.org/10.1016/j.hydromet.2020.105503
Watling HR (2014) Chalcopyrite hydrometallurgy at atmospheric pressure: 2. Review of acidic chloride process options. Hydrometallurgy 146:96–110. https://doi.org/10.1016/j.hydromet.2014.03.013
Watling HR, Collinson DM, Corbett MK, Shiers DW, Kaksonen AH, Watkin ELJ (2016) Saline-water bioleaching of chalcopyrite with thermophilic, iron(II)- and sulfur-oxidizing microorganisms. Res Microbiol 167:546–554. https://doi.org/10.1016/j.resmic.2016.05.003
Zammit CM, Mangold S, Rao Jonna V, Mutch LA, Watling HR, Dopson M, Watkin ELJ (2012) Bioleaching in brackish waters–effect of chloride ions on the acidophile population and proteomes of model species. Appl Microbiol Biotechnol 93(1):319–329. https://doi.org/10.1007/s00253-011-3731-3
Zammit CM, Watkin ELJ (2016) Adaptation to extreme acidity and osmotic stress. In: Quatrini R, Johnson DB (eds) Acidophiles. Life in extremely acidic environments. Caister Academic Press, Norfolk, UK, pp 49–62. https://doi.org/10.21775/9781910190333

Chapter 14
Metal Recovery from E-wastes

Agathe Hubau and Christopher George Bryan

Abstract Waste electrical and electronic equipment (WEEE) is an important secondary resource of metals and critical raw materials. Printed circuit boards (PCBs), which account for the major fraction of valuable metals in e-waste, are mainly recycled using pyrometallurgy, though there is currently no suitable option for low-grade PCBs and more generally a lack of PCB processing capacity. Between 2001 and 2021, over 60 publications described the technical viability of bioleaching PCBs using acidophilic microorganisms to recover base metals. While most of these have been simple laboratory-scale tests, together with more comprehensive, larger-scale work carried out the state of the art by the end of 2021 was that: (1) indirect bioleaching is applicable at commercial scale; (2) adaptation of the microbial consortia to PCB leachate and raffinate should be possible; (3) microbial colonisation of PCB leaching reactors is not only possible, but advantageous. Furthermore, iron in the PCBs can be recycled and oxidised biologically, providing sufficient oxidant for the process to be continuous. However, some issues, such as acid consumption, are still pending. A few studies are now scaling up the process and aim to optimise the operating conditions as well as to gather techno-economic and environmental data to evaluate the commercial feasibility of PCB bioleaching. The integration of the process in the whole recycling chain, with consideration of PCB pre-treatment and metal recovery from the leachate and residue, is now required.

Keywords Bioleaching · Acidophiles · Electrical waste · Electronic waste · Printed circuit boards

A. Hubau (✉) · C. G. Bryan
BRGM, Orléans, France
e-mail: a.hubau@brgm.fr; c.bryan@brgm.fr

D. B. Johnson et al. (eds.), *Biomining Technologies*,
https://doi.org/10.1007/978-3-031-05382-5_14

14.1 Waste Electrical and Electronic Equipment and E-waste

Waste electrical and electronic equipment (WEEE) has been loosely defined as discarded, surplus, or broken electrical or electronic devices and represents the fastest growing and most complex global waste stream. Electronic waste (e-waste) is a subset of WEEE, encompassing waste electronic goods such as computers, televisions, and mobile phones. Given the rapidly increasing proliferation of personal computers and other electronic equipment and their comparatively short lifespans (typically less than 3 years), e-waste is considered to contribute the majority of WEEE by mass, and the two terms are often used interchangeably.

Globally e-waste production grew from an estimated 45 Mt in 2016 to over 50 Mt in 2019 and is projected to reach nearly 75 Mt by 2030 (Forti et al. 2020). These wastes are highly diverse, considering their usage, structure, and composition. They include many different individual components, including batteries, wiring and visual displays, and printed circuit boards (PCBs). Metals comprise over 50% of WEEE by mass. Most of this is iron and steel though large amounts of copper and aluminium and a significant quantity of high-value metals such as cobalt, nickel, and precious metals are also present (UNEP 2013).

There has been a major economic and societal transition since the 1970s with the emergence of digital information and communication technologies, sustainable mobility, alternative energies, etc., which has contributed to an increase in metal consumption worldwide, in terms of both volume and variety (discussed in Chap. 1). In this context, securing supplies of critical metals has become a major issue of concern to many developed and developing countries. Primary ore grades are declining and so WEEE represents an opportunity to recover metals on a local scale and to increase metal reserves by considering both primary and secondary resources. Many of the metals found in e-wastes are of significant strategic importance and are considered to be reaching their extraction peaks. China is the world's major exporter of electrical and electronic equipment (EEE). This is an "open loop" in that it can be viewed as a net export of valuable raw materials. However, reprocessing is not generally considered during the design and production of EEE, preventing WEEE from being processed via conventional recycling streams. Moreover, the collection rate of WEEE is very low worldwide, with only 8.9 Mt reported to be collected and formally treated in 2016, meaning the fate of over 80% of generated WEEE is unknown (Forti et al. 2020). This reduces pressure to develop reprocessing options, while disposal with normal household waste and improper processing leads to significant environmental pollution.

During e-waste recycling, the refining of metals occurs after selective disassembly, sorting, and upgrading steps which use a combination of manual and automated deconstruction, size reduction, and separation processes. The wastes are usually manually sorted to remove components, such as batteries, before being shredded. This stream then passes a series of separation (such as magnetic and Eddy-current separation) and grinding steps to upgrade the material. As a result, the waste is split into high-value streams which can be sold on for profit (for example, directly to metal producers) and a low-grade stream of difficult to separate, low metal content components and reject material from physical separation steps (e.g., dusts, tailings, and mixed/unsorted material). This low-grade stream is often sold off at a loss and includes certain grades of PCB and internal wiring and the lead glass from cathode ray tubes. These components are a significant hurdle to establishing viable e-waste reprocessing operations.

14.1.1 Printed Circuit Boards

Printed circuit boards contain more than 50 elements and compounds, from toxic pollutants such as polychlorinated and polybrominated biphenyls, to highly valuable metals such as gold, and platinum group metals. They consist of a nonconductive board, predominantly made of a polymer reinforced with glass fibres or ceramics, a conductive substrate, a thin layer of copper that can be on one side, both sides, or inside the board, and electronic components. The polymers may be phenolic resins for monolayer boards, cyanate, and epoxy resins for multi-layer boards, with flame retardants and hardeners. Various ceramics and silicate fibres may be found, as well as alkali-, alkali-earth, and aluminium oxides, and others. Electronic components vary depending on the utility of the object: there maybe power supply components, filters, and also transistors, operational amplifiers, resistors, diodes, coils, etc. PCB composition is therefore highly variable but can be considered to be a mixture of plastics (approx. 30% w/w), ceramic and glasses (30%), and metals (40%).

PCBs makeup just ~3% of WEEE mass but account for the major fraction of valuable metals (e.g., UNEP 2013). They contain many metals and metalloids that are classified as critical substances by the European Union due to their economic importance and their supply risk, including tin, tungsten, cobalt, germanium, gallium, indium, and (sometimes) tantalum. The current economic value of PCBs is mainly based on their precious metal content, which can contribute up to 90% of the PCB value while making up less than 1% of their weight. Given this, and their greater complexity, PCBs have been the target of most research into WEEE reprocessing. Therefore, while there is interest in the development of bioleaching approaches to recover metals from batteries (Li, Ni, Co) and phosphors used in light-emitting diodes (LEDs) and screens (REE, In, Sn), this chapter focuses on the bioprocessing of PCBs.

14.1.2 PCB Recycling

PCBs are usually recovered as a separate stream after dismantling WEEE. The development of metal recovery routes for this stream faces several difficulties: (1) the diversity of this waste in terms of size, shape, constituents, and composition;

Table 14.1 Some pyrometallurgical PCB processing operations at industrial scale and their capacities

Company	Location	Capacity (kt $year^{-1}$)	Primary feed	E-waste in feed (%)
Aurubis A.G. (Kayser Recycling System)	Lünen, Germany	260	Copper-bearing residues, alloys, and electronic scrap	No data
Boliden (Kaldo furnace)	Rönnskår, Sweden	120	E-waste	100[a]
Dowa Eco-system	Kosaka, Japan	140	Secondary copper resources	<100
Glencore (Fonderie Horne)	Rouyn-Noranda, Canada	840	Copper and precious metal-bearing materials	14
GRM Co.	Danyang, Korea	110	Secondary copper resources	<100
Umicore (Hoboken plant)	Hoboken, Belgium	500	Precious metal-bearing industrial residues and end-of-life material	10

[a]The smelted material ("black copper") joins the main smelter feed; e-waste accounts for approximately 10% of the total input of the Rönnskår facility

(2) the presence of many different chemical elements, which makes their purification difficult compared to primary ores or concentrates (which usually contain less than 20 elements in usually lower proportions (UNEP 2013); (3) the tight bonding of the different elements; and (4) the dispersion of metals within individual PCBs.

14.1.2.1 Pyrometallurgy

The pyrometallurgical route is currently, at industrial scale, the only choice for metal refining from PCBs, and e-waste in general. This is implemented at several sites throughout the world, currently three in Europe (Table 14.1). The main advantage is the very high recovery of both copper and precious metals (UNEP 2013). Operating costs can be reduced by using plastics as reducing agents as a substitute for coke and as a source of energy and PCBs do not need to undergo heavy upstream processes to liberate the metals; the majority of metal losses during pyrometallurgical processing of PCBs are a result of pre-treatment such as shredding (UNEP 2013). However, there are many drawbacks associated with pyrometallurgical treatments, including:

- High energy consumption.
- Loss of some metals in slags (including Al, Ti, Zn, Ga, Mo, and W).
- Generation of toxic gas and dusts due to the thermal decomposition of organic materials that require additional treatment units.
- Inability to recover plastics as materials or other products such as halogens or fuels.

Smelters are generally high-capacity installations and have turned to recycling metal-bearing wastes as a means to supplement shortfalls in their typical mineral concentrate feedstocks. The disparate nature of e-waste production means complex collection and processing networks are required to channel suitable wastes to large smelter installations. Consequently, the capacities of the processing units currently in operation are insufficient to process the whole flux of scrap PCBs and a requirement for the PCBs to enter as input streams is that the concentration of precious metals should exceed a set cut-off grade. Furthermore, the decrease of Cu and precious metals content in newer PCBs (due to improved manufacturing techniques) is resulting in a stream of waste PCBs less attractive as a feed for pyrometallurgy.

14.1.2.2 Hydrometallurgy

Hydrometallurgy uses leaching solutions (lixiviants) to solubilise target metals from (typically) finely ground substrate. The metal-rich leachate must then pass through a refining process to recover metals and/or metal products as, for example, hydroxides or sulfides. Hydrometallurgical processes can offer relatively low capital expenditure (CAPEX) and are particularly suitable for small-scale installations compared to pyrometallurgy. This is important for PCB and WEEE processing as a distributed network of smaller, more flexible installations may alleviate some of the collection, transport, and logistical challenges associated with WEEE. However, to date, certain bottlenecks have limited the application of such processes to PCBs.

Hydrometallurgical processes proposed for PCB feeds are based on pre-treatment steps, with milling and concentration, followed by metal leaching and refining. Many lixiviants have been tested at laboratory and pilot scales to recover both base and precious metals from PCBs as have many refining techniques such as solvent extraction and electrowinning (SX/EW), (bio)sorption, and precipitation.

Mechanical pre-treatments to prepare PCBs (shredding, grinding, and metal concentration) are challenging. In hydrometallurgical processes, the metal accessibility for the reactants determines the kinetics of reactions, and thus the size of process equipment, which is directly linked to the capital costs. However, mechanical pre-treatments are energy-intensive, result in metal losses (particularly precious metals) and dust generation. The costs of the lixiviants contribute to the overall operating expenditure (OPEX) of hydrometallurgical options, which may also include treatment of heavily polluted by-products in special waste disposal facilities and further limit the grade of PCB that can be processed economically.

The majority of research into PCB leaching remains relatively small-scale and somewhat disparate. The generation of operational and cost data via pilot-scale tests is essential in process development. In recent years, some pilots were developed to assess the economic and environmental impact of hydrometallurgical approaches. For example, the EU FP7-funded "HYDROWEEE DEMO" project[1] developed a

[1] https://cordis.europa.eu/project/id/308549

universal containerised pilot unit (2–3 m^3) to process different e-waste fractions (lamps, cathode ray tubes, liquid crystal displays, PCBs, and Li-batteries). Two pilot units were operated, a static plant that ran continuously for 18 months, and a transportable unit which was moved to five different locations to process different WEEE feeds. Such intermediate-scale studies allow the demonstration of technological feasibility as well as initial estimates of process economics.

By the end of 2021, despite a large number of patents to develop hydrometallurgical options for PCBs (48 in July 2017; Rocchetti et al. 2018), very few companies were commercialising hydrometallurgical processes for PCBs. Umicore uses hydrometallurgy to refine the Cu and Pb bullion they obtain through pyrometallurgy to recover precious metals (Ag, Au, Pt, Pd, Rh, Ir, Ru), Cu and In, Se, Te and Pb, Bi, Sb, Sn, and As. Mint Innovation (New Zealand) have commercialised a full-scale process, and are developing two hydrometallurgical plants (one in Australia and one in the UK) to recover base and precious metals from scrap PCBs.[2] The waste is milled, and base metals leached in a first step. Following filtration, the precious metals (mainly gold) are leached from the residue before being recovered by biosorption. The Mint flowsheet uses chemical (i.e., not biogenic) lixiviants, with biological processes used only in the purification of the solubilised gold ("biorefining"). In 2018, EnviroLeach (now EnviroMetal) Technologies Inc.[3] commissioned a hydrometallurgical plant using a circumneutral pH treatment process. The plant produced an initial 21 t gold and copper concentrate from 98 t of low-grade PCBs which was further processed by a pyrometallurgical plant (Glencore's Horne Smelter).

14.2 Biohydrometallurgy

Biohydrometallurgy (bioleaching/biomining) is a hydrometallurgical process that relies on biological (normally microbial) processes to generate lixiviants (biolixiviants) and can be categorised on the basis of the type of biolixiviant produced. These are organic acids and complexing agents produced by heterotrophic microorganisms, and inorganic acids and oxidants produced by (predominantly autotrophic) acidophilic chemolithotrophs.

14.2.1 Organic Biolixiviants

Two types of organic biolixiviants are typically proposed: (1) cyanide, produced from glycine transformation by cyanogenic microorganisms; and (2) organic acids

---

[2] https://www.mint.bio/

[3] https://envirometal.com/

such as lactic acid, oxalic acid, and citric acid, which are produced from organic carbon compounds such as glucose by various heterotrophic microorganisms (bacteria and fungi). These biogenic agents react by creating acidic leach liquors and/or by chelating metals to form stable metal complexes that will enhance their dissolution, an approach collectively referred to as "heterotrophic bioleaching".

In conventional biomining, as applied to mineral resources (particularly stirred-tank bioleaching of concentrates), the economic margins are often very tight, especially for base metal operations. The biotechnology has to compete directly with alternative hydrometallurgical options. In stirred-tank bioleaching, the provision of routine nutrients (NPK) constitutes the third-largest OPEX item after power consumption and pH control (van Aswegen et al. 2007); heterotrophic bioleaching requires relatively expensive organic substrates such as molasses (market cost ~US\$150–300 t^{-1}) or simple sugars such as glucose (~\$400 t^{-1})[4] on top of these. Furthermore, in such a nutrient-rich environment, careful process control is required to prevent the growth of unwanted organisms. Both of these considerations add significantly to the cost of metal production by heterotrophic processing, often rendering it uneconomic as well as technically challenging.

In an effort to overcome some of the complications of using heterotrophic leaching to recover precious metals from PCBs, initial leaching of base metals using inorganic leaching has been proposed, with subsequent organic leaching used to dissolve the precious metals. Although the recovery of precious metals affords greater flexibility since this can support higher production costs due to their higher market value, the question arises why use a complex biological system to produce organic lixiviants in the first place rather than purchase them as reagents from chemical manufacturers. The advantage of using biology in such a system can therefore become unclear.

Heterotrophic bioleaching (of either ores and concentrates or WEEE and PCBs) is currently restricted to laboratory-scale studies that indicate insurmountable problems, such as slow reaction kinetics, low leaching yields, expensive growth substrates, and high potential for contamination by other microorganisms. Heterotrophic leaching has not been employed at an industrial scale, and there have been no pilot-scale demonstrations of proof of concept or generation of useful process data. Before work continues in this area, the techno-economic case needs to be made that such an approach is at all feasible at a commercial scale.

14.2.2 Inorganic Biolixiviants

Commercial biomining processes exploit the activities of chemolithotrophic, extremely acidophilic microorganisms to catalyse the dissolution of sulfide minerals (in ores, concentrates, or mineral wastes), for example, to liberate gold from an

[4]Typical bulk purchase market prices in 2021.

arsenopyrite matrix or to dissolve base metals directly. The primary role of these organisms is to generate and maintain the sulfuric acid and ferric iron lixiviant solution through the oxidation of soluble ferrous iron, elemental sulfur, and reduced inorganic sulfur compounds (RISCs; discussed in Chaps. 2, 3, and 5). This biolixiviant oxidises the sulfide moiety, dissolving the solid phase through a process of oxidative dissolution, and (re)generates ferrous iron to be (re)oxidised by the microorganisms. A successful bioleaching operation relies on the microbial oxidation of the ferrous iron and "reduced S" (elemental sulfur and RISCs) occurring faster than the consumption of ferric iron and protons at the mineral surface, thereby maintaining a suitable, often relatively high redox potential and low pH environment.

A benefit of this approach is that it is autocatalytic: the products of (microbially mediated) sulfide mineral oxidation (ferric iron and sulfuric acid) are reactants for the process (reducing or eliminating the need for external sources of these oxidants). While it remains a niche technology, it has advantages over other hydrometallurgical processes in that (stirred-tank) leaching can be done at relatively low (compared to pyrometallurgy) temperatures and pressures, and heap leaching can process ores otherwise considered wastes due to their extremely low target metal concentrations. The particular challenges (beyond those common to all hydrometallurgical designs) come in providing an environment conducive to microbial growth and activity: regulation of temperature, pH, oxygen, and CO_2 supply, avoiding the build-up of toxic concentrations of dissolved elements and compounds, etc.

14.3 Bioleaching of PCBs

As of July 2021, over 60 peer-reviewed articles had been published on PCB bioleaching by acidophilic microorganisms, the majority of which are simple shake flask tests that vary some basic parameters (usually pH, temperature, microbial consortia, etc.). Due to the large variability of operating conditions and uncontrolled mass transfer, it is not possible to use shake flask experiments for any form of process optimisation (despite stated aims of doing so in some studies), though they are useful for initial screening and demonstration of basic concepts and there are more and more studies using bench-top to pilot-scale reactor systems and columns. Metal leaching kinetics for PCBs are highly variable (Fig. 14.1), mostly as a consequence of the large range of operating conditions (and microbial consortia) used.

Generally, studies have used some form of ferrous sulfate medium, inoculated with one or more iron-oxidising acidophiles, to which milled PCBs are added. One parameter that has been frequently studied is the impact of adding the PCBs once initial microbial growth and iron oxidation has occurred, or is well underway (a so-called "two-step", as opposed to "one-step" approach). Broadly, two-step leaching is more rapid though, in most cases, final leaching yields are high (>90%) for the majority of metals regardless of the choice of approach. While

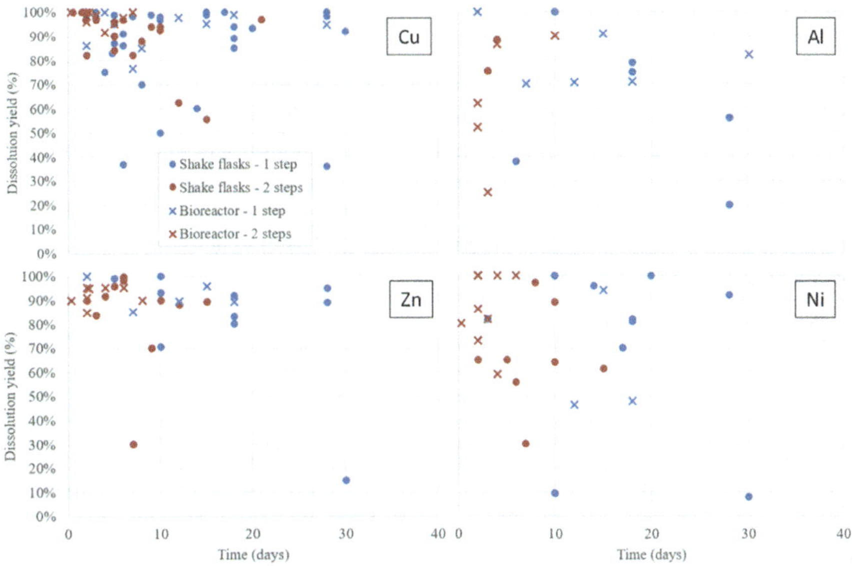

Fig. 14.1 Copper, aluminium, zinc, and nickel dissolution yields as a function of time, from 62 studies run in shake flasks (●, ●) or bioreactors (×, ×), either in one step (blue symbols) or two steps (red symbols). Time = time after PCB addition; denotes the contact time when the study was performed in batch mode, and the hydraulic residence time when the study was performed in continuous mode [Hubau (2019), and references therein]

final yields and dissolution rates have often been found to be as good in shake flasks as in reactors, this is explained by a combination of low pulp densities (usually up to 1% w/v) and rapid dissolution rates of elemental metals, as discussed below. These studies have so far demonstrated the concept, and the more detailed reactor and column studies provide some useful process data. Such scale-up is essential to determine optimal operating conditions for maximising bioleaching efficiency and to provide the necessary data to enable an initial techno-economic assessment of the concept as a whole.

The bioleaching of PCBs (and other e-wastes) may seem to be analogous to conventional bioleaching as described above: the microorganisms generate and recycle a lixiviant (ferric iron) which solubilises base metals from an inert matrix. However, there are several critical differences that present specific challenges in adapting bioleaching to PCBs, which are apparent, to a greater or lesser degree, in all these studies and can be considered as reaction kinetics, toxicity, acid consumption, and upstream and downstream linkages.

14.3.1 Reaction Kinetics

PCB bioleaching is based on two main reactions operating in tandem: metal dissolution in the presence of ferric iron (Eq. 14.1) and the biological regeneration of ferric iron (Eq. 14.2). The oxidation of the metals, often present in their zero-valent states, in PCBs is usually very fast; far more so than the oxidation of most sulfide minerals in conventional bioleaching. As a result, in PCB bioleaching microbial ferrous oxidation becomes the limiting step, and the system rapidly becomes depleted of the ferric iron oxidant.

$$M^0 + zFe^{3+} \rightarrow M^{z+} + zFe^{2+} \tag{14.1}$$

$$2Fe^{2+} + 0.5O_2 + 2H^+ \rightarrow 2Fe^{3+} + H_2O \tag{14.2}$$

Metals may also be dissolved in the presence of acid, as in Eq. (14.3) (for some metals such as Zn, Mn, Ni) or acid and oxygen, as in Eq. (14.4). These reactions may be thermodynamically favourable compared to the reaction with ferric iron [(Eq. 14.1); at least for Fe and Cu], but experimentally the reaction kinetics are much slower (e.g., Hubau 2019).

$$M^0 + zH^+ \rightarrow M^{z+} + 0.5zH_2 \tag{14.3}$$

$$M^0 + 0.5O_2 + 2H^+ \rightarrow M^{2+} + H_2O \tag{14.4}$$

Where the available metals and other oxidant-consuming reactants are in stoichiometric excess of the ferric iron, a distinctive two-phased dissolution profile is typically observed: an initial, very rapid dissolution of metals such as copper, followed by a much slower phase once the initial ferric iron has been consumed (Fig. 14.2). Where the stoichiometric ratio of ferric iron is sufficiently high, this second phase is not usually observed, which explains the apparently rapid reaction rates in two-step studies with low pulp densities. The second phase is primarily driven by the rate of proton-mediated dissolution (Eq. 14.3) and the rate of microbial (re)oxidation of ferrous iron (Eq. 14.2), although factors such as metal liberation (accessibility) and reactivity also play a role. Therefore, to maximise metal dissolution rates, ferric iron must be supplied in excess, or the net microbial iron oxidation rate must be greater than the rate of ferric iron consumption.

Multiple precipitation reactions may take place during bioleaching, and these can negatively impact metal dissolution. The predominant reaction is jarosite precipitation which reduces the availability of ferric iron. Metals such tin and lead are also prone to precipitate after their dissolution, particularly in sulfate-rich media. Cementation (electrochemical reactions between cationic metals and more electro-positive zero-valent metals) of metals such as copper may also occur, but the variability of composition of the leachates and the PCBs makes this hard to predict. Both reactions may create kinetic barriers and passivation, thus decreasing metal dissolution kinetics. Many studies use some form of modified “9K” medium (Chap. 6), which

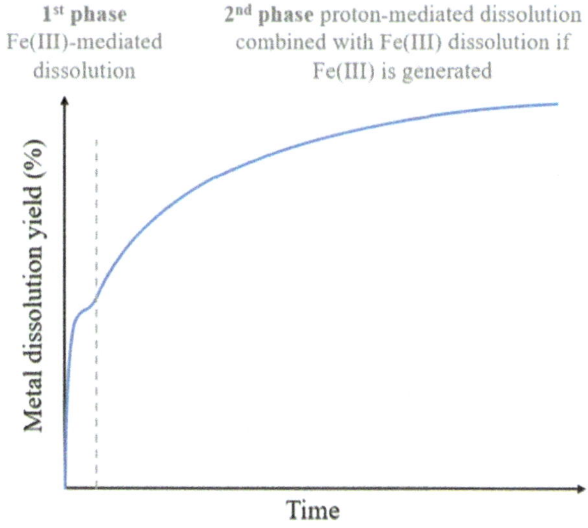

Fig. 14.2 Typical metal dissolution profile when metals are in stoichiometric excess of the Fe^{3+} concentration

generally contains excessive concentrations of ammonium which exacerbates problems linked with ammonium jarosite formation, and of phosphate, enhancing the precipitation of various metal phosphates. Indeed, it has been shown that reducing ammonium concentrations in growth media improves PCB bioleaching performance by minimising ferric iron loss through precipitation (e.g., Hubau et al. 2018). There have been no studies looking specifically at the effect of excess phosphate on the bioleaching of PCBs but it is known that ferric phosphate precipitates form readily in sulfide mineral bioleaching, reducing the available concentrations of the ferric iron oxidant.

14.3.2 *Toxicity and Inhibition*

Microbial iron oxidation rates are directly affected by factors such as gas mass transfer, temperature, and pH. Other factors such as dissolved CO_2 and nutrient concentrations indirectly affect oxidation rates through their impact on microbial growth and metabolic activity. Mass transfer can be optimised through reactor and impeller design (see Chap. 3), while temperature, pH, and nutrient supply can be selected as a compromise between the optima for the microbial consortium used (see Chaps. 5 and 7) and reaction rates and solubility. However, microbial growth is also greatly affected by the presence of toxic elements and compounds. Many studies indicate that PCB leachates are toxic, which is an issue that needs to be tackled to obtain effective dissolution kinetics. This inhibition is considered as one of the bottlenecks of bioleaching PCBs, and may be due to dissolved metals, the presence of plastics and resins (organics), other toxic elements such as bromine (from flame retardants) or, more likely, a combination of all three.

The toxicity of metal cations during PCB bioleaching has been studied using synthetic solutions containing various elements, pure or in mixture, at varying concentrations. The most commonly studied metals are copper, nickel, zinc, and aluminium, as these tend to occur in greater concentrations in PCB leachates. Some studies have also assessed the toxicity of more minor metal cations (such as Co and Cd; e.g., Hubau et al. 2020). There have been no studies published on metal anions in the PCB bioleaching context. Soluble As, which may come from GaAs chips, only exists in very low concentrations in leachates, if it is detected at all. Soluble Cr, which may come from the dissolution of steel components, is usually only detected at concentrations below 1 mg L^{-1} and probably exists as cationic Cr^{3+} under typical PCB leaching conditions. As in conventional bioleaching, mixtures of different metal ions can increase toxicity compared to the individual effect of each metal (e.g., Nurmi et al. 2009).

Most commercial bioleaching reactors operate at 40–45 °C and rely on iron-oxidising *Leptospirillum* and/or *Ferroplasma* spp. as the principal iron-oxidising organisms (Chap. 7), as do many of the published laboratory-based bioleaching studies of e-waste. PCB bioleaching poses particular challenges to the leptospirilli, as they have been shown to be sensitive to the polymetallic leachates, particularly Ni and Al and are known to be highly sensitive to soluble organic compounds which may occur in PCB leachates due to the decomposition of components such as thermosetting resins. Moreover, the highly reactive nature of the PCBs often prevents the development of a high redox environment favoured by these species.

14.3.3 Acid Consumption

When applied to PCBs and other e-wastes, the bioleaching process is not entirely autocatalytic as there is no inherent source of acid. During bioleaching, a source of protons is required to maintain low pH compatible with high microbial activity (Chap. 5) and minimise precipitation reactions, as well as forming part of the lixiviant itself. The leaching of PCBs is highly acid consuming and so significant quantities of acid are required. Based on metal content alone, theoretical acid consumption may be above 300 kg H_2SO_4 t^{-1} PCB, while experimental data imply it may be as high as 500–700 kg H_2SO_4 t^{-1} (probably due to additional acid-consuming reactions with nonmetallic components). On-site SX/EW of base metals from the pregnant leach solutions (PLS) will generate significant quantities of protons (Chap. 2), but even this acid is unlikely to fully compensate for acid consumption by reactions with PCB components and during microbial oxidation of Fe. Therefore, an external source of acid must be supplied, contributing to operating costs and negatively affecting overall economics.

14.3.4 Upstream and Downstream Linkages

The ultimate objective of R&D in this area is that PCB bioleaching becomes an established unit process option in the WEEE recycling toolbox. Apart from optimising the operating conditions, bioleaching of PCBs should be integrated into a complete recycling chain in which the pre-treatment of the waste, as well as the post-treatment of the leachate and residues, must be considered. To improve the dissolution of metals, their accessibility to the lixiviant needs to be maximised. Mechanical processes, and particularly shredding and grinding of waste PCBs, are usually the first step carried out ahead of bioleaching. However, this is not straightforward and often results in the production of large particles. To date, the influence of the upstream processes on PCB bioleaching are not fully considered or evaluated. Unlike classical bioleaching, the bioleaching of PCBs tends to show good dissolution of metals even with relatively coarse (>1 mm particle diameter) material. However, while this may minimise losses during milling (as well as potentially reduce pre-treatment costs), ensuring good mixing and preventing solids retention in a continuous system, such as a continuous stirred-tank reactor (CSTR), becomes a particular challenge.

PCB bioleaching studies have tended to focus on the dissolution of base metals, especially copper. However, in order for recycling to support a circular economy, processing must recover as much of the potential value as possible. Therefore, PCB bioleaching must be considered in the context of recovering more than just copper and readily soluble base metals, for example, how a bioleaching unit might impact downstream treatment options. This is essential because a consequence of using sulfuric acid–ferric iron biolixiviants, relatively non-soluble metals (such as Au, Pb, ad Sn) that make up significant proportions of the PCB value (as well as potential environmental impact) remain in the solid residue as undissolved or precipitated elements. Integrated flowsheets need to be developed to allow the recovery and valorisation of these metals, most likely using either non-biological hydrometallurgical or pyrometallurgical processes, yet few studies have considered this.

Some studies have proposed a two-stage inorganic and organic bioleaching approach, whereby inorganic bioleaching is used to first leach base metals, and organic/heterotrophic bioleaching is used to recover precious metals such as gold from the leach residue. However, such an approach is unlikely to be commercially viable for the reasons discussed in Sect. 14.2.1. Other studies have proposed the addition of chemical leaching units to recover gold from bioleaching residue and other approaches for the recovery of precious metals from (chemically) leached PCBs have been reviewed elsewhere (Sethurajan et al. 2019). However, such studies generally remain at the bench-scale and require further scale-up and evaluation of process data.

Hydrometallurgical and pyrometallurgical options for the recovery of precious metals from untreated or treated (chemically leached) PCBs exist at both pilot and commercial scale (as discussed in Sect. 14.1.2). However, there are few published data describing how an upstream bioleaching unit for base metal recovery may affect these processes, particularly hydrometallurgical options, nor how these may allow for the recovery of other valuable or problematic metals such as tin or lead Ultimately, it may be most appropriate to sell the precious metal-enriched bioleaching residues to existing established smelting or other processing operations, but this needs to be evaluated based on market price versus the additional capital and operating costs of recovering these metals within the process flowsheet. These options may be limited though as most smelters are primarily interested in the base metal (e.g., Cu) content, which would have been removed during bioleaching. Besides metals, PCBs also contain organic resins, bromine, and glass fibres, all of which might be recovered in a recycling process.

The recycling of process effluents is an important part of hydrometallurgical processes, either to minimise water consumption by recycling raffinate solutions or to increase target metal concentrations in intermediate leachates (Chap. 2). However, as discussed above, PCB leachates are often toxic, due to their complex nature. Without adequate purging, these inhibitory factors will become increasingly concentrated, potentially causing system failure. Therefore, further work is required to fully understand the toxicity of PCB leachates and raffinates in order to propose adequate management strategies.

14.4 PCB Bioleaching Strategies

In order to address the challenges of adapting bioleaching to PCBs, a number of solutions can be considered. Figure 14.3 gives a simplified overview of bioleaching strategies, and some of the major considerations.

14.4.1 Direct Versus Indirect Bioleaching

To try to eliminate the inhibition of microbial growth and activity associated with direct bioleaching of PCBs and to overcome reaction kinetics issues, a number of studies have opted for indirect bioleaching, in which the biological oxidation of ferrous iron is physically separated from the chemical oxidative dissolution (by ferric iron) of metals in PCBs. The simplest form considers two connected CSTRs, where ferrous iron is oxidized in the first reactor and fed to the second reactor containing ground PCBs.

Studies are increasingly moving towards using immobilised biomass systems for iron oxidation. Such an approach has a proven track record at commercial scale in the mining industry for the extraction of uranium, with indirect production of Fe^{3+} by a biofilm of *Acidithiobacillus ferrooxidans* (BacFox process, General Mining Group Laboratories, South Africa). The system was used at the Buffelsfontein uranium plant in a 10,000 t day^{-1} pachuca (a large airlift-type reactor, with a conical bottom and central draft tube) leach circuit (Ring 1980). Another commercial process based

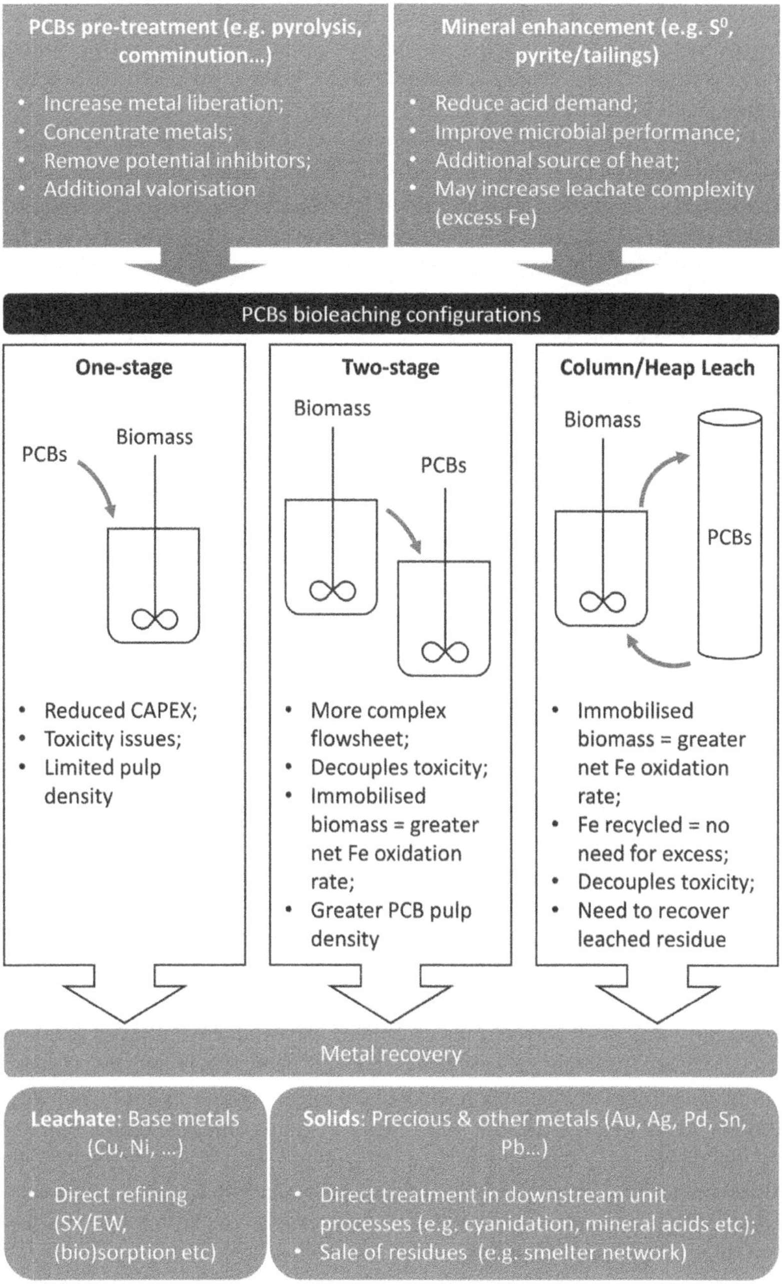

Fig. 14.3 Simplified overview of PCB bioleaching strategies

on the use of bacteria to regenerate ferric iron is currently under development for Orano in Niger (the IronBiox process), and the application of biohydrometallurgy to the processing of U ores from lab to commercial scale is reviewed elsewhere (e.g., Kaksonen et al. 2020).

The commercial success of the Bacfox process, as well as other studies on indirect mineral bioleaching, show that such an indirect bioleaching approach is technologically feasible at a commercial scale, albeit for different resources. Laboratory-scale studies have shown that very high volumetric iron-oxidation rates can be achieved using different immobilised biomass systems. Kinnunen and Puhakka (2004) reported ferrous iron oxidation rates of 26.4 g L^{-1} h^{-1} using a fluidised bed reactor though this system was sparged with 99.5% oxygen. When fed with atmospheric air, the oxidation rate was 8.2 g L^{-1} h^{-1}, which was still an order of magnitude higher than that achieved in the Bacfox process, though still in the absence of potential inhibitors introduced by recycling raffinate or intermediate leaching solutions (ILS).

An indirect PCB bioleaching process in a mini-pilot operation was run in continuous-flow mode for several months (Hubau et al. 2020). In the first stage, a bubble column (biooxidation reactor) containing immobilised biomass (activated charcoal) produced a Fe^{3+}-rich leaching solution that was delivered to a continuous stirred-tank reactor for PCB leaching. The operation of this laboratory-scale pilot demonstrated high metal dissolution yields for copper (96%), nickel (73%), zinc (85%), and cobalt (93%) with a hydraulic residence time of 48 h. Metal dissolution was further enhanced when the microorganisms, initially only found in the biooxidation stage, colonised the second stage. Simultaneous metal dissolution and Fe^{3+} generation by microbial Fe^{2+} oxidation were obtained using this configuration, demonstrating that a high bioleaching efficiency was achievable in steady-state conditions. The second stage maintained a high iron oxidation rate (~140 mg L^{-1} h^{-1}) resulting in a high redox potential ($E_H > 850$ mV). As a result, the primary biooxidation stage could be disconnected (i.e., the second stage could be run as a single-stage, direct bioleaching system) and the soluble iron contained in PCBs was sufficient for the process, and no external source of iron was required. The transitional regime, which took several weeks before achieving a high and stable microbial activity, was reduced to some hours by preparing the microbial community through successive subcultures in a PCB environment (Anaya-Garzon et al. 2021). However, raffinate recycling has not yet been tested and the system was observed to be less robust when challenged with sudden changes in operating conditions. Tests were carried out with 2% (w/v) solids, but the high redox potential shows that there was "excess" biological iron-oxidising capacity and that higher pulp densities could be achieved. Furthermore, biooxidation tests in leachates from different PCBs concentrations (<6% w/v) imply that toxicity may not be a limiting factor (Anaya-Garzon et al. 2021).

Regardless of overcoming potential toxicity issues related to high pulp densities, the system may ultimately be limited by the rate of ferric iron regeneration, and ultimately a two-stage (decoupled) approach may be the only viable option. Because of the high metal content, and reactivity of the PCBs, high ferric iron concentrations are usually required (>10 g L^{-1}), resulting in ferrous iron-rich leachates. This might be detrimental to the downstream metal recovery and purification steps. An alternative approach, demonstrated at laboratory-scale, is to use fixed bed columns in a heap leaching approach. For example, Ilyas et al. (2010) obtained high yields of metal dissolution using this approach (80% Zn, 64% Al, 86% Cu, and 74% Ni after 280 days). The recirculation of the leachate through the columns means minimal, if any, additional iron is required, and it does not need to be in stoichiometric excess as the solution is continually recycled. However, the scale-up of this approach to commercial-scale poses some challenges, such as how to recover and process the leached solid residue to valorise the precious (and other non-solubilised) metals as well as the typical challenges of heap bioleaching (Chap. 2).

To evaluate the impact of raffinate recycling on biooxidation, Looms (2014) used an immobilised biomass system (HDPE netting) with a synthetic raffinate from metals recovery from PCB leaching. The system was inoculated with a mixed consortium of *L. ferriphilum, L. ferrooxidans, Acidiferrobacter* spp., and *Acidithiobacillus* spp. and achieved an iron oxidation rate of 3.3 g L^{-1} h^{-1} with a simple ferrous iron medium (11.2 g L^{-1}; 201 mM). The addition of copper (2 g L^{-1}; 31 mM) and zinc (6 g L^{-1}; 92 mM) caused an immediate decline in iron oxidation rates, but this recovered and stabilised at approximately 1.6 g L^{-1} h^{-1}; still in excess of those reported for Bacfox. However, the addition of Al (at 9 g L^{-1}; 333 mM) had a more significant effect, indicating that further microbial adaptation would be required before the system could be used with a typical PCB leach raffinate.

Collectively, the results from these detailed studies demonstrate that indirect bioleaching is applicable at commercial scale, that adaptation to PCB leachate and raffinate should be possible and that microbial colonisation of PCB leaching reactors is not only possible, but advantageous. Furthermore, the iron in the PCBs can be recycled and oxidised biologically, providing sufficient oxidant for the process. At the same time, the accumulation of metals such as aluminium in the leach circuit as a result of recycling raffinate may prove to be a particular challenge that would need to be managed.

14.4.2 Mineral-Enhanced Bioprocessing

The work above suggests the processes can be partially autocatalytic, i.e., the PCBs can provide the source of iron in the lixiviant, though the process still lacks a source of acid. The use of pyrite as a source for the biolixiviant for PCB bioleaching was demonstrated by Bryan et al. (2015), proposing that PCBs could be co-processed with acidogenic mine wastes. Akbari and Ahmadi (2019) proposed blending PCBs with sulfidic (pyrite) tailings (from a desulfurisation plant at an iron complex) at a ratio 4:1 (tailings to PCB). At pilot scale (3×50 L continuous stirred-tank reactors), copper dissolution reached 95% with a 10 days residence time at 10% total solids. They further demonstrated successful copper extraction and recovery by SX/EW, though did not test the recycling of the raffinate or try to recover other valuable metals. This approach could be seen to some extent to mirror the way in which PCBs are blended with smelter feed, but in this case the PCBs are used as a way to augment the metal content of barren wastes. While this study demonstrated that the addition of the sulfidic tailings allowed the direct leaching of PCBs without the need to decouple iron oxidation, the feed to the reactor comprised 20% (w/w) PCB, meaning that the concentration of PCBs in the reactor was just 2% (w/v) and no economic evaluation of the process was carried out. Given that most tank bioleaching systems run at 20–25% (w/v) pulp density, it is important to see if the system could operate with greater concentrations of PCBs, and whether the presence of the sulfidic tailings results in a more robust system. When considering the whole process, one of the main disadvantages of such an approach lies in the complexity of recovering precious and other valuable insoluble metals, as the PCB residues will be blended with the leached tailings.

The CEReS co-treatment process[5] generates a lixiviant from sulfidic mining wastes by bioleaching sulfides, removing their potential to generate acid mine drainage (AMD) and allowing for their safe disposal or reuse. Concurrently, shredded PCBs are pyrolised, generating hydrocarbon fuel (which is used in cogeneration engines to power the process) from the catalytic cracking of the organic fraction, and a halogen-rich brine that can be sold. This allows greater valorisation of the PCBs and concentrates the metals in the char. Base metals are then solubilised from the char using the leachate from the bioleaching stage (to be recovered by SX/EW) while the precious and valuable undissolved metals remain in the final leach residue. An advantage of using a two-stage leaching process (leaching the mine waste in one stage, and the char in another) is that these undissolved metals become more concentrated in the leached residue).

The individual unit operations of the flowsheet were successfully demonstrated at both laboratory and pilot scale for scrap (low-grade) PCBs and sulfidic coal production wastes (Bryan et al. 2020). The study generated techno-economic feasibility data, and computer modelling and simulation demonstrated the technical viability of the process. While the economic assessment was unfavourable (the process did not create sufficient value to cover the operating costs), life cycle analysis (LCA) validated the environmental advantages of reprocessing such wastes in this way (over the current "do nothing" or "zero-option" scenario). Therefore, the value recovered from the low-grade waste PCBs off-sets the costs of permanently removing the AMD-generating potential from sulfidic wastes resulting in a durable long-term solution. The challenge is to optimise unit process performance (reduce OPEX) and maximise value recovery so that the cost per ton is less than the current best available options. For example the flowsheet design tested during the CEReS project did not include a model for the specific valorisation of precious metals but rather incorporated the sale value of the precious metal-bearing char leaching residue to a smelter. It also assumed that products made from the bioleaching residue had no market value. Nevertheless, co-processing two waste streams can provide a

[5] ceres.biohydromet.net

relatively low-cost waste treatment option resulting in demonstrable environmental benefits.

Although co-processing in this way may provide options for treating acidogenic mine wastes and tailings, it requires the co-occurrence of these two waste streams, and is not a stand-alone processing option for PCBs. Furthermore, the process contributes to overall iron concentrations in process effluents, which must be managed, with associated operating and disposal costs.

The use of elemental sulfur instead of sulfide minerals contributes a source of acid without increasing iron concentrations and may be attractive if a cheap source of elemental sulfur is available. Some studies (e.g., Liang et al. 2013) have proposed incorporating elemental sulfur in a single-step bioleaching approach, but this has not been scaled up and there are no data on the issues of toxicity as discussed above, or acid balance. Pakostova and Johnson (2019) demonstrated that a flow-through bioreactor could be used to generate sulfuric acid which could be combined with an additional (bio)lixiviant for the processing of mineral or metallic wastes such as PCBs in a two-stage process. In any case, the differences in the rates of acid consumption and biological acid generation would need to be addressed in practice.

The incorporation of sulfur or sulfide-bearing material provides an additional energy source for bioleaching microorganisms. This may facilitate greater resistance to certain toxic components, increase system biodiversity and provide additional surface for biomass formation. Moreover, certain iron-oxidising bacteria (such as *Sulfobacillus* spp.) and archaea either require or are more active in the presence of a reduced sulfur source. This is especially important where operating temperatures exceed those of iron-oxidising species such as *L. ferriphilum*.

A further effect of adding either sulfide minerals or elemental sulfur is heat generation during their oxidation. The heat of reaction for the complete oxidation of elemental copper is approximately -3.5 MJ kg^{-1} (-221 kJ mol^{-1}), whereas the heat of reaction of various sulfide minerals ranges from approximately -13 to -9 MJ kg^{-1} (-1800 to -1000 kJ mol^{-1}) and approximately -19.4 MJ kg^{-1} (-623 kJ mol^{-1}) for elemental sulfur (Chap. 3). Therefore, the addition of such materials in even modest ratios could significantly increase heat generation during processing. For example, data from the bioleaching of PCB/tailings study by Akbari and Ahmadi (2019) allow an estimation of over six times greater heat of reaction from the oxidation of the blended feed than PCB alone (and that the added pyrite would cover the total acid demand).

14.5 Conclusions

A significant number of small-scale studies have shown that biolixiviants from the oxidation of ferrous iron and reduced sulfur can facilitate the solubilisation of base metals from waste PCBs, whatever the configuration (single- or two-stage approach). The proof of concept of PCB bioleaching should be validated with further data regarding acid production, recycling of ILS and/or raffinates, and precious and base metal recovery. The scale-up of the process is required to obtain techno-economic data to ensure its possible commercial application.

References

Akbari S, Ahmadi A (2019) Recovery of copper from a mixture of printed circuit boards (PCBs) and sulphidic tailings using bioleaching and solvent extraction processes. Chem Eng Process 142:107584. https://doi.org/10.1016/j.cep.2019.107584

Anaya-Garzon J, Hubau A, Joulian C, Guezennec A-G (2021) Bioleaching of e-waste: influence of printed circuit boards on the activity of acidophilic iron-oxidizing bacteria. Front Microbiol 12: 2190. https://doi.org/10.3389/fmicb.2021.669738

Bryan CG, Watkin EL, McCredden TJ, Wong ZR, Harrison STL, Kaksonen AH (2015) The use of pyrite as a source of lixiviant in the bioleaching of electronic waste. Hydrometallurgy 152:33–43. https://doi.org/10.1016/j.hydromet.2014.12.004

Bryan CG, Williamson BJ, Całus-Moszko J, van Haute Q, Guezennec AG, Gaydardzhiev S, Wavrer P, Frączek R (2020) CEReS—co-processing of coal mine & electronic wastes: novel resources for a sustainable future. Hydrometallurgy 197:105444. https://doi.org/10.1016/j.hydromet.2020.105444

Forti V, Balde CP, Kuehr R, Bel G (2020) Quantities, flows and the circular economy potential. United Nations University/United Nations Institute for Training and Research, International Telecommunication Union, and International Solid Waste Association, Bonn, Geneva and Rotterdam

Hubau A (2019) Conception d'un procédé de biolixiviation pour la valorisation des métaux contenus dans les déchets de cartes électroniques (Design of a bioleaching process for the recovery of metals contained in spent Printed Circuit Boards). PhD. Université Paris Sciences et Lettres

Hubau A, Minier M, Chagnes A, Joulian C, Perez C, Guezennec AG (2018) Continuous production of a biogenic ferric iron lixiviant for the bioleaching of printed circuit boards (PCBs). Hydrometallurgy 180:180–191. https://doi.org/10.1016/j.hydromet.2018.07.001

Hubau A, Minier M, Chagnes A, Joulian C, Silvente C, Guezennec A-G (2020) Recovery of metals in a double-stage continuous bioreactor for acidic bioleaching of printed circuit boards (PCBs). Sep Purif Technol 238:116481. https://doi.org/10.1016/j.seppur.2019.116481

Ilyas S, Ruan C, Bhatti HN, Ghauri MA, Anwar MA (2010) Column bioleaching of metals from electronic scrap. Hydrometallurgy 101(3–4):135–140. https://doi.org/10.1016/j.hydromet.2009.12.007

Kaksonen AH, Lakaniemi A-M, Tuovinen OH (2020) Acid and ferric sulfate bioleaching of uranium ores: a review. J Clean Prod 264:121586. https://doi.org/10.1016/j.jclepro.2020.121586

Kinnunen PHM, Puhakka JA (2004) High-rate ferric sulfate generation by a *Leptospirillum ferriphilum*-dominated biofilm and the role of jarosite in biomass retainment in a fluidized-bed reactor. Biotechnol Bioeng 85(6):683–694. https://doi.org/10.1002/bit.20015

Liang G, Tang J, Liu W, Zhou Q (2013) Optimizing mixed culture of two acidophiles to improve copper recovery from printed circuit boards (PCBs). J Hazard Mater 250-251:238–245. https://doi.org/10.1016/j.jhazmat.2013.01.077

Looms J (2014) The study of a mixed culture biofilm reactor and the effect of high dissolved metal content and low pH on the rate of ferric iron regeneration for application in printed circuit board bioleaching. MSc, University of Exeter

Nurmi P, Özkaya B, Kaksonen AH, Tuovinen OH, Puhakka JA (2009) Inhibition kinetics of iron oxidation by *Leptospirillum ferriphilum* in the presence of ferric, nickel and zinc ions. Hydrometallurgy 97(3):137–145. https://doi.org/10.1016/j.hydromet.2009.02.003

Pakostova E, Johnson DB (2019) Microbial generation of sulfuric acid from granular elemental sulfur in laboratory-scale bioreactors. Hydrometallurgy 190:105152. https://doi.org/10.1016/j.hydromet.2019.105152

Ring RJ (1980) Ferric sulphate leaching of some Australian uranium ores. Hydrometallurgy 6(1): 89–101. https://doi.org/10.1016/0304-386X(80)90010-9

Rocchetti L, Amato A, Beolchini F (2018) Printed circuit board recycling: a patent review. J Clean Prod 178:814–832. https://doi.org/10.1016/j.jclepro.2018.01.076

Sethurajan M, van Hullebusch ED, Fontana D, Akcil A, Deveci H, Batinic B, Leal JP, Gasche TA, Ali Kucuker M, Kuchta K, Neto IFF, Soares HMVM, Chmielarz A (2019) Recent advances on hydrometallurgical recovery of critical and precious elements from end of life electronic wastes—a review. Crit Rev Environ Sci Technol 49(3):212–275. https://doi.org/10.1080/10643389.2018.1540760

UNEP (2013) In: Reuter MA, Hudson C, van Schaik A, Heiskanen K, Meskers C, Hagelüken C (eds) Metal recycling: opportunities, limits, infrastructure. A report of the working group on the global metal flows to the international resource panel

van Aswegen PC, van Niekerk J, Olivier W (2007) The BIOX™ process for the treatment of refractory gold concentrates. In: Rawlings DE, Johnson DB (eds) Biomining. Springer, Berlin, pp 1–33. https://doi.org/10.1007/978-3-540-34911-2_1

Chapter 15
Reductive Mineral Bioprocessing

Ana Laura Santos and Axel Schippers

Abstract While biomining is currently restricted to reduced (sulfide) ores, many commercially valuable metals can be found in significant concentrations in oxidised ores. These comprise laterites, polymetallic marine nodules, and ores from oxidation of sulfide deposits. Currently, these oxide ores are processed using pyro- or hydrometallurgical techniques, but these can have several drawbacks which have restricted their exploitation. This chapter summarises reductive bioprocessing options for metal oxide ores and focuses chiefly on laterites for recovery of nickel and cobalt. Over the past 40 years, several laboratory studies have demonstrated the possibility of bioleaching saprolitic and limonitic laterite ores, as well as tailings, using organic acids generated by heterotrophic bacteria or fungi. However, pilot-scale tests have not been reported and the viability of this approach is questionable. The anaerobic reductive dissolution of iron and manganese oxy-hydroxide minerals coupled to the oxidation of elemental sulfur is catalyzed by acidophilic, chemolithotrophic bacteria such as *Acidithiobacillus ferrooxidans* and has shown to be a more promising approach, especially for bioprocessing of limonitic laterite ores, as an integral part of the *Ferredox* process. Aerobic reductive dissolution of laterites with *Acidithiobacillus* species has also been demonstrated at low pH (<1). These promising bioprocessing options for limonitic laterites are currently awaiting full process development.

Keywords Reductive bioleaching · Laterites · Cobalt · Nickel · *Ferredox* process · *Acidithiobacillus*

A. L. Santos
Science Group, Natural History Museum, London, United Kingdom
e-mail: a.santos@nhm.ac.uk

A. Schippers (✉)
Federal Institute for Geosciences and Natural Resources (BGR), Hannover, Germany
e-mail: axel.schippers@bgr.de

D. B. Johnson et al. (eds.), *Biomining Technologies*,
https://doi.org/10.1007/978-3-031-05382-5_15

15.1 Introduction

Biomining is a well-established global biotechnology which uses aerobic, acidophilic microorganisms to catalyze the oxidative dissolution of sulfide minerals present in ores, concentrates, and mining wastes. Mineral bioprocessing at commercial scale is currently restricted to sulfide ores where the target metal is either surrounded by sulfide minerals obstructing its extraction (e.g., refractory gold ores) or present within the structure of host minerals (most base metals). In both cases, mineral dissolution occurs via oxidative processes in extremely acidic conditions and in the presence of acidophilic prokaryotes, whose main role is to generate ferric iron and sulfuric acid. Many commercially valuable metals are also, or sometimes exclusively, found in (iron and manganese-rich) oxide ore bodies, and in some cases their mineral reserves are more extensive (and accessible) than their sulfide counterparts (e.g., nickel in laterites).

While oxidative bioleaching of sulfide ores has been studied extensively and applied commercially for over 50 years, bioprocessing of silicate and oxide ores falls behind in theory and application. Oxide ores comprise laterite ores, polymetallic marine nodules, and oxidised ores from sulfide deposits. Currently, such oxide ores are processed, if at all, using pyro- or hydrometallurgical techniques (Stanković et al. 2020). Apart from environmental issues, these have been associated with high capital and operational costs due to high demand for energy (e.g., smelters and high-pressure acid leaching—HPAL) or slow leaching rates (e.g., heap leaching). The mining industry is therefore seeking alternative processes to cope with both continuous depletion of high-grade sulfide ore reserves and tightening environmental regulations.

The use of microbially mediated reductive processes to extract economically valuable metals from oxide (and oxy-hydroxide-containing) ores is a novel biotechnology that is currently receiving considerable attention. This chapter provides an overview of the mechanisms and recent applications of bioreductive dissolution of metal oxide ores with a focus on laterite ores for recovery of nickel and cobalt.

15.2 (Bio)Hydrometallurgical Processing of Oxide Ores

15.2.1 Laterite Ores

Laterites are supergene ore bodies (i.e., they occur relatively near the surface) originating from chemical and mechanical weathering processes of ultramafic rocks in tropical and subtropical areas. Chemical weathering processes mobilise the most soluble elements (Mg, Ca, and Si) and concentrate the least soluble elements (most transition metals as well as aluminium). These iron-rich ore deposits contain several types of iron oxides, hydroxides, and oxy-hydroxides in which atoms of iron (mostly ferric iron) are linked to oxygen and/or hydroxyl groups. Laterite ore

Fig. 15.1 Laterite ore deposits. (**a**) Zoned profile of the Çaldag mine (Turkey). A green saprolite zone at the base is covered with a thick brown limonite layer (goethite-rich zone) and a white silica-rich cap; (**b**) ancient weathering profile at the Treni mine (Albania). At the base of the pit is a thin saprolite zone, which is overlain by goethite-rich, partly reworked limonite zone. The weathered profile is covered by later sedimentary deposits (sandstones and limestones)

deposits follow a similar weathering profile (Fig. 15.1) which is generically divided into three main layers. At the bottom, there are clay silicates with dominant nickel-smectites, above which there is a silicate-rich saprolitic layer, often rich in magnesium-nickel hydrous silicates. On top, there is an oxide zone of limonitic laterite, consisting of iron oxides dominated by goethite (α-FeO·OH) or limonite (FeOOH·nH_2O), and manganese oxides such as asbolane ($(Ni,Co)_xMn(O,OH)_{4n}H_2O$) and lithiophorite ($(Al,Li)MnO_2(OH)_2$; Butt and Cluzel 2013).

Although complex in structure, laterite deposits may contain exploitable reserves of nickel in one or more of these layers, and therefore they have been commercially defined as "Ni-laterite ores". In addition to nickel, laterite deposits may also include appreciable amounts of cobalt, copper, and scandium and, in some cases, rare earth elements. Nickel in limonitic ores is typically associated with ferric iron minerals (e.g., goethite) whereas cobalt is associated with Mn(IV) minerals, such as asbolane. Rare earth elements are generally associated with phosphate minerals present in lateritic deposits (Ñancucheo et al. 2019). Laterite ores usually contain between 0.8–3% nickel and 0.05–0.2% cobalt, though geochemical and mineralogical characteristics may differ significantly even amongst neighbouring deposits (Table 15.1).

Production of nickel from laterite deposits in New Caledonia began in 1875. However, with the discovery of sulfide deposits containing nickel and copper in Canada in the late 1800s, the focus diverted towards processing of sulfide ores, and by the 1950s, approximately 90% of the nickel was produced from the latter. Nickel laterites account for 72% of the world's nickel reserves and are mostly found in equatorial regions (e.g., Southeast Asia, Northern Brazil, Northern Australia, and Cuba), though some of these deposits occur in nontropical areas, such as Greece in Europe, the Urals in Russia, Turkey and Kazakhstan in Asia, and USA (Oregon, California and North Carolina; US Geological Survey 2020).

Complex and heterogeneous mineralogy, costly energy requirements, and remoteness from processing and distribution facilities are amongst the reasons why laterite processing was overlooked in the past (Marrero et al. 2020). However, due to a greater demand for nickel (and cobalt) in the last few decades, the development of new processing technologies as well as the rapid depletion of sulfide ore deposits, nickel production from laterite ores increased to 46% by 2008, exceeded 50% of global production in 2010 and it is expected to reach 72% by 2022 (Oxley et al. 2016).

Nickel laterites currently contribute 20–30% of total global supply of cobalt. This metal is mostly obtained as a co-product of copper and nickel sulfide ore processing, with the Democratic Republic of Congo currently being the world's leading source. The global demand for cobalt has increased exponentially over the past 30 years, reflecting its increased use in high-tech materials (e.g., rechargeable batteries) as society moves towards a more sustainable economy. In 2011, the European Commission added cobalt and other materials to the list of "*critical raw materials (CRMs) for the European Union economy*", which are fundamental to industry, essential for enabling technological development, and in need of reliable and sustainable supply.

Currently, nickel extraction from laterite ores is mainly performed using pyrometallurgical techniques for the production of ferro-nickel and matte smelting, though this is only suitable for saprolite zones of lateritic ores, and limonite layers are often not utilised. Hydrometallurgical processing for nickel and cobalt recovery

Table 15.1 Geochemical composition of different limonitic laterite deposits

	Çaldag, Turkey	Piaui, Brazil	Nkamouna, Cameroon	Acoje, Philippines	Shevchenko, Kazakhstan	Kastoria mine, Greece	Evia mine, Greece	Ag Ioannis mine, Greece	Penamax, New Caledonia	Tiebaghi, New Caledonia
SiO_2[a]	35.4	48.8	12.4	14.2	42.7	30.6	36.2	26.9	2.15	1.36
Fe_2O_3[a]	39.3	27.5	40.8	45.3	25.3	22.9	30.7	49.4	69.6	47.8
Cr_2O_3[a]	1.5	0.86	1.01	2.27	0.96	0.99	1.52	2.69	2.95	1.85
Al_2O_3[a]	4.6	2.37	20.9	7.99	4.33	0.99	3.92	5.49	5.5	18.6
MnO[a]	0.4	0.41	6.98	0.61	1.72	0.30	0.28	0.29	1.19	7.34
MgO[a]	1.71	7.63	0.36	4.31	8.37	17.4	6.99	3.46	0.47	0.71
Ni[a]	1.16	1.78	0.87	0.99	1.47	1.02	0.56	0.82	1.35	0.75
Co[b]	710	902	10850	626	2700	334	274	516	1640	12450
Cu[b]	60	1710	553	108	35	17	23	38	56	85
Sc[b]	40	13	43	58	24	19	24	47	59	42
As[b]	163	<5	6	8	8	<5	17	7	<5	<5
V[b]	117	45	15	271	128	66	122	197	157	128

[a]wt. %
[b]$mg\ kg^{-1}$

Table 15.2 Summary of hydrometallurgical techniques used for extraction of nickel from laterites

Process	HPAL	Caron Process	Heap Leaching	Acid leaching	Direct Nickel	Neomet process
Ore type	Limonite	Limonite	Saprolite	Limonite and Saprolite	Limonite and Saprolite	Limonite and Saprolite
Lixiviant	H_2SO_4	NH_3-$(NH_4)_2CO_3$	H_2SO_4	H_2SO_4	HNO_3	HCl
Leaching time	90 min	n.a.	120–150 days	12 h	2–4 h	n.a.
Temp (°C)	245–250	850	Ambient	95	105	100–110
Ni and Co extraction (%)	90–95	80–85	70–80	85–95	> 90	> 95

Modified from Stanković et al. (2020)

can, however, be applied for both, saprolitic and limonitic layers of laterite ores. Table 15.2 summarises existing metallurgical technologies developed for extraction of nickel from laterites. All these methods require high energy and/or reagent consumption, expensive capital equipment costs, and incur several technical and environmental challenges. In addition, hydrometallurgical processing of limonitic ores results in the co-dissolution of gangue minerals, increasing the complexity and cost of recovering valuable metals as well as treatment and disposal of wastes. For this reason, in most existing mines, limonitic laterite ores are currently being stockpiled as mining waste.

Although it is widely acknowledged that processing limonite can be a major solution to meet the future demand of Ni and contribute to the supply of Co, Cu, Sc, and V, there is still a lack of novel and sustainable robust processing routes allowing reduced energy and reagent inputs and producing non-polluting residues. Biohydrometallurgy has a potentially major role in this context.

15.2.2 Biological Processing of Ni–Co Laterites

Although still mostly studied at laboratory scale, the biological processing of lateritic ores has recently received more attention due to the increasing demands for nickel and cobalt. Over the past 40 years, several studies have demonstrated the use of acid bioleaching of both saprolitic and limonitic laterite ores, as well as laterite tailings by organic acids generated by heterotrophic bacteria or fungi (e.g., Bosecker 1977; Nasab et al. 2020). Metal dissolution by heterotrophic microorganisms generally involves an indirect process with microbial production of organic acids, such as citric, oxalic and gluconic, as metabolic by-products (Bosecker 1986). Solubilisation of metals occurs by direct displacement of metal ions from the ore matrix by protons and by the formation of soluble metal complexes and chelates. Bioprocessing of laterites using filamentous fungi of the genera *Aspergillus* and *Penicillium* with the production of organic acids and other metabolites had been demonstrated to be effective in previous studies (Bosecker 1986; Coto et al. 2008). *Bacillus* spp. have been shown to solubilise nickel from a low-grade nickel saprolite ore at circum-neutral pH values (Giese et al. 2019). However, heterotrophic (fungal and bacterial) approaches have a number of issues which may impact bioleaching operations and downstream processing, such as large biomass production, cost of growth substrates, prevention of growth of undesired microorganisms, stability of metal–organic acids complexes, adsorption of metals by fungal biomass and relative low dissolution rates.

Many species of acidophilic bacteria are well known for their ability to catalyze the oxidative dissolution of metal sulfide minerals. Some of them, including species of both heterotrophic and autotrophic acidophiles, are facultative anaerobes and have been shown to be able to catalyze the dissimilatory reduction of soluble ferric iron to ferrous iron [reviewed in Marrero et al. (2020)] and, in some cases, to mediate the reductive dissolution of ferric iron minerals (Bridge and Johnson 1998). Brock and Gustafson (1976) first reported that the chemolithotrophic prokaryotes *At. ferrooxidans*, *At. thiooxidans*, and *Sulfolobus acidocaldarius* were able to reduce soluble ferric iron when growing on elemental sulfur as an energy source, but it was not confirmed that these acidophiles could actually respire on ferric iron. Pronk et al. (1992) later demonstrated that *At. ferrooxidans* was able to grow by using ferric iron as an alternative electron acceptor to oxygen, and Bridge and Johnson (1998, 2000) reported that moderately thermophilic iron-oxidising bacteria and *Acidiphilium* SJH (an obligately heterotrophic and mesophilic acidophile) were able to solubilise a range of ferric iron-containing minerals (e.g., goethite and magnetite) under anaerobic conditions. Subsequent to this, Hallberg et al. (2011) screened four pure cultures of acidophilic bacteria for their ability to accelerate the reductive dissolution of a low-grade Ni-laterite ore using relatively low temperatures (< 30–45 °C) and acidic conditions ($pH < 2$). The acidophilic heterotroph *Acidicaldus organivorus* (using glycerol as electron donor) and the chemolithotroph *At. ferrooxidans* (using elemental sulfur) were able to solubilise nickel present in the ore under anaerobic conditions.

Reductive mineral dissolution requires the provision of an extraneous electron donor since the mineral itself does not contain the energy supply to promote growth of the microorganisms. Both organic and inorganic substrates can be provided depending on the energy requirements of the microorganisms driving iron reduction. For heterotrophic iron-reducing prokaryotes, small molecular weight organic compounds (such as glucose and glycerol) are often the substrate of choice, though complex organic carbon compounds, such as those from agricultural or food industries (e.g., sugar beet molasses) might also be considered. These substrates can, however, significantly increase operational costs, and contamination by undesirable bacteria and fungi is also highly likely.

In the case of chemolithotrophic microorganisms, such as the facultative anaerobe *At. ferrooxidans*, the oxidation of inorganic compounds (e.g., H_2 and elemental sulfur) can be coupled to the reduction of ferric iron in the absence of oxygen. In extremely acidic conditions (pH <2), chemolithotrophic iron-reducing acidithiobacilli use ferric iron as electron acceptor when oxygen is absent, since (1) ferric iron tends to be more bioavailable due to its greater solubility, and (2) the high redox potential of the Fe(II)/Fe(III) couple (~700 mV at pH 2, in sulfate-rich liquors) makes ferric iron a thermodynamically attractive alternative electron acceptor to molecular oxygen.

Reductive dissolution of ferric iron oxy-hydroxides, such as goethite, is highly consumptive of protons, though when sulfur is used as electron donor less acid is required per mol of goethite (Eq. 15.1) than when reductive dissolution is coupled to, for example, glucose oxidation (Eq. 15.2). In addition, the direct acid dissolution of goethite (Eq. 15.3) consumes almost twice as many protons than sulfur-enhanced reductive bioleaching.

$$6\,FeO \cdot OH + S^0 + 11\,H^+ \rightarrow 6\,Fe^{2+} + HSO_4^- + 8\,H_2O \quad (15.1)$$

$$6\,FeO \cdot OH + 0.25\,C_6H_{12}O_6 + 12\,H^+ \rightarrow 6\,Fe^{2+} + 1.5\,CO_2 + 10.5\,H_2O \quad (15.2)$$

$$6\,FeO \cdot OH + 18\,H^+ \rightarrow 6\,Fe^{3+} + 12\,H_2O \quad (15.3)$$

In addition, using elemental sulfur as an electron donor for iron reduction has other advantages: (1) it is a more cost-effective alternative to organic electron donors, (2) elemental sulfur is produced in vast quantities as a secondary product, for example, in removing hydrogen sulfide from natural gas reserves, (3) elemental sulfur oxidation (coupled to oxygen (Eq. 15.4) or soluble ferric iron reduction) generates sulfuric acid which helps to maintain the pH at suitable levels for acidophilic bacteria and retaining metals in solution, and (4) since most acidophiles that oxidise elemental sulfur are autotrophic, using sulfur for enhancing bioleaching consumes rather than produces CO_2.

$$S^0 + 1.5\,O_2 + H_2O \rightarrow HSO_4^- + H^+ \quad (15.4)$$

Coto et al. (2008) compared the use of organic and inorganic bio-acids on the recovery of cobalt and nickel from laterite tailings. Sulfuric acid was biologically generated by the oxidation of elemental sulfur by *At. thiooxidans* in aerobic conditions. Results showed that production of sulfuric acid by sulfur oxidation was more effective in extracting nickel than organic acids produced by fungi. In this study, 80% Co and over 99% Mn and Ni present in the tailings were solubilised in a period of 15 days. The advantage of biologically produced sulfuric acid as leaching agent in comparison to hydrometallurgical sulfuric acid leaching is that the cost of sulfur is less than that of sulfuric acid, and generating sulfuric acid biologically at a remote mine site can reduce hazards and costs involved in its transportation, though an additional bioreactor would be required.

While acid leaching seems to be the only (indirect) bioprocessing route for saprolitic laterites, du Plessis et al. (2011) developed an innovative biohydrometallurgical approach to extract valuable metals from limonitic laterites

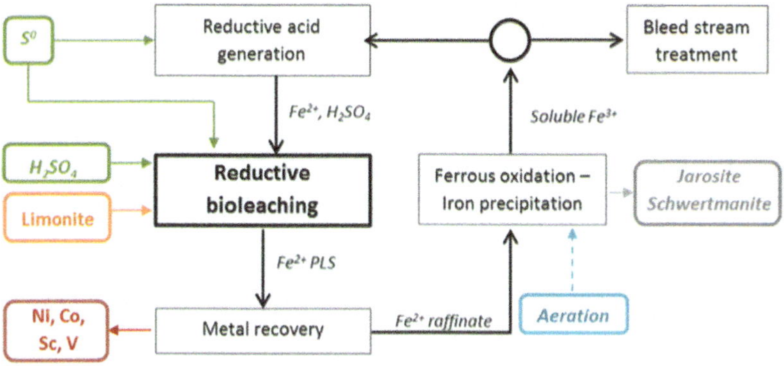

Fig. 15.2 Bioprocessing of laterite ores based on *Ferredox* concept (du Plessis et al. 2011, modified)

using mild pH and temperature. The *Ferredox* process was firstly designed to treat limonitic ores for nickel and cobalt recovery by means of anaerobic reductive dissolution of iron and manganese oxy-hydroxide minerals. Overall the proposed process consists of four components: (1) acid-consuming reductive leaching of limonitic laterite ores, (2) recovery of valuable metals (nickel, cobalt, and copper) by direct sulfide precipitation or solvent extraction or ion exchange, (3) aerobic oxidation and precipitation of ferric iron as jarosites or schwertmannite, and (4) a reductive generation of sulfuric acid which can then be used to assist the leaching step (Fig. 15.2; du Plessis et al. 2011).

The anaerobic reductive bioleaching stage of the *Ferredox* process was described previously by Hallberg et al. (2011). This consisted of two stages: an aerobic phase in which oxidation of sulfur generated acidity and promoted bacterial growth, and an anaerobic mineral leaching stage. The abiotic dissolution of ferric iron minerals proceeds via proton attack which releases ferric iron (Eq. 15.3). The abiotic rate of ferric iron solubilisation by acid is relatively slow, and is related to the crystallinity of the mineral. In the presence of iron-reducing acidophiles, soluble ferric iron is reduced relatively rapidly which causes a disequilibrium between ferric iron in the mineral phase and that in solution thereby supporting the chemical dissolution of the ferric mineral.

Another important reaction in the anaerobic reductive dissolution of laterite ores is the dissolution of manganese oxy-hydroxide minerals. As mentioned previously, cobalt (and some nickel) is chiefly concentrated in manganese oxy-hydroxides such as asbolane and lithiophorite. The sulfur-enhanced reductive dissolution of a generic manganese oxy-hydroxide mineral can occur either directly via biological dissimilatory reduction of Mn(IV) to Mn(II) (Eq. 15.5) or indirectly via manganese reduction by ferrous iron (Eq. 15.6) derived from goethite dissolution (Eq. 15.1). The resulting soluble ferric iron may then be biologically reduced (du Plessis et al. 2011).

$$Mn_3O_3(OH)_6 + S^0 + 5\,H^+ \rightarrow 3\,Mn^{2+} + 5\,H_2O + HSO_4^- \quad (15.5)$$

$$Mn_3O_3(OH)_6 + 6\,Fe^{2+} + 12\,H^+ \rightarrow 3\,Mn^{2+} + 6\,Fe^{3+} + 9\,H_2O \quad (15.6)$$

Anaerobic sulfur-enhanced reductive bioprocessing of limonitic laterite, laterite overburden, and processing residues using pure cultures or consortia of acidophilic bacteria has been investigated at laboratory scale for extracting different primary target metals, including nickel, cobalt (Johnson et al. 2013; Marrero et al. 2020), and copper (Ñancucheo et al. 2014). Table 15.3 shows examples of studies on metal extraction from laterite ores and processing residues using acidophilic, sulfur-oxidising bacteria.

As an alternative approach to anaerobic processing, aerobic reductive dissolution of laterites with *Acidithiobacillus* species has been demonstrated at low pH (<1), including the use of pure cultures of *At. thiooxidans* (Marrero et al. 2015, 2017). Since dissimilatory reduction of ferric iron has not been described for *At. thiooxidans* the question of the mechanism of how this occurs arises. It is possible that intermediary sulfur compounds, such as thiosulfate, formed during enzymatic oxidation of elemental sulfur to sulfuric acid, serve as a chemical reductant for iron and manganese oxides, as suggested for enhanced dissolution of seafloor manganese nodules in aerobic bioleaching experiments with *At. thiooxidans* (Kumari and Natarajan 2001), though this has not been proven. Aerobic reductive bioleaching has some potential advantages over anaerobic reductive dissolution of laterites, including a lower requirement for acid and the lack of a requirement to ensure oxygen-free conditions. Aerobic reductive bioleaching using *At. thiooxidans* was reported to be more efficient in extracting total iron, ferrous iron, manganese, and cobalt than the anaerobic process using *At. ferrooxidans* (Marrero et al. 2015, 2017, 2020). The downsides include the relatively slow abiotic reduction of ferric iron in cultures of sulfur-oxidising acidophiles such as *Acidithiobacillus caldus* and the fact that in a commercial operation it would hardly be feasible to exclude iron-oxidising bacteria that are also active at pH ~1, such as *Leptospirillum* spp., that would probably regenerate ferric iron and thereby counteract the reductive bioleaching process.

15.2.3 Biological Reductive Dissolution of Other Oxide Minerals

Besides iron and manganese oxides in limonitic laterites, reductive bioleaching could be applied for dissolution of other oxide minerals or as a pre-treatment step for refractory ores such as gold, platinum group element (PGE) oxide ores (Hedrich et al. 2020) or rare earth elements (Ñancucheo et al. 2019).

There is considerable potential for applying reductive bioleaching to extract metals from mining and industrial wastes. For example, a combination of oxidative and reductive bioleaching was shown to be highly effective in extracting Cu from tailings (Falagán et al. 2017). Extraction of Al and rare earth elements from red mud

Table 15.3 Examples of laboratory-scale studies on bioleaching of laterite ores and processing residues using acidophilic, sulfur-oxidising bacteria

	Coto et al. (2008)	Hallberg et al. (2011)	Marrero et al. (2015)	Marrero et al. (2017)	Smith et al. (2017)	Santos et al. (2020)
Sample type	Laterite tailings	Limonite	Laterite tailings	Laterite overburden	Limonite	Limonite, filter dust, and slag
Aerobic/anaerobic	Aerobic	Anaerobic	Aerobic/Anaerobic	Aerobic	Anaerobic/aerobic	Anaerobic
Organism	*At. thiooxidans*	*At. ferrooxidans*	*At. thiooxidans*, *At. ferrooxidans*	*At. thiooxidans*	Acidophilic consortium	Acidophilic consortium
Pulp density (% w/v)	10	5	10	5	5	5
Leaching time (days)	12–15	20–30	7	7	22	25–30
Temp (°C)	30	30	30	30	35	35
pH	1.5	1.8	1.8	0.8	1.8	1.5
% Co extracted	~70	~50	55–60	85	99/70	39–49
% Ni extracted	80	70	53–57	16	70/20	37–68

has been demonstrated by a two-stage aerobic and anaerobic bioleaching process. In the anaerobic stage *Acidianus manzaensis* dissolved jarosites via ferric iron reduction coupled with sulfur oxidation (Zhang et al. 2020). Reductive dissolution of jarosite, schwertmannite, and other ferric iron-containing minerals by heterotrophic acidophiles such as *Acidiphilium* species have also been demonstrated (Bridge and Johnson 2000).

Deep-sea polymetallic deposits such as manganese nodules and crusts also represent an important resource of metals, including Co, Cu, Ni, V, and Mo. They consist mainly of manganese and iron oxides with valuable metals incorporated within the structure of the host minerals. Conventional pyro- and hydrometallurgical techniques as well as bioleaching can be applied to process marine nodules (Kumari and Natarajan 2001). Since the minerals in the nodules are present in their oxide form, acidophilic bacteria are able to reduce iron and manganese oxides via sulfuric acid production both aerobically and anaerobically. Heller and Schippers (2015) reported preliminary results of aerobic reductive bioleaching of manganese nodules using a mixed culture of acidophilic chemolithotrophic iron- and sulfur-oxidising bacteria (*At. thiooxidans*, *At. ferrooxidans*, *L. ferrooxidans*, *L. ferriphilum*) and *A. cryptum*. Data showed that 40% Ni, 25% Cu, 1.2% Mn, 0.3% Co, 1% Fe, 70% Zn, and 70% Zr were leached from the Mn-nodules after 56 days. Chemical anaerobic reductive dissolution was also tested in this study by incremental addition of soluble ferrous iron. Ferrous iron was capable of reducing Mn (IV) to Mn (II) with up to 82% Ni, 98% Co, 68% Cu, and 97% Mn leached in these experiments.

15.3 Summary

This chapter has reviewed the fundamentals of mineral reductive bioprocessing and highlighted its challenges and potential advantages over existing pyro- and other hydrometallurgical techniques. The bioprocesses described have the potential to increase metal recovery in existing mines and to transform the categorisation of some unexploited ores, limonite stockpiles, and tailings from laterite ore processing, as well as deep-sea nodules, into valuable resources. Most studies are, however, still at the laboratory stage and additional pilot-scale operations are required to explore the technical (e.g., bioleaching in heaps, ponds, or large tank reactors) and economic potential of this new development in biomining technology.

Acknowledgements Ana Laura Santos is grateful for the financial support from the European Union Framework Programme for Research and Innovation Horizon 2020 (CROCODILE project; grant reference 776473) and the Natural Environment Research Council (UK) under its Security of Minerals Supply programme (grant reference NE/M010899/1). Axel Schippers thanks for financial support from the German BMBF (CLIENT II project BioProLat, FKZ 033R271A). Also, we would like to thank Prof. Richard Herrington (Natural History Museum, London, UK) for providing the photographs and Dr. Paul Schofield and Dr. Agnieszka Dybowska (Natural History Museum, London, UK) for sharing the data on geochemical composition of laterites.

References

Bosecker K (1977) Studies in the bacterial leaching of nickel ores. 1977. In: Schwartz W (ed) Conference bacterial leaching 1977. GBF Braunschweig, Verlag Chemie, Weinheim, pp 139–144

Bosecker K (1986) Leaching of lateritic nickel ores with heterotrophic microorganisms. In: Salley J, McCready RGL, Wichlacz P (eds) Biohydrometallurgy international symposium proceedings. CANMET, Ottawa, pp 15–24

Bridge TAM, Johnson DB (1998) Reduction of soluble iron and reductive dissolution of ferric iron containing minerals by moderately thermophilic iron-oxidising bacteria. Appl Environ Microbiol 64:2181–2186. https://doi.org/10.1128/AEM.64.6.2181-2186.1998

Bridge TAM, Johnson DB (2000) Reductive dissolution of ferric iron minerals by *Acidiphilium* SJH. Geomicrobiol J 17:193–206. https://doi.org/10.1080/01490450050121161

Brock TD, Gustafson J (1976) Ferric iron reduction by sulphur- and iron-oxidizing bacteria. Appl Environ Microbiol 32:567–571. https://doi.org/10.1128/AEM.32.4.567-571.1976

Butt CRM, Cluzel D (2013) Nickel laterite ore deposits: weathered serpentinites. Elements 9:123–128

Coto O, Galizia F, Hernández I, Marrero J, Donati E (2008) Cobalt and nickel recoveries from laterite tailings by organic and inorganic bio-acids. Hydrometallurgy 94:18–22. https://doi.org/10.1016/j.hydromet.2008.05.017

du Plessis CA, Slabbert W, Hallberg KB, Johnson DB (2011) Ferredox: a biohydrometallurgical pro-cessing concept for limonitic nickel laterites. Hydrometallurgy 109:221–229. https://doi.org/10.1016/j.hydromet.2011.07.005

Falagán C, Grail BM, Johnson DB (2017) New approaches for extracting and recovering metals from mine tailings. Miner Eng 106:71–78. https://doi.org/10.1016/j.mineng.2016.10.008

Giese EC, Carpen HL, Bertolino LC, Schneider CL (2019) Characterization and bioleaching of nickel laterite ore using *Bacillus subtilis* strain. Biotechnol Progress 35:e2860. https://doi.org/10.1002/btpr.2860

Hallberg KB, Grail BM, du Plessis CA, Johnson DB (2011) Reductive dissolution of ferric iron minerals: a new approach for bio-processing nickel laterites. Miner Eng 24:620–624. https://doi.org/10.1016/j.mineng.2010.09.005

Hedrich S, Kraemer D, Junge M, Marbler H, Bau M, Schippers A (2020) Bioprocessing of oxidised platinum group element (PGE) ores as pre-treatment for efficient chemical extraction of PGE. Hydrometallurgy 196:105419. https://doi.org/10.1016/j.hydromet.2020.105419

Heller C, Schippers A (2015) Geobiotechnical recovery of metals from manganese nodules: first experiments. Adv Mater Res 1130:201–204. https://doi.org/10.4028/www.scientific.net/AMR.1130.201

Johnson DB, Grail BM, Hallberg KB (2013) A new direction for biomining: extraction of metals by reductive dissolution of oxidised ores. Fortschr Mineral 3:49–58. https://doi.org/10.3390/min3010049

Kumari A, Natarajan KA (2001) Electrobioleaching of polymetallic ocean nodules. Hydrometallurgy 62:125–134. https://doi.org/10.1016/S0304-386X(01)00192-X

Marrero J, Coto O, Goldmann S, Graupner T, Schippers A (2015) Recovery of nickel and cobalt from laterite tailings by reductive dissolution under aerobic conditions using *Acidithiobacillus* species. Environ Sci Technol 49:6674–6682. https://doi.org/10.1021/acs.est.5b00944

Marrero J, Coto O, Schippers A (2017) Anaerobic and aerobic reductive dissolutions of iron-rich nickel laterite overburden by *Acidithiobacillus*. Hydrometallurgy 168:49–55. https://doi.org/10.1016/j.hydromet.2016.08.012

Marrero J, Coto O, Schippers A (2020) Metal bioleaching: fundamentals and geobiotechnical application of aerobic and anaerobic acidophiles. In: Lee NM (ed) Biotechnological applications of extremophilic microorganisms, Life in extreme environments 6. De Gruyter, Berlin, pp 261–287. isbn:978-3-11-042773-8

Ñancucheo I, Grail BM, Hilario F, du Plessis C, Johnson DB (2014) Extraction of copper from an oxidised (lateritic) ore using bacterially-catalysed reductive dissolution. Appl Microbiol Biotechnol 98:6297–6305. https://doi.org/10.1007/s00253-014-5687-6

Ñancucheo I, Oliveira G, Lopes M, Johnson DB (2019) Bioreductive dissolution as a pretreatment for recalcitrant rare-earth phosphate minerals associated with lateritic ores. Fortschr Mineral 9: 136. https://doi.org/10.3390/min9030136

Nasab MH, Noaparast M, Abdollahi H, Amoozegar MA (2020) Kinetics of two-step bioleaching of Ni and co from iron rich-laterite using supernatant metabolites produced by *Salinivibrio kushneri* as halophilic bacterium. Hydrometallurgy 195:105387. https://doi.org/10.1016/j.hydromet.2020.105387

Oxley A, Smith ME, Caceres O (2016) Why heap leach nickel laterites? Miner Eng 88:53–60. https://doi.org/10.1016/j.mineng.2015.09.018

Pronk JT, de Bruyn JC, Bos P, Kuenen JG (1992) Anaerobic growth of *Thiobacillus ferrooxidans*. Appl Environ Microbiol 58:2227–2230. https://doi.org/10.1128/AEM.58.7.2227-2230.1992

Santos AL, Dybowska A, Schofield PF, Herrington RJ, Johnson DB (2020) Sulfur-enhanced reductive bioprocessing of cobalt-bearing materials for base metals recovery. Hydrometallurgy 195:105396. https://doi.org/10.1016/j.hydromet.2020.105396

Smith SL, Grail BM, Johnson DB (2017) Reductive bioprocessing of cobalt-bearing limonitic laterites. Miner Eng 106:86–90. https://doi.org/10.1016/j.mineng.2016.09.009

Stanković S, Srećko S, Sokić M, Marković B, Friedrich B (2020) Review of the past, present and future of the hydrometallurgical production of nickel and cobalt from lateritic ores. Metall Mater Eng 26(2):199–208. https://doi.org/10.30544/513

U.S. Geological Survey (2020) Mineral commodity summaries 2020. U.S. Geological Survey, 200 p. https://doi.org/10.3133/mcs2020

Zhang DR, Chen HR, Nie ZY, Xia JL, Li EP, Fan XL, Zheng L (2020) Extraction of Al and rare earths (Ce, Gd, Sc, Y) from red mud by aerobic and anaerobic bi-stage bioleaching. Chem Eng J 401:125914. https://doi.org/10.1016/j.cej.2020.125914

Chapter 16
Biological Removal and Recovery of Metals from Waste Streams and Process Waters

Sabrina Hedrich and Päivi Hanna-Maarit Kinnunen

Abstract Waste streams and process waters frequently contain metals of economic importance. Recovery of these resources would help to secure the supply of metals for industry and domestic consumption. Contaminants are also often a problem for further treatment of these waste streams and effluents, and need to be destroyed or removed to overcome potential pollution hazards. The development and optimisation of methods in biohydrometallurgy have opened new opportunities for the processing of low-grade waste materials and metal-containing effluents with potential for metal recovery. Bioleaching has been successfully applied in full-scale operations for reprocessing mine tailings and removing contaminants from concentrates, and promising results have been obtained for metallurgical waste and industrial residues. Biological treatment of process waters via bioprecipitation and biosorption has been used successfully in laboratory to full-scale operations to recover transition metals and remove contaminants from both acid mine drainage waters and mineral-processing effluents. Biological methods can also be selective and applicable for mine waters where metal concentrations are relatively small. The biological systems described in this chapter are flexible and can be integrated with chemical systems to overcome issues of metal and salt/solute toxicity. The combination of biological and chemical methods for metal recovery from secondary resources can help to secure metal supply and support the principle of zero-waste generation.

Keywords Tailings · Biohydrometallurgy · Bioleaching · Critical metals · Bioprecipitation

S. Hedrich (✉)
TU Bergakademie Freiberg, Institute of Biosciences, Freiberg, Germany
e-mail: sabrina.hedrich@bio.tu-freiberg.de

P. H-M. Kinnunen
VTT Technical Research Centre of Finland Ltd, Tampere, Finland
e-mail: paivi.kinnunen@vtt.fi

D. B. Johnson et al. (eds.), *Biomining Technologies*,
https://doi.org/10.1007/978-3-031-05382-5_16

16.1 Introduction

The mining and metallurgical industries create large amounts of solid and liquid waste streams that often contain elevated concentrations of valuable and toxic metals and metalloids. A number of metals in these waste streams are of significant economic value and are considered critical by major global economies, causing a potential risk to the supply stream. In addition to the economic and environmental drivers to transform wastes into valuable material streams there is political pressure to increase metals supply from secondary materials sources.

The vast majority of mine wastes are relic deposits, generated when the recovery of metals from primary ores was less efficient than is possible with more recent developments. Most technologies used in the mining and (hydro-)metallurgical industry have been developed for high-grade primary ores and effluents with large concentrations of transition metals and are not efficient at recovering metals from low-grade waste materials and effluents that contain trace amounts of metals. Consequently, metals of low commercial value or present in small concentrations in process waters are often removed by chemical precipitation and the sludge is deposited in designated landfill sites.

In addition to the lost metal value, waste deposits, and tailings storage facilities often cause negative environmental impacts, additional costs, and long-term liabilities to the companies. Waste disposal costs vary in different countries depending also on landfill taxation (Fig. 16.1). For example, in the European Union the landfill taxation varied (in 2021) between 5 €/ton for municipal solid waste in Lithuania to 267.55 €/ton for a mix of hazardous and non-hazardous waste in Belgium. In some

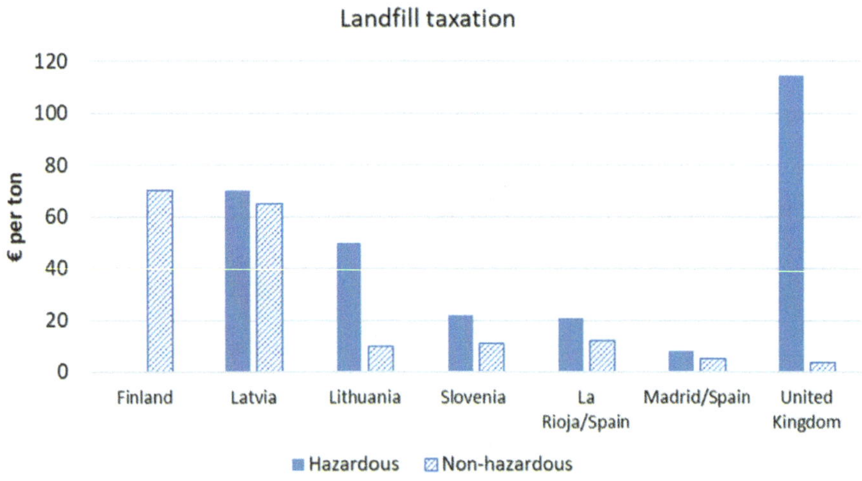

Fig. 16.1 Landfill taxation levels in some European countries for hazardous and non-hazardous waste in 2021 [data based on CEWEP (2021)]. In Finland, there is no taxation for hazardous waste. In the UK, the lower rate applies to non-hazardous waste streams with low greenhouse gas emissions and polluting potential. UK pound conversion 1 £ = 1.18 €

countries, landfill taxation may apply only to non-hazardous waste and in other countries only to hazardous waste. Many countries do not impose landfill taxes. Often there is an exemption of landfill tax for tailings, whereas metallurgical residues are subject to landfill taxation. The divergence of landfill costs and gate fees between countries and waste categories means that the economic assessment of re-mining projects can vary, based on the location of the operation and type of waste. These disposal costs could be avoided if the materials were recycled and used for other purposes.

Often, metal and sulfide contents or particle size of solid wastes and metal content in process waters hinders the possibility to use these for other purposes, such as in the construction industry or water recycled and used for other processes. The recovery of metals from waste streams also removes the pollution hazard and enables the use of the material in other applications, thereby adhering to the zero-waste principle. This has resulted in the need to develop new and modify conventional processing methods. With further development of processing methods, such as biohydrometallurgical technologies that are suitable for processing low-grade materials and selective metal recovery from process waters, these streams have now the potential to be transformed from waste to value. The recovery of metals from waste streams is one solution to help boost the supply of metals, and it also follows the principle of the circular economy. When the solid waste materials are already of small particle size, the energy needs, and costs associated with crushing and grinding compared to conventional primary mining of metals can be avoided. The following sections outline where biohydrometallurgy has already been applied for metal recovery or where the greatest potential has been highlighted for future developments.

16.2 Reprocessing Solid Wastes for Value Recovery and Environmental Benefit

This section describes how bioprocessing has been tested and assessed for recovering metals from a variety of materials, including mineral tailings, pyrometallurgical slags, and incinerated municipal wastes. Bioleaching of electronic waste is described in detail in Chap. 14.

16.2.1 Tailings Reprocessing and Removal of Impurities

Tailings are produced as a waste stream in ore processing, where target minerals are enriched and separated by flotation and physical means, and account for one of the largest global waste streams. Billions of tonnes of tailings waste have been already produced by mining mineral resources and disposed of in ponds or mine sites. When production rates increase and the ore grades decrease, the relative amounts of tailings generated increase further. Tailings typically contain carbonates, silicates and sulfides, but also often economically-viable concentrations of metals, especially in the case of historic tailings. Since tailings have already been mined, crushed, and ground, the related costs can be avoided. In addition, tailings deposits can generate acidity and leaching of heavy metals if not properly managed, and therefore pose an environmental risk. Recovery of metals from these challenging secondary metal-containing material streams can be mediated by various approaches, with bioleaching being one of the more applicable technologies (Kinnunen and Kaksonen 2019). Technology development has been considered not only as a driver to the tailings valorisation but also as a bottleneck, since further technology development is required.

The first commercial tailings reprocessing operation using bioleaching technology was operated between 1999 and 2013 in Kasese, Uganda, for the recovery principally of cobalt, with nickel and arsenic as secondary products (Morin and d'Hugues 2007). The processed tailings contained mostly pyrite (80%) and 1.1–1.4% cobalt associated with sulfide minerals. The dominant bioleaching microorganisms in the 42 °C bioleaching reactors were *Leptospirillum ferriphilum*, *Acidithiobacillus caldus*, and *Sulfobacillus benefaciens*. Almost 80% extraction of cobalt was obtained in approximately 6 days of retention time (Morin and d'Hugues 2007). Also, recent bioleaching laboratory pilots in Finland have reached almost 90% cobalt recovery from sulfide tailings, where cobalt is in Co-pentlandite and in the pyrite structure, in 10 days and the process is being further optimised and piloted (Mäkinen et al. 2020).

Some other larger projects have shown that bioleaching technology itself may function well for metal recovery, but the whole process flow sheet affects the potential use of the technology. For example, bioleaching was assessed for remediating stockpiled arsenopyrite tailings at Snow Lake (Canada). The main obstacles to take the process to a full-scale operation were the relatively low price of gold at the time and the costs of precipitation and neutralisation reagents due to the geographical location of the deposit. This highlights the need to evaluate re-mining of tailings deposits case-by-case.

Bioleaching of tailings with a mixed acidophilic culture has also been tested, at laboratory scale, as a pre-treatment method to leach nickel, cobalt, zinc, and iron prior to the chemical chloride leaching of copper and gold. In all chloride leaching experiments, highest metal yields were obtained when bioleaching was used as a pre-treatment method (Altinkaya et al. 2018). The combination of bioleaching with chemical leaching may be an alternative to improve the overall leaching yields from low-grade materials.

Typically, bioleaching has been considered to be suitable for low-grade ores and wastes where other process options are too expensive or not technically possible. Existing commercial bioleaching applications treat ore in heaps or use biooxidation of gold-containing material prior to chemical leaching. When concentrates are considered, one commercial bioleaching process already exists where nickel sulfide is leached and arsenic impurities removed from a nickel concentrate (generated as a by-product of talc production) in order to produce a higher value product from the nickel side stream. In 2015, Mondo Minerals constructed the world's first nickel sulfide concentrate bioleaching plant to treat approximately 12,000 tons of nickel concentrate annually at full operation with the target to recover 93% nickel with a residence time of 7 days. The nickel plant process consists of pre-treatment, bioleaching in primary and secondary reactors, precipitation of iron and arsenic into the waste, and precipitation of nickel and cobalt end product. This operation is described in detail in Chap. 12.

16.2.2 Metallurgical Wastes

When metal sulfide concentrates are smelted, metal-containing matte and slag are produced as a waste. Significant amounts of valuable metals may be present as impurities in the slag, depending on the original mineral and the selected smelting process. Slags are typically characterised as hazardous wastes, and metal recovery from them using pyrometallurgical, (bio)hydrometallurgical, or hybrid technologies could potentially transform the waste status, bringing economic and environmental benefits in addition to the value of the recovered metal. The challenge with bioleaching is that slags are already oxidised and may not contain sufficient reduced sulfur for microbial oxidation and acid generation. However, even relatively low leaching yields (e.g., 50%) may transform the categorisation of slags from hazardous to non-hazardous and allow their use in other applications. Sulfuric acid has proven to be a successful leaching agent in the chemical leaching of slag. Since sulfur-oxidising microorganisms can oxidise acid-labile metal sulfides and elemental sulfur and produce sulfuric acid, bioleaching has potential to decrease the need to add sulfuric acid to the slag leaching process. Comparable metal yields have been obtained for bioleaching and chemical leaching from copper slag, with decreased acid consumption with the former due to biological elemental sulfur oxidation. Bioleaching has therefore been considered as a potential technology to treat slag. Copper slag bioleaching yields in laboratory studies have varied depending on the nature of the slag and bioleaching conditions. Pulp density has been found to have a major influence on metal yields. For example, copper yields of 93–97% were obtained with 7–10% pulp density, compared to 73–77% when the pulp density was increased to 15–25%, in tests carried out in shake flasks (Panda et al. 2015). In a 1-L bioreactor experiment, 55% Cu, 37% Co, and 41% Zn recoveries were obtained at 10% pulp density within 5 days (Kinnunen et al. 2020).

In addition to biological sulfur oxidation, acidic solutions of biologically produced ferric iron have been used to bioleach copper slag (Carranza et al. 2009). However, biological iron oxidation and the subsequent increase in redox potential have also resulted in decreased metal yields either due to the precipitation of leached metals or due to the metal leaching not being based on redox reactions (Kinnunen et al. 2020). When the metals have been removed by leaching, the processed slag can be used in value-added products, for example, in the construction industry.

16.2.3 Bottom Ash

Incinerated municipal solid waste is thermally converted to produce heat and power. Significant amounts of municipal solid waste are treated by this means; in the European Union alone 58 million tonnes were estimated by Eurostat (ec.europa.eu/eurostat) to have been incinerated in 2018. The incineration process produces bottom ash and fly ash as residues. Since the municipal waste is highly variable, the bottom ash from the incineration process is also a very heterogeneous material stream, which contains, for example, glass, unburned organic matter, and a variety of metals. Also, the particle sizes vary from very fine particles to much larger (several centimetres diameter) fragments, with metal wires, for example, often still being visible. The metals are differently distributed between the fines and larger particles and also between different mineralogical fractions. Metal oxides are the most dominant form in the ash after incineration, but also metal sulfides and metallic phases occur. The heterogeneous properties of the bottom ash have made technology development for processing them highly challenging, and residual metals could potentially be recovered by (bio)leaching. One of the main targets in the treatment of bottom ash is to remove sufficient amounts of metals in order to be able to use the remaining material for other purposes. Therefore, even lower residual metal leaching yields could enable the valorisation of bottom ash. At the same time, solubilisation of iron needs to be minimised to avoid iron removal costs in the later steps of the process.

The proposed technologies for the treatment of bottom ash include hydrometallurgical leaching with acids, bioleaching, washing after the aging of the ash or direct utilisation as a solid material, when possible. Municipal solid waste incineration (MSWI) ash is alkaline, and some residual toxic metals are challenging for acidophilic bioleaching microorganisms. Metals, such as chromium, which can be present as oxy-anions, are particularly toxic to acidophiles, and the alkaline nature of fly ash requires higher acid production to maintain the pH at the appropriate level for these microorganisms. Bioleaching of MSWI fly ash with acidophilic iron- and sulfur-oxidisers has been reported to leach target metals from the ash in comparable amounts with chemical leaching (>85% yields for Al, Cu, Mg, Mn, and Zn), but bioleaching has shown selectivity towards certain elements (Ni and Pb) and lower need for additional sulfuric acid addition (Funari et al. 2017). Bioleaching is one option to be studied further for the recovery of metals from fly ash, but so far results are from only laboratory tests and currently no commercial plants exist.

16.3 Biological Metal Recovery from Process Waters

16.3.1 Source and Chemistry of Process Waters

Natural and anthropogenic processes can result in the generation of large volumes of wastewaters (such as acid mine drainage; AMD) that are enriched in metals, metalloids, and other contaminants. AMD resulting from the microbially catalysed dissolution of metal sulfides is found worldwide and is commonly characterised by high loads of sulfate, low pH, and elevated concentrations of many metals and metalloids. The most abundant sulfide mineral affected by microbial attack and responsible for acid mine drainage formation is pyrite (Eq. 16.1), which occurs as a major gangue mineral in actively mined ore bodies and also, like other iron sulfides, in tailings dumps. Other major sources of AMD are flooded, abandoned mines that generate waters contaminated with, for example, Fe, As, Cu, U, and Sn. Whereas Eq. (16.1) describes AMD formation under oxic conditions, soluble ferric iron can also cause further chemical dissolution of the pyrite (Eq. 16.2) in anaerobic zones of, for example, waste dumps, thereby enhancing the process.

$$FeS_2 + 3.75\,O_2 + 0.5\,H_2O \rightarrow Fe^{3+} + HSO_4^- + SO_4^{2-} \qquad (16.1)$$

$$FeS_2 + 6\,Fe^{3+} + 3\,H_2O \rightarrow 7\,Fe^{2+} + S_2O_3^{2-} + 6\,H^+ \qquad (16.2)$$

The main industrial processes causing discharge of metal-rich effluents include mining, electroplating industry, tannery, battery, and printed circuit board manufacturing, paper and fertiliser industries. All of these effluents are considered as potential environmental pollutants and risks to human and animal health due to enrichment of transition metals such as Fe, Zn, Cu, Ni, and Mn that can occur in relatively large concentrations and other contaminants that usually occur in smaller concentrations (e.g., As, Hg, Pb, Mo, Cr, Sb, and Se) in water bodies and sediments, causing severe toxicity to living organisms. Depending on the source of these waters, acid mine drainage, or the industrial process, these waters have different chemistries (Table 16.1).

Whereas process waters can also be considered as an environmental threat, most of them contain valuable metals in concentrations that are economically feasible for recovery. Commercial downstream processing of process waters can involve solvent extraction coupled with electrowinning, ion exchange, membranes, or other alternative methods. However, biological methods have proven to be more efficient and selective in overall metal recovery and more suitable when metal contaminants are present in relatively low concentrations.

Table 16.1 Examples of mine drainage and process water chemistries

	pH	As	Zn	Ni	Cu	Cr	Fe	Co	Cd	Al	Other metals	
Mining PLS	–	<20	4000–7500	1500–3500	500	<20	1000–2000	<200	<20	3000–9000	U: 10–30	Talvivaara (2012, unpublished)
Acid mine drainage	2.2–7.1	1.3	0.8–464	–	<0.01–483	–	<0.01–1072	–	–	0.98–132	Mn: 0.95–55 U: 0.28–4.82	Johnson and Santos (2020)
Electroplating wastewater	2.2	–	239	28	0.92	47	–	–	–	–	–	Rene et al. (2017)
Battery factory	3.8–5.8	–	28.3	0.07–0.38	<0.38	<0.08	0.02–20		0.02–0.12	0.2–7.3	Pb: 4–13	Rene et al. (2017)
Copper smelting	0.64	1979	455.6	12	164.48	2.3	88	0.04	76.05	–	Bi: 85; Sb: 1.5; Pb: 4.6	Rene et al. (2017)

Concentrations shown are $mg\ L^{-1}$

16.3.2 Biological Methods for Recovering Metals from Waste Liquid Streams

Biological methods described for metal recovery mainly from AMD comprise biosorption, bioprecipitation, constructed wetlands, permeable reactive barriers, phytomining, green walls and roofs, etc., which have been extensively reviewed elsewhere (e.g., Rene et al. 2017). Recent lab-scale methods also comprise the use of nanoparticle production and application of phages (Kaksonen et al. 2020).

Even though mine drainage and process waters often display extreme chemistries, they are almost always populated by chemolithotrophic and other microorganisms. On the one hand, these are responsible for accelerating the release of metals from sulfidic and other minerals, but their ability to, e.g., catalyse redox reactions or adsorb metals can, in turn, be used for the recovery of metals and metalloids if stimulated under the right conditions. Some microorganisms that are indigenous to mine-related environments and process waters have remarkable traits for application in metal recovery and process water treatment.

Biosorption using microbial cells or other biological materials has shown some promising results for the removal and recovery of metals from process waters. Different mechanisms are involved to entrap metals and the efficiency of the process depends on various parameters. However, biosorption has not as yet been applied at an industrial scale due to various factors (Chap. 17) such as the selectivity of metal recovery.

Bioprecipitation is one of the most intensively studied and industrially applied biohydrometallurgical processes, and is based on the conversion of soluble metals and metalloids into insoluble precipitates by microbially catalysed redox and other reactions. The most common precipitates are iron hydrous oxides, hydroxides, and hydroxysulfates (e.g., jarosites, schwertmannite, and ferrihydrite), metal sulfides, and phosphates and carbonates. Bioprecipitation for iron recovery and metal sulfide precipitation have been applied in pilot and technical scale and is discussed in detail below.

16.3.3 Biological Iron Recovery

Iron is the most abundant metal on planet Earth, and also in many mines and other process waters as it occurs in many primary minerals and residues. Depending on pH, oxygen concentration, and presence of iron-oxidising microorganisms or oxidising agents it occurs either as Fe^{2+} or Fe^{3+}. Although iron is a low-value metal, it often requires removal prior to downstream processing due to its elevated concentrations in most process waters and interference with the recovery of other metals. In acidic liquors, Fe^{2+} can be the more prevalent form of soluble iron and requires oxidation to less soluble Fe^{3+} to be removed by precipitation.

Conventional iron removal approaches applied by most mining companies involve the addition of lime or limestone to raise the pH in combination with aeration, to achieve iron oxidation and precipitation. During this process, a mixed-metal sludge is produced, which requires deposition in designated areas. Recent biological iron oxidation and removal approaches allow the selective precipitation of iron at low pH, avoiding the production of waste sludge and promote the subsequent recovery of valuable metals and removal of contaminants (e.g., Hedrich and Johnson 2014; Johnson and Santos 2020).

Iron-oxidising microorganisms catalyse the conversion of Fe^{2+} to Fe^{3+} at acidic pH in an oxygen-requiring and proton-consuming reaction (Eq. 16.3).

$$2\,Fe^{2+} + 0.5\,O_2 + 2\,H^+ \rightarrow 2\,Fe^{3+} + H_2O \tag{16.3}$$

Iron oxidation efficiency in bioreactors is mostly determined by oxygen mass transfer rather than ferrous iron oxidation kinetics. The resulting ferric iron precipitates subsequently in an acid-producing reaction, depending on the type of iron mineral being formed [Eq. (16.4) for jarosite, where "M" can be K, Na, H, or NH_4, and Eq. (16.5) for schwertmannite].

$$3\,Fe^{3+} + M^+ + 2\,HSO_4{}^- + 6\,H_2O \rightarrow MFe_3(SO_4)_2(OH)_6 + 8\,H^+ \tag{16.4}$$

$$8\,Fe^{3+} + SO_4{}^{2-} + 14\,H_2O \rightarrow Fe_8O_8\,(SO_4)(OH)_6 + 22\,H^+ \tag{16.5}$$

The type of secondary iron mineral being formed is determined by pH, temperature, and available cations in solution. Common secondary iron minerals formed after biological oxidation are jarosites schwertmannite (Fe_8O_8 $(SO_4)(OH)_6$) or ferrihydrite ($Fe_{10}O_{14}(OH)_2$).

Biological iron-oxidising systems for process water treatment using, for example, acidophilic iron-oxidising *Acidithiobacillus* and *Leptospirillum* spp., have been intensively studied depending on inflow pH, iron concentration, and accompanying metals [reviewed in Rene et al. (2017)]. Two other studies reported that the more recently discovered iron-oxidisers "*Ferrovum myxofaciens*" and *Acidithrix ferrooxidans* were the most efficient to oxidise iron from acidic waters in continuous flow bioreactors without the requirement of support material, as these bacteria were able to form filamentous structures attaching to the bioreactor walls.

Iron oxidation and precipitation at various redox potentials from acidic waters containing 15 g L^{-1} Fe^{2+} at ambient temperatures were demonstrated in a two-stage bioreactor system housing the mesophilic iron-oxidisers *At. ferrooxidans* and *L. ferriphilum* (Kaksonen et al. 2014). The system achieved 94% iron oxidation and 31% iron removal mainly as well-settling jarosite with minor co-precipitation of Ni and Cu. A modular continuous bioreactor system housing "*Fv. myxofaciens*" achieved >90% oxidation of iron from AMD (pH 1.9; 0.28 mg L^{-1} Fe^{2+}) also containing Al, Cu, Mn, and Zn and subsequent precipitation of schwertmannite in a separate vessel by alkali addition without co-precipitation of other metals (Hedrich and Johnson 2012). However, only a few microbial iron-oxidising systems that

Fig. 16.2 (**a**) Microbial iron oxidation and precipitation plant treating mine waters from a lignite mine in Nochten (Germany) (left picture). (**b**) Aeration tank of the plant with untreated mine water (right picture). (images provided by D.B. Johnson)

facilitate the removal of soluble iron from process waters have been implemented beyond laboratory scale. One example was a pilot plant operation by the company GEOS at a lignite mine in Nochten (Germany), where Fe^{2+}-rich (0.63 g L^{-1}) groundwater pH 4.9 was treated in a continuous aerated bioreactor (Janneck et al. 2010). The 10.5 m^3 pilot plant started its operation in 2007 and consists of an oxidation basin (8.14 m^3) with removable growth carriers and an aeration and precipitation tank (Fig. 16.2). The microbial consortium in the plant was dominated by the iron-oxidiser "*Fv. myxofaciens*", which oxidises ferrous iron, facilitating the removal of soluble iron by precipitation of schwertmannite, as the pilot plant is operated at pH around 3.0 (Hedrich and Johnson 2012). While iron is conventionally precipitated together with other metals and deposited, schwertmannite has potential application as a sorbent for oxy-anions to remove As, Se, or Mo from process waters or as a pigment in, for example, ceramic tiles, though this has not yet been realised commercially.

16.3.4 Biological Metal Sulfide Precipitation

The dissimilatory bacterial reduction of sulfate and elemental (zero-valent) sulfur to hydrogen sulfide has been applied in various configurations for the removal and recovery of metals from mine and process waters as active or passive system (Johnson and Santos 2020). The technology harnesses sulfate- and/or sulfur-reducing bacteria (SRB) which are capable of converting sulfate or sulfur to sulfide. Biosulfidogenesis is fuelled by organic substrates (e.g., ethanol, methanol, or glycerol) or hydrogen, which allows for a wide range of low-cost waste material of various production routes. The sulfide generated reacts with soluble chalcophilic metals in solution to precipitate insoluble metal sulfides. Under pH-controlled conditions metals can be selectively precipitated, due to their different solubility products, and later recovered from the precipitate (Johnson and Sanchez-Andrea 2019).

Biosulfidogenesis has also the advantage of removing sulfate and increase solution pH, as sulfate reduction is a proton-consuming reaction at pH values above ~6.5 (Eq. 16.6). The microorganisms can therefore be used to recover metals and at the same time remediate these waters by raising the pH and lowering the sulfate concentration to avoid environmental pollution.

$$2\,H^+ + SO_4{}^{2-} + CH_2O \rightarrow H_2S + H_2CO_3 \tag{16.6}$$

Commercial-scale processes applying biosulfidogenesis to recover metals from process waters typically use neutrophilic bacteria, which are sensitive to low amounts of metals and acidity, requiring sulfidogenesis and metal precipitation to occur in separate vessels. The first of these commercial processes include the Biosulphide® process by the company BQE Water (previously BioteQ Environmental Technologies Inc. Canada) and various processes implemented by the Dutch company Paques. Both companies utilise neutrophilic SRB to generate hydrogen sulfide from either sulfate (Paques) or elemental sulfur (BioteQ). The systems allow selective metal sulfide precipitation and removal of contaminants by controlling pH and sulfide concentration in separate precipitation tanks containing the AMD or process water. BQE Water has successfully installed and run their first Biosulphide® process plant at the Caribou Mine (Canada) to remove Cu, Zn, Cd, and Pb from AMD as metal sulfides applying elemental sulfur as an electron acceptor for the SRB. Further installations are at the Copper Queen Mine (North America) to recover copper from a copper concentrator wastewater (decommissioned in 2015) and the Raglan Nickel mine (Canada) to recover nickel.

While Biosulphide® is a single-step bioprocess for H_2S production and subsequent metal sulfide precipitation, Paques' SULFATEQ™ process involves a second biological step for the oxidation of excess H_2S to elemental sulfur by sulfur-oxidising bacteria in a separate, aerated bioreactor (Boonstra et al. 1999). The SULFATEQ™ technology using sulfate as electron donor has been implemented at various sites, with the oldest one at the Nyrstar zinc refinery (Netherlands) installed in 1992 for the recovery of zinc from a metal- and sulfate-rich groundwater. The plant comprises an upflow anaerobic sludge blanket reactor (UASB) for biosulfidogenesis and a submerged fixed-film reactor for aerobic oxidation of hydrogen sulfide to zero-valent sulfur. The second biogenic metal sulfide precipitation process developed by Paques (THIOTEQ™ Metal) possesses only one biological step for H_2S production from elemental sulfur and only aims at metal recovery, e.g., copper from an acidic process water at the Pueblo Viejo Gold Mine in the Dominican Republic.

The development of laboratory-scale low pH sulfidogenic bioreactors utilising acidophilic and acid-tolerant SRB allowed simultaneous sulfate reduction from acidic, metal-rich waters, and selective metal sulfide precipitation (reviewed in Johnson and Santos 2020). The organisms used in the process were enriched from

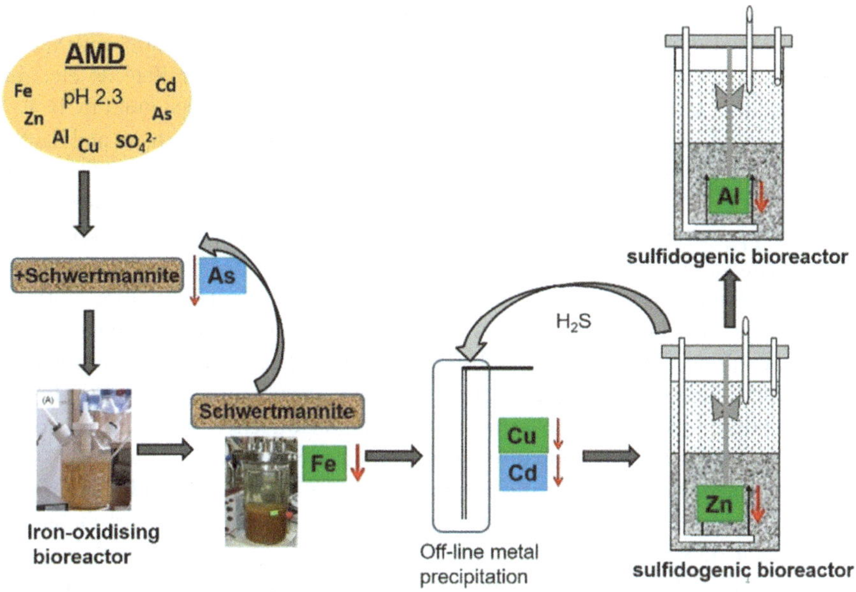

Fig. 16.3 Integrated scheme for biological recovery of valuable metals (green) and contaminant removal (blue) from acid mine drainage [modified from Hedrich and Johnson (2014)]

mine-impacted environments and tolerate low pH and elevated concentrations of transition metals, which allows them to get in contact with acidic, metal-rich waters, and sulfidogenesis and metal precipitation to occur in a single vessel. The system, which has proven to be very robust to various water chemistries, operates as a continuous bioreactor and pH is maintained by control of the inflowing acidic waters, as the bacterial sulfate reaction is proton-consuming at low pH. Depending on the chemistry of the wastewater, metal sulfide precipitation can either be achieved within the bioreactor or off-line, by stripping the H_2S produced. The sulfate-reducing consortium can operate over a wide pH range by changing its composition and has been applied for the recovery of metals from highly and moderately acidic wastewaters. The technology was applied in an integrated approach to recover Zn (0.46 g L^{-1}) and Fe (0.40 g L^{-1}) from AMD at the Maurliden mine (Sweden) and at the same time removing contaminants (As, Cd, Al, Cu, Mn, Fig. 16.3). In a first step As was removed by adsorption onto schwertmannite, followed by microbial iron oxidation and schwertmannite precipitation (far in excess of that required to remove As). The third and fourth stages comprised offline removal of Cd and Cu by biogenic H_2S produced in the fourth stage, a sulfidogenic bioreactor (pH 4.0) fuelled with glycerol to produce H_2S from the mine water (pH 2.3) and precipitate Zn in the same vessel. The setup allowed selective precipitation of ZnS without coprecipitation of other metals or metalloids (Fig. 16.3).

Although aluminium, which is also often present in elevated concentrations in acidic mine waters, does not form a sulfide phase, the increase in pH associated with sulfidogenesis at low pH can facilitate the precipitation of aluminium hydroxysulfate minerals [hydrobasaluminite and felsőbányaite; reviewed in Johnson and Santos (2020)]. A modified system using both elemental sulfur and sulfate as electron acceptors was used to remove Zn and Cu, present in small concentrations in two circum-neutral AMDs from abandoned Zn/Pb mines in the UK (Johnson and Santos 2020). The acidophilic SRB system has also proven suitable for the selective recovery of Cu, Zn, and Cd/Ni from an acidic, highly sulfate- and metal-rich effluent from a copper concentrate bioleaching process, after partial chemical recovery of Fe and Cu and sulfate.

Biosulfidogenesis has also recently been used to recover selenium from synthetic wastewaters in a promising process combining microbial selenite reduction and selenium sulfide formation with subsequent bioreduction of the selenium sulfide and crystallisation of hexagonal selenium. Paques also developed the BIOMETEQ™ technology to achieve precipitation of, e.g., Se, U, Mo, and Cr present in μg–mg L^{-1} concentration in contaminated waters.

Wastewaters from the electroplating industry are often acidic and contain elevated metal concentrations, and have therefore been considered to be suitable for biological metals sulfide precipitation.

Biological metal recovery systems are flexible and can be adjusted according to the water being treated and combined with chemical recovery methods. Sulfidogenic bioreactor systems have the overall advantage of controlled and consistent performance while recovering valuable metals for recycling from wastewater and additionally allowing sulfate removal and alkalinity production for water treatment. Furthermore, excess hydrogen sulfide produced by these systems can be converted to elemental sulfur, which can serve as a substrate for other biotechnological processes or be used as a fertiliser. The major costs are related to system construction, maintenance, and electron donor for the SRB, which are currently pure-grade chemicals (e.g., ethanol, methanol, or hydrogen), but have the option to be replaced by lower-cost waste materials.

The biological iron oxyhydroxide and metal sulfide precipitation systems described provide a more environmentally benign alternative to lime treatment, with low operational costs, near-zero waste production, consistent performance, efficient iron and sulfate reduction, and production of potentially saleable end products (schwertmannite, metal sulfides, and elemental sulfur). As shown for the integrated biological approach in Fig. 16.3, metal products are recovered from process waters in combination with contaminant removal for waste reduction and environmental protection (Hedrich and Johnson 2014). Both systems have been operated and investigated in terms of efficiency and economy either in laboratory (2 L tank reactor) or pilot scale and recommended for technical use.

16.3.5 Biological Removal of Other Contaminants (As, U, Cr, and Se)

Increased processing of complex ores and industrial processes, generating metal-rich wastewaters, also causes the solubilisation of accompanying toxic contaminants, such as As, U, Cr, and Se, which require removal. Microorganisms are able to either contribute directly or indirectly to the removal of many of these. However, biological systems for the recovery and removal of minor contaminants from process waters are not yet as advanced and efficient as the biological iron oxyhydroxide and metal sulfide precipitation systems described in the previous section.

Arsenic occurs either as As(III) (H_3AsO_3) or As (V) ($H_2AsO_4^-$), often in the mg L^{-1}-range in AMD but sometimes >10 g L^{-1} total As, in some mine process waters. Arsenic removal by sulfidogenic bacteria was reported in the 1990s already using the neutrophile *Desulfotomaculum auripigmentum*, which reduced both arsenate and sulfate, causing the yellow mineral orpiment (As_2S_3) to precipitate. Orpiment is, however, less stable at higher pH values and in the presence of high dissolved sulfide concentrations and may result in the formation of FeAsS in the presence of ferrous iron. Battaglia-Brunet et al. (2012) demonstrated how the formation of orpiment could be controlled by choice of the electron donor and control of pH and thereby sulfate-reduction rate in a fixed bed sulfidogenic bioreactor. When running the bioreactor at pH 5 with glycerol as electron donor, hydrogen sulfide production was low, most likely due to acetate formation and inhibition of the SRB, but favoured the formation of orpiment achieving 100% As removal. In contrast, when using hydrogen as electron donor the rate of sulfate-reduction increased and resulted in higher pH and elevated concentration of H_2S, which lead to the formation of soluble thioarsenic species and mobilisation of arsenic from the bioreactor.

The microbial iron precipitation pilot plant process described in Sect. 16.3.3 produces the positively charged ferric iron mineral schwertmannite with a large surface area and a high sorption potential for oxy-anions. The mineral has been shown to aid in the removal of arsenate, selenite, and molybdate.

A similar approach has been shown with iron oxide and siderite precipitate formed by a ferric citrate-reducing microbial community. The precipitate could adsorb up to 0.4 mmol of metals and metalloids g^{-1} from dilute solutions (Kaksonen et al. 2020).

The formation of scorodite ($FeAsO_4 \cdot 2H_2O$) from arsenic-containing wastewaters has been intensively studied in recent years as arsenic contamination has become a major problem, e.g., effluents from the processing of arsenic bearing minerals (enargite and tennantite). The coprecipitation of arsenate As(V) with ferric iron requires As (III) to be oxidised, which may be biologically or chemically mediated, which requires a catalyst at temperatures <50 °C. The thermo-acidophilic iron-oxidiser *Acidianus sulfidivorans* was shown to form scorodite at pH 1 and 80 °C when grown on 13 mM ferrous iron and 13 mM arsenate, even without the presence of any minerals or seed crystals, and Okibe et al. (2013) reported biogenic scorodite crystallisation at 70 °C in cultures of *Acidianus brierleyi* at low concentrations of arsenic (III) (3.3–20 mM). This process was, however, driven by initial microbial iron oxidation and subsequent reduction of the ferric iron coupled to the oxidation of arsenic (III) to arsenate and proposed to occur in the extracellular polymeric substance (EPS) compartment near the microbial surface. The accumulation of small amounts of arsenic in the EPS compartment allows the microorganism to form scorodite even at low arsenic (III) concentrations compared to the pure chemical process and highlights the advantage of biogenic metal precipitation. The Dutch company Paques has also developed a technology (THIOTEQ™ Scorodite) forming bioscorodite, which is based on microbial iron oxidation and optional arsenic (III) oxidation resulting in highly stable bioscorodite crystals. The liquids for this process required arsenate concentrations of at least 1 g L^{-1}, which are much greater than those used in laboratory experiments.

Uranium contamination is often a problem associated with mining, and this metal has a severe impact on human health. Some microorganisms are capable of reducing soluble uranium (VI) to insoluble uranium (IV), thereby contributing to the removal of uranium from waste streams mostly under anaerobic conditions and circum-neutral pH. In situ uranium removal has been widely studied as a low-cost process with a reduced risk of exposure to radiation than using above-ground bioreactors. While the latter allows better control of the process, they incur higher capital costs and require measures for protection from radiation.

Chromium has also become a major concern as a contaminant in wastewaters related to its sometimes intensive use in industrial applications. The most common oxidation states in these waters are Cr (VI), which is highly mobile in the environment, and the less soluble Cr(III). Although Cr(VI) is generally highly toxic to microorganisms, some can catalyse the enzymatic reduction of Cr(VI) to Cr(III) and subsequent immobilisation of chromium, Sulfate-reducers can also indirectly contribute to Cr(VI) reduction by producing H_2S which reduces Cr (VI) to Cr (III), and also by alkalinity production which promotes Cr (III) precipitation. Efficient Cr (VI) reduction under aerobic conditions and removal of 94% Cr(VI) has been shown by a *Bacillus* spp.-dominated mixed culture (Rene et al. 2017). Iron-reducing acidophiles have also been shown to indirectly reduce Cr(VI) by their ability to reduce Fe(III) to Fe(II), which then reacts with Cr(VI) to form Cr(III). Most of the studies achieved complete Cr(VI) reduction, but no efficient removal of soluble Cr (III). The removal of chromium from acidic wastewater in a sulfidogenic system fuelled with ethanol achieved 98% total chromium removal after bioreduction and pH increase to >7.3.

Elevated concentrations of selenium discharged into the global water cycle by industrial processes can also cause environmental pollution. Biological selenium removal can be achieved via phytoremediation, constructed wetlands, or microbial reduction and volatilisation of SeO_4^{2-}. Microbial selenium reduction for selenium removal from effluents is efficient even at low selenium concentrations and offers a cost-effective alternative to chemical methods. Microbial selenium removal is most efficient under anaerobic conditions, where soluble selenium oxyanions are converted into elemental selenium. The efficiency of biological selenium recovery has been demonstrated in various types of bioreactors at laboratory and pilot scale

(Tan et al. 2016). However, the efficiency of selenium reduction and recovery using actual process waters still needs improvement. Once this is achieved, the production of selenium nanoparticles through microbial reduction could have promising technical applications.

16.4 Conclusions

Biohydrometallurgical methods have proven to be attractive for the treatment of various low-grade solid wastes, and drainage and process waters, since investment and operation costs are often relatively low and biological processes can be efficient and effective. Although bioprocessing of slags, ashes, and end-of-life consumer waste is challenging due to most of them being oxidised, containing toxic elements, with capacity to neutralise acidity and deficient of reduced sulfur content, research has shown some promising approaches where the waste materials are converted into saleable products that could be used in the construction industry, and possibly others. Adaptation of microorganisms and washing of the material can be used as methods to overcome these challenges. Addition of iron and sulfur to the process creates additional costs, which need to be compared to the costs of using chemical reagents and other consumable materials alone. Sustainable metal recovery and contaminant removal are enhanced by the application of biological methods to process water treatment, as biological methods allow metal recovery even from both dilute and chemically complex solutions. Toxicity of metals and high salt/solute contents are also sometimes an issue with concentrated process waters and can be overcome by coupling chemical and biological methods for efficient metal recovery and contaminant removal. The bioleaching and metal recovery systems described in this chapter are flexible and can be integrated with other technologies depending on the waste stream treated. Oxidative and reductive biological processes investigated in the laboratory with potential for upscaling to industrial scale have the potential to help secure the supply of strategic and critical metals. The major factors to consider for upscaling and application of these processes are the environmental impacts and operational costs as well as revenue from valuable metals recovered, compared to conventional processes.

References

Altinkaya P, Mäkinen J, Kinnunen P, Kolehmainen E, Haapalainen M, Lundström M (2018) Effect of biological pretreatment on metal extraction from flotation tailings for chloride leaching. Miner Eng 129:47–53. https://doi.org/10.1016/j.mineng.2018.09.012

Battaglia-Brunet F, Crouzet C, Burnol A, Coulon S, Morin D, Joulian C (2012) Precipitation of arsenic sulphide from acidic water in a fixed-film bioreactor. Water Res 46:3923–3933. https://doi.org/10.1016/j.watres.2012.04.035

Boonstra J, van Lier R, Janssen G, Dijkman H, Buisman CJN (1999) Biological treatment of acid mine drainage. Process Metal 9:559–567. https://doi.org/10.1016/S1572-4409(99)80145-X

Carranza F, Romero R, Mazuelos A, Iglesias N, Forcat O (2009) Biorecovery of copper from converter slags: slags characterization and exploratory ferric leaching tests. Hydrometallurgy 97:39–45. https://doi.org/10.1016/j.hydromet.2008.12.012

CEWEP, Confederation of European Waste-to-Energy Plants (2021) Landfill taxes and bans. https://www.cewep.eu/wp-content/uploads/2021/08/Landfill-taxes-and-bans-overview.pdf. Last update 20 Aug 2021

Funari V, Mäkinen J, Salminen J, Braga R, Dinelli E, Revitzer H (2017) Metal removal from municipal solid waste incineration fly ash: a comparison between chemical leaching and bioleaching. Waste Manag 60:397–406. https://doi.org/10.1016/j.wasman.2016.07.025

Hedrich S, Johnson DB (2012) A modular continuous flow reactor system for the selective bio-oxidation of iron and precipitation of schwertmannite from mine-impacted waters. Bioresour Technol 106:44–49. https://doi.org/10.1016/j.biortech.2011.11.130

Hedrich S, Johnson DB (2014) Remediation and selective recovery of metals from acidic mine waters using novel modular bioreactors. Environ Sci Technol 48:12206–12212. https://doi.org/10.1021/es5030367

Janneck E, Arnold I, Koch T, Meyer J, Burghardt D, Ehinger S (2010) Microbial synthesis of schwertmannite from lignite mine water and its utilization for removal of arsenic from mine waters and for production of iron pigments. In: Wolkersdorfer C, Freud A (eds) Mine Water and Innovative thinking. Proceedings of the International Mine Water Association symposium 2010. Cape Breton University Press, Sydney

Johnson DB, Sanchez-Andrea I (2019) Dissimilatory reduction of sulfate and zero-valent sulfur at low pH and its significance for bioremediation and metal recovery. Adv Microb Physiol 75: 205–231. https://doi.org/10.1016/bs.ampbs.2019.07.002

Johnson DB, Santos AL (2020) Biological removal of sulfurous compounds and metals from inorganic wastewaters. In: Lens PNL (ed) Environmental technologies to treat sulphur pollution: principles and engineering. IWA, London, pp 215–246

Kaksonen AH, Morris C, Hilario F, Rea SM, Li J, Usher KM, Wylie J, Ginige MP, Cheng KY, Du Plessis C (2014) Iron oxidation and jarosite precipitation in a two-stage airlift bioreactor. Hydrometallurgy 150:227–235. https://doi.org/10.1016/j.hydromet.2014.05.020

Kaksonen AH, Deng X, Bohu T, Zea L, Khaleque HN, Gumulya Y, Boxall NJ, Morris C, Cheng KY (2020) Prospective directions for biohydrometallurgy. Hydrometallurgy 195:105376. https://doi.org/10.1016/j.hydromet.2020.105376

Kinnunen P, Kaksonen A (2019) Towards circular economy in mining: opportunities and bottlenecks for tailings valorization. J Clean Prod 228:153–160. https://doi.org/10.1016/j.jclepro.2019.04.171

Kinnunen P, Mäkinen J, Salo M, Soth R, Komnitsas K (2020) Efficiency of chemical and biological leaching of copper slag for the recovery of metals and valorization of the leach residue as raw material in cement production. Fortschr Mineral 10:654. https://doi.org/10.3390/min10080654

Mäkinen J, Salo M, Khoshkhoo M, Sundkvist J-E, Kinnunen P (2020) Bioleaching of cobalt from sulfide mining tailings; a mini-pilot study. Hydrometallurgy 196:105418. https://doi.org/10.1016/j.hydromet.2020.105418

Morin D, d'Hugues P (2007) Bioleaching of a cobalt-containing pyrite in stirred reactors. In: Rawlings DE, Johnson DB (eds) Biomining. Springer, Berlin, pp 35–54

Okibe N, Koga M, Sasaki K, Hirajima T, Heguri S, Asano S (2013) Simultaneous oxidation and immobilization of arsenite from refinery waste water by thermoacidophilic iron-oxidizing archaeon, *Acidianus brierleyi*. Miner Eng 48:126–134. https://doi.org/10.1016/j.mineng.2012.08.009

Panda S, Mishra S, Rao DS, Pradhan N, Mohapatra U, Angadi S, Mishra BK (2015) Extraction of copper from copper slag: mineralogical insights, physical beneficiation and bioleaching studies. Korean J Chem Eng 32:667–676. https://doi.org/10.1007/s11814-014-0298-6

Rene ER, Sahinkaya E, Lewis A, Lens PNL (2017) Sustainable heavy metal remediation. Springer International, Cham

Tan LC, Nancharaiah YV, van Hullebusch ED, Lens PNL (2016) Selenium: environmental significance, pollution, and biological treatment technologies. Biotechnol Adv 34(5):886–907. https://doi.org/10.1016/j.biotechadv.2016.05.005

Chapter 17
The Future of Biomining: Towards Sustainability in a Metal-Demanding World

Anna Henriikka Kaksonen and Jochen Petersen

Abstract Increasing demand for metals, declining ore grades, and the need for improved sustainability pose challenges to the mining sector. Biomining offers solutions to alleviate these challenges, enabling value recovery from wastes and mineral resources that have previously been considered subeconomic, and mitigating harmful environmental impacts of mining processes. This chapter provides an outlook of the future of biomining in a metal-demanding world, starting with a brief review on implications of the kinetic mechanisms of bioleaching for process design. This is followed by a discussion on new avenues for biomining to enable the extraction of unexploited mineral resources, such as continental deep sub surface and deep sea minerals, extraterrestrial minerals, as well as metal-containing wastes. Unconventional and emerging biotechnologies for extracting and recovering metals, including unconventional biolixiviants, bioelectrochemical leaching, biosorption, bioaccumulation, phytomining, biobeneficiation of minerals, and upcycling of metals through biomineralisation, are discussed. While there are many interesting potential routes for sustainable metal extraction, it is important to evaluate them critically in terms of what is feasible within techno-economic limitations.

Keywords Bioaccumulation · Biobeneficiation · Bioleaching · Biolixiviant · Biomineralisation · Biomining · Biosorption · Emerging technologies · Phytomining · Unconventional resources

---

A. H. Kaksonen (✉)
Commonwealth Scientific and Industrial Research Organisation (CSIRO), Land and Water, Floreat, Western Australia, Australia
e-mail: anna.kaksonen@csiro.au

J. Petersen
Department of Chemical Engineering, University of Cape Town, Rondebosch, Cape Town, South Africa
e-mail: Jochen.Petersen@uct.ac.za

D. B. Johnson et al. (eds.), *Biomining Technologies*,
https://doi.org/10.1007/978-3-031-05382-5_17

17.1 Introduction

Metals are essential for economic growth and modern society. Global resource use is expected to double in the period 2010–2030, while the quality and grade of ores have declined over time (Kinnunen and Kaksonen 2019). Biomining has enabled the extraction of value from low-grade resources, the utilisation of which would not necessarily be feasible through traditional pyrometallurgical or hydrometallurgical processing. However, many techno-economic challenges associated with established biomining processes remain and need to be understood and addressed, not only to consolidate biomining as a technology of choice in the conventional mining context, but also to facilitate its successful extension into novel applications and resources.

As conventional mineral resources are becoming increasingly depleted and/or more complex, there has been growing interest in the exploration of deposits in the continental deep subsurface, deep sea, and outer space. Moreover, metal-containing wastes are increasingly considered valuable secondary resources that have not yet been extensively exploited, and the utilisation of which would improve the sustainability of mining industry (Kaksonen et al. 2020).

The application of biomining to new types of mineral resources may also expand the range of biotechnologies being considered for metal extraction and recovery. Novel biolixiviants can provide more environmentally friendly alternatives to chemical leaching agents, whereas "phytomining" may facilitate both metal extraction and remediation of mine sites. Techniques such as bioflotation, biosorption, and biomineralisation are gaining renewed interest as potentially sustainable techniques to complement bioleaching. This chapter aims to offer an outlook towards future biomining, starting with a brief review of the mechanisms and limitations of bioleaching processes and from there expanding into an overview of new avenues.

17.2 From Understanding the Rate Limitations of Bioleaching Mechanisms to Improved Process Design

The various sub-processes that govern metal sulfide bioleaching and their complex interactions in various bioleaching processes have been discussed in Chaps. 2–5 and are also summarised in Petersen and van Staden (2019) and Petersen (2010a). Figure 17.1 offers an illustration of this network of interactions. The overall dynamics of a leaching system are determined by the relative dynamics of the individual steps and sub-processes in such a way that the overall rate (or "speed") of a given bioleaching process is governed by the slowest step in the network of interactions.

The design of bioleaching processes is concerned with sizing of equipment and drawing up of operating schedules to achieve maximum extraction of elements of economic interest at the fastest rate. In the applied context, however, considerations of costs to build and operate a given process and the size of the operation, critically

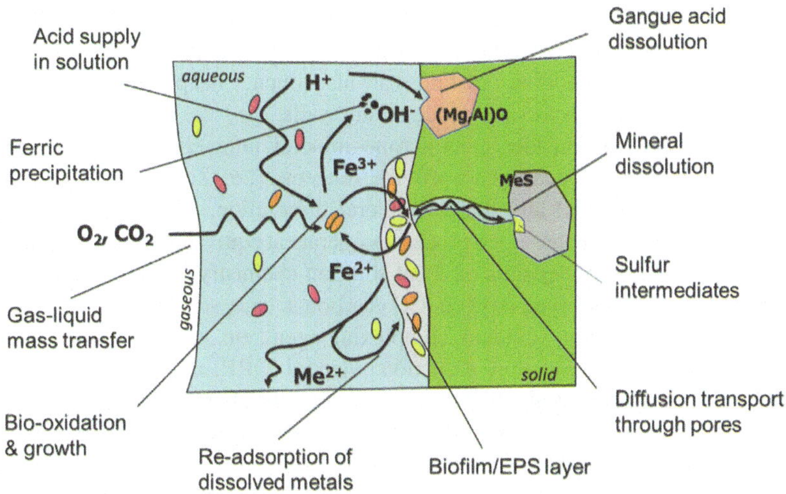

Fig. 17.1 Schematic view of physical/chemical and biological interactions in metal sulfide bioleaching. EPS = extracellular polymeric substances

determine what is economically feasible compared to what is technologically possible.

Bioleaching processes tend to be significantly slower than their conventional chemical counterparts, and it is this aspect, and the subsequent requirement for larger tankage to achieve the same throughput as a competing process, which have commonly limited the broader uptake of bioleaching technology in industrial practice. It is, therefore, critical to understand the factors affecting bioleaching rates and yields in order to maximise cost-effectiveness in a given processing context for the future establishment of bioleaching technologies.

A key rate-limiting step in bioleaching is the transfer of oxygen and CO_2 from the gas phase into aqueous solutions. In tank bioreactors this is achieved through the injection of air bubbles and vigorous agitation, creating a large gas–liquid interface. However, this comes at the cost of energy input into agitators and compressors. Also, due to hydrodynamic stress experienced by microorganisms at vigorous agitation, there is a need to maintain relatively dilute slurries in tank reactors, meaning the volumes of water handled per unit weight of mineral feed are relatively large, affecting the size of both primary and downstream equipment for metal recovery.

The effective solid-to-liquid ratio is much more favourable in heaps, but here the channelling of gas through the wet ore bed is relatively difficult to achieve and control. This is primarily due to the significant non-homogeneity of the ore packing and solution distribution. Even under the most optimal conditions, the gas–liquid interface is relatively small, resulting in significant rate limitations of the process (Petersen 2010b).

Inner-particle diffusion through narrow pore spaces is a slow process, and the time to diffuse increases with the square of the particle radius. In heap leaching this sets up a complex trade-off in terms of the particle size distribution of the ore being leached—crushing finer will reduce diffusion time but risks creating a more compact bed through which gas and solution channel less easily, while also incurring a higher cost for ore preparation. Fine crushing can also impact heap stability, leading to engineering failures such as subsidence of sidewalls.

Further factors to consider are microbe–mineral interactions and the relative kinetics of microbial processes to provide reagents (i.e., sulfuric acid and ferric iron for sulfide minerals) and that of mineral dissolution, which consumes these reagents. At steady state these two processes operate at equivalent rates. The position of this balance strongly depends on local solution chemistry which influences both processes (effect of dissolved oxygen, CO_2/carbon source, substrate, toxic ions, etc., on the biological reactions, oxidant, complexing agent, etc., on the mineral reaction, as well as pH, solution oxidation–reduction potential (ORP) and temperature on both reactions). As with the supply of oxygen, control of the chemical conditions is more easily achieved to be at their optimum in tank reactor operation, whereas in heaps any sort of local control is essentially impossible, making it all the more critical that external operating parameters (such as irrigation and aeration rates and modes; Chap. 2) are chosen and manipulated such that optimal operating conditions are maintained in the heap throughout and over the entire period of its operation. This necessitates the use of a comprehensive mathematical model of the process (Petersen and van Staden 2019; Chap. 2).

The time to complete extraction has a critical influence on process costing for the treatment of primary ores since the cost for producing the feed material must be borne upfront, whereas revenue is generated only after extraction and recovery. Thus, material held up in the process represents an inventory cost, which can make heap leaching uncompetitive, if alternative processing options are feasible, despite the much lower capital and operating costs of its operation. Similarly, the need for large tanks to accommodate long residence times in relatively dilute slurries limits the viability of the process.

Future technological development of biomining, therefore, needs to address the key issues that limit the rate of the process, such as supply of oxygen and CO_2, process intensification in tanks, and packing/permeability/particle size in heaps. In the light of this, agitator design for tank leaching continues to improve, whereas the design and operation of heaps (particularly copper sulfides) have moved to using smaller top particle size in more carefully stacked heaps, controlled intermittent irrigation to achieve uniform wetting and permeability in heaps, as well as rigorously designed aeration systems. Temperature control to achieve within the heap conditions suitable for thermophilic acidophiles for effective bioleaching of chalcopyrite has been shown to be possible but requires a systematic understanding of the thermal interactions within the heap based on comprehensive mathematical models.

The "speeding up" of a biomining process often involves financial trade-offs, as improved extraction does not necessarily justify the means required to achieve it. In an industrial bioleaching context, the cost of processing is ultimately limited by the value of the metal extracted. This trade-off can be successful for biomining, especially in the context of gold: its high value in relatively small volume operations creates a favourable niche for biological extraction. Similarly, carefully designed and operated heap bioleaching offers favourable opportunities for the extraction of low-grade minerals, especially chalcopyrite, complex minerals such as shales, or enargite and other minerals containing toxic elements that make them unsuitable for conventional processing. It is probable that future developments of biomining technology will focus on these areas.

Process economics are somewhat different in bioremediation applications, where there is less of a short-term incentive but more of a long-term benefit through the destruction of negative value associated with disposal of the untreated material. In this context, the slowness of heap leaching would not present an inventory cost and therefore make it an interesting option to consider more systematically.

A hybrid scenario exists in the context of biomining waste materials from mining and mineral processing, where the focus is on recovering remaining metal value and simultaneously remediating the residual waste. There is little cost associated with preparing the feed material, as its mining from surface deposits is usually straightforward, and it is already available in a granular form. Copper heap and dump bioleaching in their current form fall into this category, but similar opportunities exist for many concentrator tailings materials from base metal sulfide ores, providing the challenge of agglomerating the finely ground tailings onto stable "particles" suitable for heap leaching can be addressed.

Many other options for biomining for metal production in the future are discussed in the following sections, but these still need to be assessed in light of the processing challenges outlined above. The success of harnessing new mineral resources for biomining, employing new bioleaching chemistries, and implementing new biotechnologies critically hinges on their industrial application offering a competitive opportunity to provide resources in a sustainable manner.

17.3 Biomining Unexploited Mineral Resources

Industrial-scale bioleaching processes have so far mainly targeted relatively shallow sulfide mineral deposits that can be easily accessed by open cut or underground mining. The depletion of these mineral ores has triggered the search for alternative resources that could be amenable for metal extraction. Among these are ore deposits in deep continental subsurface and deep sea environments, extraterrestrial bodies as well as metal-containing mining, metallurgical and postconsumer wastes (Kaksonen et al. 2020).

17.3.1 In Situ Biomining

Conventional mining processes rely on drilling, blasting, excavating, and hauling ores to the surface before reducing particle size through crushing, and in some cases grinding, before metal extraction. This ore preprocessing has been estimated to represent 5–7% of global energy consumption, and the recovered metals often account for less than 1% of the excavated material (Johnson 2015). Deposits at depths of one or more kilometres are typically not considered to be accessible or economical for extraction using conventional mining methods, with the exception, perhaps, of precious metals. In situ recovery aims to extract metals from deep underground deposits without bringing ore to the land surface, thus reducing energy consumption, costs, above-ground footprint, and mine waste generation. Some form of in situ recoveries of copper has been practiced since medieval times, but the role of microorganisms in metal solubilisation was not known until the 1950s. The intentional use of microbial catalysts for in situ recovery has been explored since the 1970s, first for uranium and later for base metal extraction (Chap. 1).

The term deep in situ biomining (DISB) has been used to describe an emerging in situ biomining approach that targets fractured ore bodies at depths of 1–2 km. The approach utilises spatial separation of unit processes for (1) chemical underground leaching of metals from sulfide ore under saturated conditions with acidic ferric lixiviant delivered through boreholes, (2) above-ground metal recovery from pregnant leach liquor, and (3) biological lixiviant regeneration in an above-ground bioreactor (Fig. 17.2a; Johnson 2018). DISB has been recently explored in the European Union-funded BIOMOre project, which targeted a copper-containing saline and calcareous "kupferschiefer" deposit in Poland. The extraction of copper from the deposit required a three-step approach in which first water and then acid leaching was used to remove excess salinity (which would otherwise inhibit microbial activity) and acid-consuming materials, respectively, before bioleaching with biogenic ferric lixiviant (Fig. 17.2c; Johnson 2018). The use of biogenic ferric lixiviant could also be used for oxidising refractory gold ores in an in situ environment before leaching the gold with thiosulfate or iodine-based lixiviants (Fig. 17.2d; Kaksonen et al. 2014b; Kaksonen et al. 2020).

An alternative in situ biomining approach is based on forming a subsurface ore bed with sufficient permeability to operate it as an unsaturated trickle-bed bioreactor that allows in situ regeneration of the biogenic ferric iron lixiviant with underground aeration (Fig. 17.2b). Free-face blasting and underground galleries (stope leaching) have been proposed to enable the partial removal of the ore for this type of configuration. Based on modelling, this approach would allow the leaching of taller ore beds than the saturated leaching that relies on above-ground ferric lixiviant regeneration. Moreover, this approach avoids the need for an above-ground bioreactor, although the ore body pre-treatment costs are higher than in saturated leaching because of the need to remove a fraction of the ore (Vargas et al. 2020). An in situ aeration concept has also been explored for the biological oxidation of refractory sulfidic gold ores before possible chemical in situ leaching of gold (Kaksonen et al. 2014b). The simulated in situ aeration enhanced pyrite oxidation during leaching of the pyrite with biogenic ferric lixiviant. One advantage of using microbial catalysts for in situ leaching as compared to chemical ferric leaching is the ability to remove passivating sulfur layers from the mineral surfaces through microbial oxidation as some bioleaching microorganisms can oxidise reduced sulfur compounds with ferric iron as the electron acceptor (Kaksonen et al. 2014b).

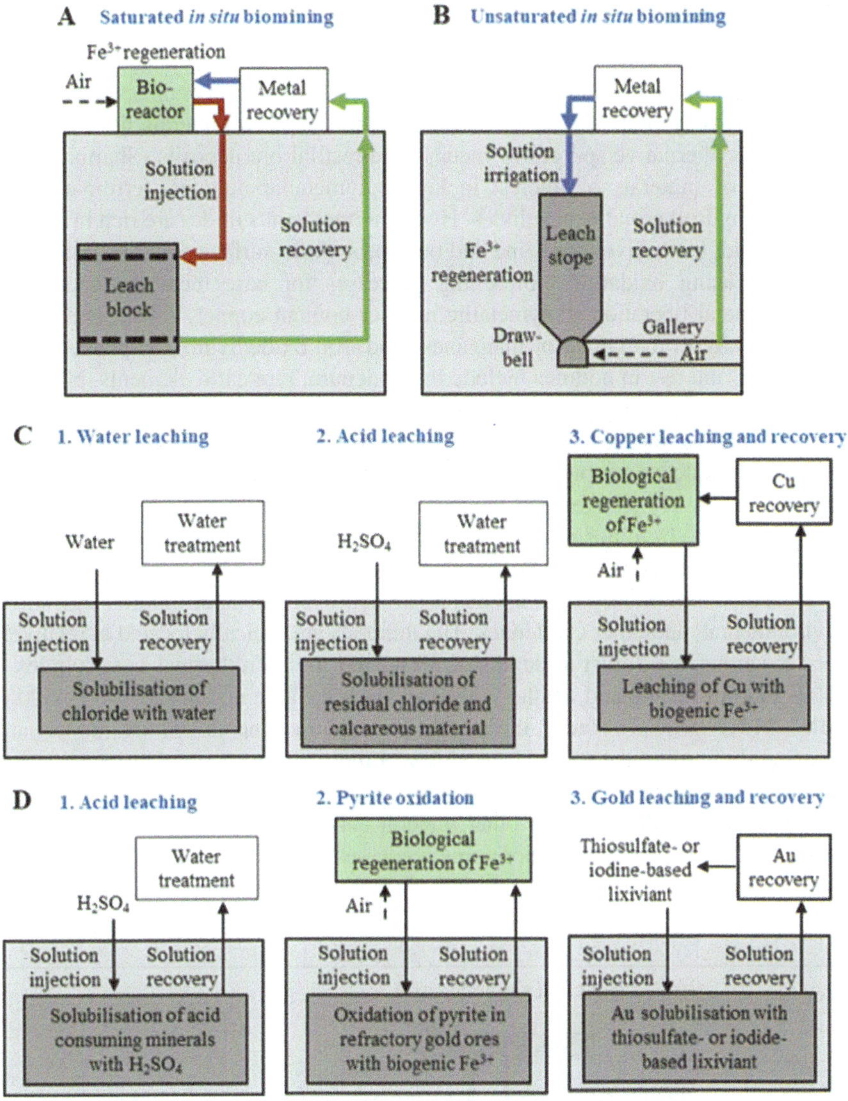

Fig. 17.2 Various in situ biomining approaches: (**a**) Deep in situ biomining under saturated conditions utilising biogenic ferric iron regenerated above ground; (**b**) in situ biomining under unsaturated conditions utilising in situ aeration and biogenic ferric regeneration in an underground stope; (**c**) in situ biomining of copper from a saline calcareous copper sulfide ore; (**d**) in situ leaching of gold from refractory sulfidic deposit after biological pre-treatment [adapted from Kaksonen et al. (2020) and Vargas et al. (2020)]

17.3.2 Deep Sea Biomining

Deep sea minerals were first discovered in 1873, but their economic potential for mining was only proposed in 1965. Since then, deep sea minerals have attracted attention as alternative sources of metals to terrestrial ore deposits (Sharma 2017). Examples of minerals of interest include polymetallic nodules, ferromanganese crust, and hydrothermal vent sulfides. Hydrothermal vent sulfides are rich in barium, copper, gold, lead, silver, and zinc and once brought to surface could potentially be processed using oxidative bioleaching processes for base metal extraction and precious metal liberation. Polymetallic nodules contain copper, cobalt, nickel, and zinc, typically in the lattice of manganese and iron oxide/hydroxide phases. Other elements of interest in nodules include molybdenum, rare earth elements, platinum, and tellurium (Kaksonen et al. 2020). The bioleaching of these minerals requires reductive bioleaching (Chap. 15). Manganese nodules have also been explored as an oxidant for chalcopyrite bioleaching (Kaksonen et al. 2020). Ferromanganese crust contains negatively charged Mn-oxyhydroxides that are bound to hydrated cations (Ca, Ni, Zn, Pb) or to positively charged Fe-hydroxides, which are complexed with anionic forms of As, P, V, and other elements (Wang and Müller 2009).

The utilisation of deep sea minerals is subject to technical, economic, legislative, environmental, and other challenges. The minerals are typically located at relatively deep locations, e.g., crust at depths of 800–2400 m, hydrothermal vent sulfides at 1000–2000 m (Wang and Müller 2009), and polymetallic nodules at 4000–6500 m (SPC 2013; Fig. 17.3). Hence, the mining of deep sea minerals requires underwater robots and pumping the minerals as slurry to the surface for metals extraction and recovery (Kaksonen et al. 2020). The profitability of deep sea mining operations is influenced by the discovery of other mineral deposits which impacts metal prices (SPC 2013). An advantage of biomining processes is that they can be easily used at a small scale and hence may be economically more feasible for small deposits than

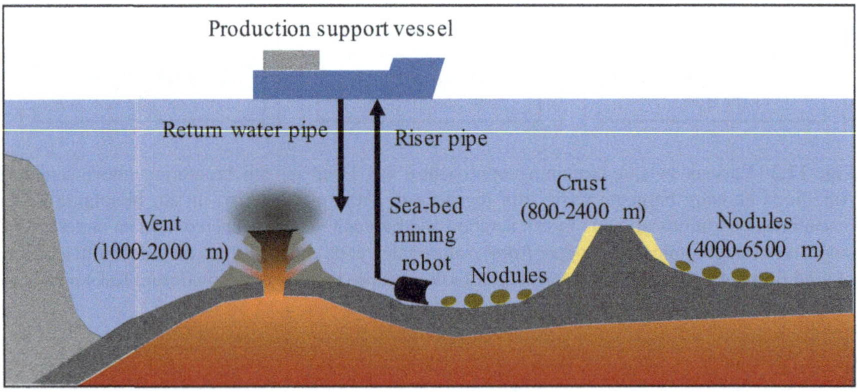

Fig. 17.3 Mining of minerals from deep sea vents, nodules, and crust [adapted from Wang and Müller (2009) and SPC (2013)]

conventional metallurgical processes. As many of the deposits are located in international waters, their utilisation is regulated under the United Nations Convention on the Law of the Sea through the International Seabed Authority (Sharma 2017). The utilisation of deep sea minerals has also raised concerns about the potential impact of the activities on marine ecosystems (Kaksonen et al. 2020).

17.3.3 Space Biomining

In situ resource utilisation (ISRU), the use of local resources for production and maintenance, is being explored to support space missions and human establishment in space. Large-scale inhabitation of space and space-based industries would need to rely on ISRU as resupply of resources from Earth is not practical nor economically feasible (Klas et al. 2015). Metals extracted from space minerals could be used for 3D printing structures, electronics, and spacecraft components to decrease the dependence of space activities on resources shipped from Earth and support human establishment beyond the lower Earth orbit. Lunar and Martian regoliths have been reported to contain SiO_2, FeO, Al_2O_3, and MgO, and additionally TiO_2 and Ca have been detected in Lunar regolith. Hydrated minerals have also been identified on both the Moon and Mars, suggesting a possible indirect source of water for in situ biomining operations (Bishop 2005; Hand 2009). Asteroids have been shown to contain platinum group metals (iridium, palladium, platinum, and rhodium) and 44 other "endangered" and "critical" elements that will face supply limitations in future years (Kaksonen et al. 2020).

The feasibility of using biomining as an enabling technology for ISRU is attracting interest. Biomining approaches proposed for space applications include, for example, oxidative and reductive bioleaching of metals for downstream recovery, bioaccumulation of iron in cells, for example as modified ferritin or magnetosomes to allow magnetic iron recovery, and partial biological iron oxidation or reduction to generate magnetite for magnetic recovery (Kaksonen et al. 2020; Volger et al. 2020a). Microorganisms investigated for space biomining include the ferrous iron- and sulfur-oxidising and ferric iron-reducing acidophile *Acidithiobacillus ferrooxidans*, ferric iron-reducing *Shewanella oneidensis* (Kaksonen et al. 2020), magnetotactic, magnetosome forming heterotrophic *Magnetospirillum gryphiswaldense* and genetically modified heterotrophic *Escherichia coli* which overexpresses a modified ferritin complex, and has an improved iron import mechanism and dysfunctional iron export mechanism (Fig. 17.4; Volger et al. 2020a).

Space environments differ from Earth conditions in terms of gravity, pressure, radiation, temperature ranges, chemical composition, as well as water and nutrient availability (Klas et al. 2015). Space biomining studies have evaluated, e.g., the ability of biomining microorganisms to survive and grow on, and extract metals from, Lunar and Martian regoliths (Kaksonen et al. 2020; Volger et al. 2020a), the effect of microgravity on biomining microorganisms (Kaksonen et al. 2020), the

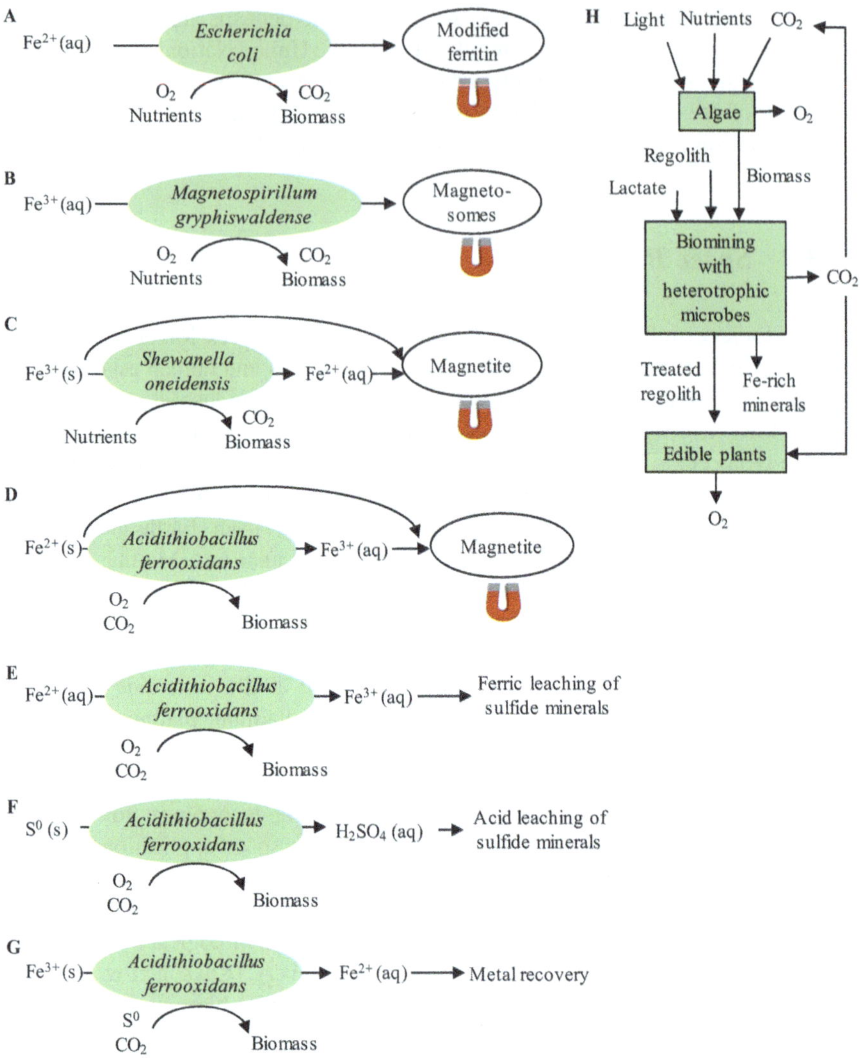

Fig. 17.4 Various approaches proposed for biomining of minerals in space (**a–g**) and an example of integrating biomining processes into other life-sustaining processes in space (**h**) [adapted from Kaksonen et al. (2020) and Volger et al. (2020a, b)]

effect of magnesium perchlorate, which is abundant in Mars, on biomining microorganisms (Volger et al. 2020a), bioreactor and process flowsheet development (Volger et al. 2020b) and payback times of biomining infrastructure (Volger et al. 2020a). Engineering challenges related to space biomining include, for example, the delivery and establishment of microbial communities in space environments, operational maintenance of microbial activities, processing, and refining of extracted resources and delivery of the products to end users (Klas et al. 2015). Robots will be essential for the implementation of ISRU in hostile space environments (Klas et al. 2015; Kaksonen et al. 2020). The use of natural resources in space is governed by the *Treaty on Principles Governing the Activities of States in the Exploitation and Use of Outer Space, including the Moon and Other Celestial Bodies* (*Outer Space Treaty*; Klas et al. 2015).

17.3.4 Biomining Waste Materials

Mining and mineral processing generate in the order of 100 billion tons of solids wastes annually. Examples of mining and metallurgical wastes include tailings, slags, converter sludges, and pyritic ashes. Biomining has been shown to be feasible for extracting metal values from minerals that could be considered as wastes due to grades that are sub-economic for traditional metallurgical processes. This enables the extension of mine life and the utilisation of deposits that would not otherwise be exploited. Bioprocessing can also make waste more amenable for final disposal, reducing risks to humans and the environment (Kaksonen et al. 2020).

Post-consumer wastes are another waste stream that can supplement declining primary ore grades and reserves and thereby support the circular economy (Chap. 14). Wastes that contain valuable metals include batteries, spent catalysts, electronic equipment (e-waste), magnets, light products, sewage sludge, municipal solid waste fly ash, and powerplant fly ash (Srichandan et al. 2019; Kaksonen et al. 2020; Yu et al. 2020). Depending on the products, these may contain precious, base, and critical metals. The complexity of the wastes, presence of hazardous substances, and relatively small waste volumes pose challenges for the use of traditional metallurgical approaches for value recovery, and biohydrometallurgy is increasingly being considered as a sustainable alternative for extracting metal values locally from these waste streams. The economic feasibility of biotechnical, and non-biological, metal extraction and recovery methods depends on waste volumes, their content of metals and impurities, geographic location, transport distances, capital and operating costs, environmental impacts, and regulatory framework including landfill and export bans, and incentives for resource recovery (Reuter et al. 2018; Kaksonen et al. 2020). Therefore, the competitiveness and sustainability of biomining approaches for waste processing need to be evaluated on a case-by-case basis.

A number of laboratory- and pilot-scale studies have been conducted to explore biological extraction and recovery of metal values from wastes [reviewed in Srichandan et al. (2019), Yu et al. (2020) and elsewhere]. Approaches evaluated include, e.g., biodismantling printed circuit boards (Monneron-Enaud et al. 2020), bioleaching with various biolixiviants, bioelectrochemical systems, biosorption, and biomineralisation (Yu et al. 2020). One-step, two-step, and spent medium bioleaching have been explored to alleviate toxicity of some wastes to microorganisms and to identify possible leaching mechanisms. Moreover, due to the complexity of wastes, the integration of various biological, chemical, and/or physical unit processes may be required for value recovery (Kaksonen et al. 2020). An example of the integration of chemical and biological unit processes is the pilot-plant of Mint Innovation, a start-up company in New Zealand, that leaches metals from e-waste using chemical leaching and recovers gold from leach liquors through biosorption with *Cupriavidus metallidurans* (Kaksonen et al. 2020).

17.4 Unconventional and Emerging Biotechnologies for Extracting and Recovering Metals

17.4.1 Bioleaching with Unconventional Lixiviants

Biogenic Organic Acids

While much of biohydrometallurgy has focused on the biologically facilitated oxidation of sulfide minerals, many metals are derived from oxide-type minerals through acid leaching. This acid could be generated from the bioleaching of acid-generating minerals, such as pyrite, elemental sulfur oxidation, or through the production of biogenic organic acids, such as acetic, citric, oxalic, or polyphenolic acids (Ilyas et al. 2018). While organic acids do not deliver the same acid strength as inorganic acids, they can form organic complexes with metal ions, often rendering them soluble in solution where an inorganic acid would not. A key example is lead, which is insoluble in sulfate systems, but well soluble in the form of lead acetate [and similar for lead citrate and oxalate]; reaction (17.1):

$$PbO(s) + 2\,CH_3COOH\,(aq) \rightarrow Pb(CH_3COO)_2\,(aq) + H_2O \tag{17.1}$$

Organic acids are well known to be produced as metabolic by-products by various microorganisms, such as *Aspergillus niger* and *Penicillium* spp., and numerous studies have been published describing these with a large range of metal sources, both primary and secondary (Anjum et al. 2010; Ilyas et al. 2018).

A key drawback of microbial organic acid producers is that they require an organic carbon source, which adds to both the operating cost and the risk of competing microorganisms. On the positive side, through their local action at the target surface, the biogenic leach reactions can be facilitated without the need for large concentrations of acid in the bulk feed which could be consumed by side reactions with non-target minerals. Thus, metal recovery through this route is of particular interest in the passive treatment of metal-bearing waste materials in heaps (Ilyas et al. 2018).

Biogenic Cyanide

Despite being much maligned for its perceived toxicity, cyanide remains the lixiviant of choice for gold leaching due to its superior properties as a complexing agent, relative stability, and low cost [reaction (17.2)]:

$$4\ Au + 8\ NaCN + O_2 + 2\ H_2O \rightarrow 4\ \left[Au(CN)_2\right]^- + 8\ Na^+ + 4\ OH^- \quad (17.2)$$

What is often forgotten is that cyanide is produced naturally by many microorganisms and plants, for defence and competitive advantage. Some innovative work has focused on harnessing the cyanide produced by certain microorganisms, in particular *Chromobacterium violaceum* (Campbell et al. 2001) and *Pseudomonas fluorescens* (Reith et al. 2007), to directly leach gold from ores, concentrates, and wastes.

A key advantage of generating cyanide in situ is that gold dissolution can be achieved with much smaller concentrations of cyanide than in conventional processes, thus potentially reducing the hazardousness of the leach solution. A major drawback of cyanide-producing organisms is that they generally grow slowly, which limits their usefulness in practical industrial applications (Zammit et al. 2012). The source of nitrogen also impacts the amount of cyanide formed; glycine is typically the best precursor but if other forms of nitrogen are available, cyanide generation tends not to be a preferred reaction (Blumer and Haas 2000). Microorganisms have also found useful application in the biological destruction of cyanide present in waste streams emanating from the mining industry (Zammit et al. 2012).

Iodide-Oxidising Microorganisms

Because of the concerns of the environmental impacts of cyanide, alternative biolixiviants are being explored for gold bioleaching. Kaksonen et al. (2014a) proposed the use of iodide-oxidising microorganisms to regenerate iodide-iodine lixiviant for gold leaching, which has shown considerable promise. Some microorganisms, such as *Roseovarius* spp. can oxidise I^- to I_2 with oxygen as the electron acceptor [reaction (17.3); Kaksonen et al. 2014a]:

$$4\ I^- + O_2 + 4\ H^+ \rightarrow 2\ I_2 + 2\ H_2O \quad (17.3)$$

Iodide (I^-) reacts chemically with iodine (I_2) to form triiodide (I_3^-) according to reaction (17.4); (Kaksonen et al. 2014a):

$$I^- + I_2 \rightarrow I_3^- \quad (17.4)$$

Gold can be solubilised according to reactions (17.5) and (17.6) (Kaksonen et al. 2014a):

$$2\ Au + I_3^- + I^- \rightarrow 2\ [AuI_2]^- \quad (17.5)$$

$$2\ Au + 3I_3^- \rightarrow 2\ [AuI_4]^- + I^- \quad (17.6)$$

Khaing et al. (2019) showed that iodide-oxidising bacteria were able to solubilise gold from sulfidic ore. Further research is needed to explore other applications of iodide-oxidising microorganisms, flow sheet development and optimisation of iodine-based bioleaching processes.

17.4.2 Bio-electrochemical Leaching

The oxidative leaching of sulfide minerals is an electrochemical process, similar to corrosion. Electron transfer is facilitated through a redox intermediary, principally Fe^{3+}/Fe^{2+} as shown in Fig. 17.1. At the mineral surface electrons are transferred from the mineral (mostly the sulfur species) to the (bio)oxidative reaction system (cathodic reaction) and the mineral dissolves (anodic reaction). The ratio Fe^{3+}/Fe^{2+}, which is linked to the ORP, critically determines the rate of oxidative dissolution at the mineral surface. Whereas the high ORP achieved during bioleaching is desirable for most sulfide minerals, in the case of chalcopyrite, the most common copper mineral, so-called passivation phenomena inhibit the reaction at high ORP, mandating a lower ORP environment (Tanne and Schippers 2017).

A form of ORP control during chalcopyrite bioleaching can be achieved through an electrolysis system in which an anode and a cathode are added to a tank reactor through which current is injected into the solution to facilitate the reduction of excess Fe^{3+} generated by microbes to Fe^{2+} thus maintaining the ORP at the desired level. Keeping the ORP at intermediate levels has been found to stimulate microbial growth due to the increased availability of Fe^{2+} (Natarajan 1992).

A further effect of the added electrolysis system is that direct mineral reduction can take place at the cathode, especially in the context of chalcopyrite reduction to chalcocite, which is much more readily leachable by bioleaching under moderate conditions than the refractory chalcopyrite [reactions (17.7) and (17.8)]:

$$2\,CuFeS_2 + 6\,H^+ + 2\,e^- \rightarrow Cu_2S + 2\,Fe^{2+} + 3\,H_2S \tag{17.7}$$

$$CuFeS_2 + 3\,Cu^{2+} + 4\,e^- \rightarrow 2\,Cu_2S + Fe^{2+} \tag{17.8}$$

Ahmadi et al. (2011) demonstrated that such a system leaches a chalcopyrite concentrate much more efficiently than bioleaching on its own. Such an electro-assisted bioleaching system may have limited implementation and uptake at large scale, but it remains a promising technique.

17.4.3 Biosorption, Bioaccumulation, and Phytomining

Almost any biomass, living or dead, can act as an adsorbent for soluble metal ions. This is facilitated through the abundance of hydroxyl, carboxyl, amino, and other functional groups at the surface of biomacromolecules which readily form complexes with the metals and thus bind them to the organic phase. Living organisms can adsorb and actively sequester metals within their cell structures, a process referred to as bioaccumulation, as a defence mechanism against metal toxicity (de Freitas et al. 2019). Physical separation of biomass loaded with metals enables their removal from aqueous solution, which makes biosorption and bioaccumulation very appealing in the context of mine wastewater treatment. However, for many types of biomass (cells and macromolecules) easy separation by physical means would require some form of supporting substrate on which the biomass is anchored.

A fundamental concern remains with the subsequent treatment of the loaded biomass to stabilise the sequestered metals so as to prevent their release once the biomass decomposes. Metal recovery from the loaded biomass is therefore a key second step. Metal desorption is usually achieved by treating the loaded biomass with strong acid or base and/or at elevated temperatures to force the adsorbed metals back into aqueous solution, often at elevated concentrations, which allows subsequent recovery through conventional hydrometallurgical techniques. However, while a given biomass may be more selective to certain types of metals than others, biosorption generally shows limited selectivity towards a mixed-metal solution. Therefore, the eluted solution would still require further purification, and the biosorption step can at best be considered an up-concentration of metals from a dilute solution. Another drawback is that the adsorbing biomass tends to become destroyed in the desorption process by the aggressive eluents used, requiring continuous new biomass supply for an ongoing process. Consequently, artificial adsorbents such as ion exchange resins and functionalised porous minerals such as zeolites remain the preferred route for selective and efficient metal removal from aqueous solution. Nonetheless, biosorption remains a feasible option for the passive treatment of metal-containing wastewaters in wetlands and barrier systems with the potential for occasional metal harvesting.

Bioaccumulation of metals in larger plants growing in a metal-rich environment has given rise to the concept of "phytomining". Certain plant species are known to accumulate metals (often quite selectively) from soils or wetland environments that are abundant in these metals. Examples are *Brassicaceae* (cruciferous plants), *Lamiaceae* (herbal plants), and *Cunoniaceae*, and it is potentially possible to selectively breed and adapt certain plants to maximise their accumulative capacity and selectivity (Sheoran et al. 2009). Such plants can be actively planted and harvested and processed further to extract the accumulated metal value. The harvested plant matter is incinerated or bio-digested, yielding energy and/or useful by-products, whereas the metals are concentrated in the ash/residue to such an extent that they can be directly fed to a conventional process for metal extraction, possibly fuelled by the energy/by-products from the primary step. While again the process is unlikely to find application in primary mining due to its slow pace/large expanse of operations required (phytomining could be perceived as a form of "metal farming"), it is an interesting option for the ameliorating of mine wastewaters in wetlands and in the treatment of contaminated soils.

17.4.4 Biobeneficiation of Minerals

Biobeneficiation of ores through bioflotation and bioflocculation for the separation of valuable minerals from gangue materials have been explored as environmentally friendly alternatives to conventional froth flotation and chemical flocculation. Moreover, bioleaching of impurities, such as phosphorus from ore has been considered as a potential approach to remove penalty elements (Kaksonen et al. 2020). Bioflotation is based on the ability of microbial cells or their metabolites to change the surface properties and hence hydrophobicity or hydrophilicity of minerals. Hydrophobicity increases the floatability of minerals whereas hydrophilicity results in depression of minerals (Behera and Mulaba-Bafubiandi 2017). Bioflocculants are biopolymers which facilitate the agglomeration of fine mineral particles by forming bridges (Kinnunen et al. 2020).

A number of microbial species and their metabolites have been evaluated for their potential for bioflotation and bioflocculation as mineral surface modifiers, activators, depressants, and collectors. The behaviour of microorganisms during bioflotation is influenced by the cell wall and membrane composition and solution pH. The structure and composition of microbial cell walls and membranes differ between archaea and bacteria and between Gram-positive and Gram-negative cells. The composition of cell walls and membranes affects the surface charge and hydrophobicity of the cells. Lipids make cells more hydrophobic resulting in flocculation of cells and adhesion of cells to solids and air bubbles (Behera and Mulaba-Bafubiandi 2017; Kinnunen et al. 2020).

Solution pH affects the surface charge of minerals and microbial cells. Repulsive forces hinder adsorption of cells to minerals when cells and minerals have the same surface charge. Microbial cells are negatively charged when the solution pH is above the microbial isoelectric point (IEP) and positively charged when the pH is below the IEP. Microbial growth substrate, growth phase, culture adaptation, and presence of minerals also influence the surface charge of microbial cells (Kinnunen et al. 2020).

Microbial metabolites evaluated for bioflotation include biosurfactants, proteins, polysaccharides, and nucleic acids. Biosurfactants are surface-active amphiphilic compounds that have both hydrophobic and hydrophilic domains and hence decrease the surface tension of solution and adsorb to mineral surfaces acting as mineral collectors. Single-stranded DNA also has an amphipathic nature with a hydrophilic phosphate backbone and hydrophobic aromatic nitrogenous bases. Polysaccharides present in extracellular polymeric substances (EPS) impart hydrophilicity to mineral surfaces, whereas hydrophobic amino acids of EPS proteins confer hydrophobicity. Iron- and sulfur-oxidising microorganisms can also induce chemical changes to mineral surfaces through biologically catalysed oxidation and reduction processes, which have been utilised, e.g., for pyrite depression (Behera and Mulaba-Bafubiandi 2017). So far bioflotation has only been used in laboratory-scale investigations (Kinnunen et al. 2020). Further research is required for the optimisation of biobeneficiation processes and the evaluation of their techno-economic feasibility.

17.4.5 Upcycling of Metals Through Biomineralisation

Microorganisms are known to play a role in the biogeochemical cycling of metals through biosolubilisation and intra- and extracellular precipitation (Reith et al. 2007). The ability of prokaryotes to precipitate metals, such as sulfides, jarosite, and scorodite, has been utilised in biotechnical mine water and hydrometallurgical water treatment (Kaksonen et al. 2018, 2020). The physicochemical and morphological properties of metals can be improved through biomineralisation and formation of biological nanoparticles (1–100 nm diameter). Metallic nanoparticles have unique antimicrobial, catalytic, electronic, magnetic, and optical properties that can be utilised for a variety of applications, such as catalysis, separation, pharmaceutical, and medical applications (e.g., antibacterial agents, cancer treatment, gene therapy, and targeted drug delivery), imaging (e.g., magnetic resonance imaging and transmission electron microscopy), fuel cells, biosensors, electronics, photonics (e.g., quantum dots) and environmental clean-up (Edmundson et al. 2014; Kaksonen et al. 2020).

The current chemical and physical methods for synthesising metallic nanoparticles typically require high temperatures and/or pressures, and hence are energy intensive and costly. Moreover, the feedstocks for these methods often need to be very pure and are therefore expensive, although biological nanoparticle formation can be carried out at lower temperatures and pressures and with less pure feedstocks, thereby saving costs and enabling the upcycling of metals from waste streams (Edmundson et al. 2014; Kaksonen et al. 2020).

A number of microorganisms have been shown to produce metallic nanoparticles, including both bacteria (e.g., *Cupriavidus metallidurans*; Cd, Cu, Co, Ge, Ni, Pb, Pd, Y, Zn; *Desulfovibrio* sp.; Au, Cr, Pd, Pt; *Magnetospirillum gryphiswaldense*; Fe; *Pseudomonas* sp.; Ag, Co, Fe, Li, Ni, Pd, Pt, Rh, Ru; *Shewanella* sp.; Fe), and fungi (e.g. *Fusarium oxysporum*; Ag, Au, Cd, Pb, Pt, Ti, Zr, and *Phoma* sp.; Ag; Edmundson et al. 2014). Synthetic biology has been proposed as a tool to functionalise and produce metallic nanoparticles. Through the ability to swap genetic modules in and out of chassis organisms, synthetic biology would allow the production of nanoparticles with various elements and post-production modifications (Edmundson et al. 2014). Further research is required to increase the understanding of and optimise the microbial genetic and metabolic elements that enable biological nanoparticle synthesis (Kaksonen et al. 2020).

17.5 Conclusions

As this concise overview has shown, there are many opportunities for biomining to play a continued, and expanding, role in providing metals for future demand. The exploitation of natural biological processes to this end contributes to a sustainable approach towards metal production, especially if it increasingly includes the recovery of metals from wastes substituting for the large-scale exploitation of primary resources. The optimisation of biomining processes requires consideration of multiple simultaneously occurring chemical, physical, and biological sub-processes. In researching new resources, chemistries, and technologies it is important, therefore, to fully appreciate the techno-economic limitations of bioleaching and judge critically whether, or to what extent, a given approach will yield an industrially feasible solution for the resource of interest in a given context.

Acknowledgements The support from CSIRO Land and Water, CSIRO Mineral Resources, CSIRO Synthetic Biology Future Science Platform (FSP), CSIRO Space Technology FSP, and New South Wales Government through its Environmental Trust to AHK are gratefully acknowledged. JP wishes to acknowledge the National Research Foundation of South Africa (NRF) through their funding of the SARChI Chair in Minerals Beneficiation (Grant UID 64829). Denis Shiers and Jason Wylie from CSIRO are thanked for reviewing the manuscript.

References

Ahmadi A, Schaffie M, Petersen J, Schippers A, Ranjbar M (2011) Conventional and electrochemical bioleaching of Sarcheshmeh chalcopyrite concentrate by moderately thermophile bacteria at high pulp density. Hydrometallurgy 106:84–92. https://doi.org/10.1016/j.hydromet.2010.12.007

Anjum F, Bhatti HN, Asgher M, Shahid M (2010) Leaching of metal ions from black shale by organic acids produced by *Aspergillus niger*. Appl Clay Sci 47:356–361. https://doi.org/10.1016/j.clay.2009.11.052

Behera SK, Mulaba-Bafubiandi AF (2017) Microbes assisted mineral flotation a future prospective for mineral processing industries—a review. Miner Process Extr Metall Rev 38(2):96–105. https://doi.org/10.1080/08827508.2016.1262861

Bishop JL (2005) Hydrated minerals on Mars. In: Tokano T (ed) Advances in astrobiology and biogeophysics. Springer, Berlin, pp 65–96. https://doi.org/10.1007/978-3-540-31538-4_4

Blumer C, Haas D (2000) Mechanism, regulation, and ecological role of bacterial cyanide biosynthesis. Arch Microbiol 173:170–177. https://doi.org/10.1007/s002039900127

Campbell SC, Olson GJ, Clark TR, McFeters G (2001) Biogenic production of cyanide and its application to gold recovery. J Ind Microbiol Biotechnol 26:134–139. https://doi.org/10.1038/sj.jim.7000104

de Freitas GR, da Silva MGC, Vieira MGA (2019) Biosorption technology for removal of toxic metals: a review of commercial biosorbents and patents. Environ Sci Pollut Res 26:19097–19118. https://doi.org/10.1007/s11356-019-05330-8

Edmundson MC, Capeness M, Horsfall L (2014) Exploring the potential for metallic nanoparticles within synthetic biology. New Biotechnol 31:572–578. https://doi.org/10.1016/j.nbt.2014.03.004

Hand E (2009) Water on the moon? Nature, 18 Sept 2009. https://doi.org/10.1038/news.2009.931

Ilyas S, Kim M-s, Lee J-c (2018) Integration of microbial and chemical processing for a sustainable metallurgy. J Chem Technol Biotechnol 93:320–332. https://doi.org/10.1002/jctb.5402

Johnson DB (2015) Biomining goes underground. Nat Geosci 8:165–166. https://doi.org/10.1038/ngeo2384

Johnson DB (2018) The evolution, current status, and future prospects of using biotechnologies in the mineral extraction and metal recovery sectors. Fortschr Mineral 8:343. https://doi.org/10.3390/min8080343

Kaksonen AH, Mudunuru BM, Hackl R (2014a) The role of microorganisms in gold processing and recovery—a review. Hydrometallurgy 142:70–83. https://doi.org/10.1016/j.hydromet.2013.11.008

Kaksonen AH, Perrot F, Morris C, Rea S, Benvie B, Austin P, Hackl R (2014b) Evaluation of submerged bio-oxidation concept for refractory gold ores. Hydrometallurgy 141:117–125. https://doi.org/10.1016/j.hydromet.2013.10.012

Kaksonen AH, Boxall NJ, Gumulya Y, Khaleque HN, Morris C, Bohu T, Cheng KY, Usher KM, Lakaniemi A-M (2018) Recent progress in biohydrometallurgy and microbial characterisation. Hydrometallurgy 180:7–25. https://doi.org/10.1016/j.hydromet.2018.06.018

Kaksonen AH, Deng X, Bohu T, Zea L, Khaleque HN, Gumulya Y, Boxall NJ, Morris C, Cheng CY (2020) Prospective directions for biohydrometallurgy. Hydrometallurgy 195:105376. https://doi.org/10.1016/j.hydromet.2020.105376

Khaing SY, Sugai Y, Sasaki K (2019) Gold dissolution from ore with iodide-oxidising bacteria. Sci Rep 9:4178. https://doi.org/10.1038/s41598-019-41004-8

Kinnunen PH-M, Kaksonen AH (2019) Towards circular economy in mining: opportunities and bottlenecks for tailings valorization. J Clean Prod 228:153–160. https://doi.org/10.1016/j.jclepro.2019.04.171

Kinnunen P, Miettinen H, Bomberg M (2020) Review of potential microbial effects on flotation. Fortschr Mineral 10:533. https://doi.org/10.3390/min10060533

Klas M, Tsafnat N, Dennerley J, Backmann S, Osborne B, Dempster AG, Manefield M (2015) Biomining and methanogenesis for resource extraction from asteroids. Space Policy 34:18–22. https://doi.org/10.1016/j.spacepol.2015.08.002

Monneron-Enaud B, Wiche O, Schlömann M (2020) Biodismantling, a novel application of bioleaching in recycling of electronic wastes. Recycling 5:22. https://doi.org/10.3390/recycling5030022

Natarajan KA (1992) Effect of applied potentials on the activity and growth of *Thiobacillus ferrooxidans*. Biotechnol Bioeng 39:907–913. https://doi.org/10.1002/bit.260390905

Petersen J (2010a) Modelling of bioleach processes: connection between science and engineering. Hydrometallurgy 104:404–409. https://doi.org/10.1016/j.hydromet.2010.02.023

Petersen J (2010b) Determination of oxygen gas-liquid mass transfer rates in heap bioleach reactors. Miner Eng 23:504–510. https://doi.org/10.1016/j.mineng.2010.01.006

Petersen J, van Staden PJ (2019) Copper heap leaching—process, principles and practical considerations. In: Wassink B, Asselin E (eds) Copper hydrometallurgy: principles and practice. Canadian Institute of Mining, Metallurgy and Petroleum (CIM), pp 52–78. isbn:978-1-926872-45-2

Reith F, Lengke MF, Falconer D, Craw D, Southam G (2007) The geomicrobiology of gold. ISME J 1:567–584. https://doi.org/10.1038/ismej.2007.75

Reuter MA, van Schaik A, Ballester M (2018) Limits of the circular economy: Fairphone pushing the limits of ewaste recycling. World of Metallurgy – ERZMETALL 71 (2018) No. 2 pp 68–79

Sharma R (2017) Deep-sea mining: current status and future considerations. In: Sharma R (ed) Deep-sea mining. Resource potential, technical and environmental considerations. Springer, Cham, pp 65–141. https://doi.org/10.1007/978-3-319-52557-0_3

Sheoran V, Sheoran AS, Poonia P (2009) Phytomining: a review. Miner Eng 22:1007–1019. https://doi.org/10.1016/j.mineng.2009.04.001

SPC (2013) In: Baker E, Beaudoin Y (eds) Deep sea minerals: manganese nodules, a physical, biological, environmental, and technical review, vol 1B. Secretariat of the Pacific Community (SPC). isbn:978–82–7701-120-2

Srichandan H, Kumar Mohapatra R, Kumar Parhi P, Mishra S (2019) Bioleaching approach for extraction of metal values from secondary solid wastes: a critical review. Hydrometallurgy 189:105122. https://doi.org/10.1016/j.hydromet.2019.105122

Tanne CK, Schippers A (2017) Electrochemical applications in metal bioleaching. In: Harnisch F, Holtmann D (eds) Bioelectrosynthesis. Advances in biochemical engineering/biotechnology, vol 167. Springer, Cham. https://doi.org/10.1007/10_2017_36

Vargas T, Estay H, Arancibia E, Díaz-Quezada S (2020) In situ recovery of copper sulfide ores: alternative process schemes for bioleaching application. Hydrometallurgy 196:105442. https://doi.org/10.1016/j.hydromet.2020.105442

Volger R, Petterson GM, Brouns SJJ, Rothschild LJ, Cowley A, Lehner BAE (2020a) Mining moon & mars with microbes: biological approaches to extract iron from lunar and Martian regolith. Planet Space Sci 184:104850. https://doi.org/10.1016/j.pss.2020.104850

Volger R, Timmer MJ, Schleppi J, Haenggi CN, Meyer AS, Picioreanu C, Cowley A, Lehner BAE (2020b) Theoretical bioreactor design to perform microbial mining activities on mars. Acta Astronaut 170:354–364. https://doi.org/10.1016/j.actaastro.2020.01.036

Wang X, Müller WEG (2009) Marine biominerals: perspectives and challenges for polymetallic nodules and crusts. Trends Biotechnol 27(6):375–383. https://doi.org/10.1016/j.tibtech.2009.03.004

Yu Z, Han H, Feng P, Zhao S, Zhou T, Kakade A, Kulshrestra S, Majeed S, Li X (2020) Recent advances in the recovery of metals from waste through biological processes. Bioresour Technol 297:122416. https://doi.org/10.1016/j.biortech.2019.122416

Zammit CM, Cook N, Brugger J, Ciobanu CL, Reith F (2012) The future of biotechnology for gold exploration and processing. Miner Eng 32:45–53. https://doi.org/10.1016/j.mineng.2012.03.016

The manufacturer's authorised representative in the EU is Springer
Nature Customer Service Centre GmbH, Europaplatz 3, 69115 Heidelberg,
Germany. If you have any concerns regarding our products, please
contact ProductSafety@springernature.com

Printed and bound by CPI Group (UK) Ltd, Croydon, CR0 4YY
28/09/2026
02224071-0012